Undergraduate Lecture Notes in Physics

Undergraduate Lecture Notes in Physics (ULNP) publishes authoritative texts covering topics throughout pure and applied physics. Each title in the series is suitable as a basis for undergraduate instruction, typically containing practice problems, worked examples, chapter summaries, and suggestions for further reading.

ULNP titles must provide at least one of the following:

- An exceptionally clear and concise treatment of a standard undergraduate subject.
- A solid undergraduate-level introduction to a graduate, advanced, or non-standard subject.
- A novel perspective or an unusual approach to teaching a subject.

ULNP especially encourages new, original, and idiosyncratic approaches to physics teaching at the undergraduate level.

The purpose of ULNP is to provide intriguing, absorbing books that will continue to be the reader's preferred reference throughout their academic career.

Series editors

Neil Ashby
Professor Emeritus, University of Colorado, Boulder, CO, USA

William Brantley
Professor, Furman University, Greenville, SC, USA

Matthew Deady
Professor, Bard College Physics Program, Annandale-on-Hudson, NY, USA

Michael Fowler
Professor, University of Virginia, Charlottesville, VA, USA

Morten Hjorth-Jensen
Professor, University of Oslo, Oslo, Norway

Michael Inglis
Professor, SUNY Suffolk County Community College, Long Island, NY, USA

Heinz Klose
Professor Emeritus, Humboldt University Berlin, Berlin, Germany

Helmy Sherif
Professor, University of Alberta, Edmonton, AB, Canada

More information about this series at http://www.springer.com/series/8917

Jan-Markus Schwindt

Conceptual Basis
of Quantum Mechanics

 Springer

Jan-Markus Schwindt
Dossenheim
Germany

ISSN 2192-4791   ISSN 2192-4805   (electronic)
Undergraduate Lecture Notes in Physics
ISBN 978-3-319-24524-9   ISBN 978-3-319-24526-3   (eBook)
DOI 10.1007/978-3-319-24526-3

Library of Congress Control Number: 2015951765

Springer Cham Heidelberg New York Dordrecht London

Printed on acid-free paper

Springer International Publishing AG Switzerland is part of Springer Science+Business Media (www.springer.com)

Preface

Quantum mechanics (QM) is still quite a mysterious theory. In contrast to the theory of relativity, for example, it did not evolve out of some basic idea, some physical principle. Only reluctantly, physicists had to accept its claims which are completely counterintuitive. Many aspects regarding its interpretation are still a matter of debate nowadays.

I tried to write a textbook I would have liked to read as a student; a book that avoids many misunderstandings, unclear and vague ideas, and the questions that have settled in my mind for quite a while, at that time; or confirms that some of my questions belong to the issues of interpretation which are considered unclear by the entire physics community and are under heavy debate. The resulting book differs in some points from most of the other QM textbooks.

First, the mysterious properties of the theory are emphasized and discussed in detail. For that reason, it starts with Bell's inequalities as a foretaste of what we embark on. Other QM books often treat QM as something that became common-place and self-evident, just as a part of the physics canon. But this is not the case. The conceptual basis of QM has been under steady research and fiery debates throughout all the decades from 1900 until today (in contrast to special and general relativity whose foundations have been understood for 90 years), with new and surprising insights coming up again and again. By now, there is a whole bunch of interpretations of the theory which fundamentally differ in their world view. A separate chapter is dedicated to these interpretations.

Second, the book develops its matter from the general to the specific. At first, the general postulates of QM are presented and discussed in detail. The wave function as a special case of a quantum state follows later. This has the advantage that the main hurdle is taken right in the beginning, and a double run-up is avoided. For many books develop the matter twice: first by means of Schrödinger's wave mechanics and later again under the title of "abstract formulation." From my point of view, it is better to begin with the abstract part and thereby avoid the misunderstanding that QM is only a theory of wave functions.

Third, the general postulates and basic notions are all explained by means of the simplest nontrivial quantum system: the two-dimensional state space of the electron spin, also called qubit.

Also for wave functions we stick to our principle of the simplest example. Instead of following most books which directly jump from one to three dimensions, we take a stopover in two dimensions. For there it is easier to explain the separation of variables and in particular some features of a rotation symmetric potential (namely without spherical harmonics).

Fourth, there is a strong focus on the clarity of notions and understanding of the mathematical background. This clarity helps to avoid misunderstandings from the beginning.

For example, a large part of QM takes place in the infinite-dimensional space of wave functions. Many theorems known from linear algebra for finite-dimensional vector spaces are no longer valid there. These peculiarities are discussed in detail.

Tensor products also get enough space, since they play an enormously important role in QM, e.g., for the notion of an entangled state, and for the combination of wave function and spin state.

Fifth, the exercises are not gathered at the end of each chapter, but are embedded in the main text. This is supposed to encourage the reader to think about and to recalculate things during his read. The solutions to the exercises are gathered at the end of the book.

The focus of this book is on the general postulates of QM, its interpretation, its basic notions, and its mathematical formulation. The first and most extensive part of the book is dedicated to this topic.

In the second part, an important special case is treated: the QM of wave functions in one, two, and three spatial dimensions, under the assumption that the Hamiltonian operator consists only of a kinetic term and a time-independent potential. Here, the most important examples are the harmonic oscillator and the hydrogen atom. Scattering theory is also discussed within this context.

The third part encompasses further topics which belong to the canonical material of a lecture on QM: combination of spin and angular momentum, QM with electromagnetism, perturbation theory, and systems with many particles. Here we only develop the basic ideas and methods, as well as some simple examples. For applications like the fine and hyperfine structure of hydrogen, or the theory of atomic transitions, we refer the reader to the literature. Finally, we provide a short explanation of the notion of a path integral and discuss the relativistic theory of the electron (Dirac equation).

The target audience of this book is, of course, in the first place students of physics who study QM in the context of a course on theoretical physics. But due to the axiomatic, deductive approach and the detailed discussion of the mathematical background, it is also very well suited for mathematicians who want to understand QM. In some places we explicitly try to overcome the "cultural barrier" between mathematicians and physicists.

Some topics which are exciting but not mandatory for the further reading (and for exams) are placed in a dedicated text environment—the so-called "nerd's corner"—for curious readers.

Acknowledgments

This book did not result from my work alone. I want to express my gratitude to the people who have enabled, supported, and contributed to this project:

- Vera Spillner at Springer for the opportunity to write this book; for the inspiring discussions; for the supply with reading material; and for proofreading, a crucial contribution to the quality of this book;
- Bianca Alton at Springer for the organizational supervision of the project; she has helped me with many questions and formalities;
- Claus Ascheron at Springer for the opportunity to translate and publish this book also in English;
- Kristin Riebe for the creation of two images, once again shining light on her graphical talent;
- Bernhard Brosda for the nice idea with the cartoons;
- Anja Stemme and Jörg Kügler for reading material about Bell's inequalities;
- Andreas Rüdinger for mentioning the Gelfand triple;
- Michael Doran for pointing me to Shankar;
- My mother for her moral support.

The fact that I am able to write such a book I owe to the many teachers, friends, and sources of inspiration on my way. At this point I want to mention two of them explicitly: I thank

- My math teacher Hanspeter Eichhorn, who a long time ago nurtured my interest in mathematics and gave it, so to speak, legs to walk.
- My Ph.D. supervisor Prof. Christof Wetterich for inspiration and support over many years.

Frankfurt, June 2013
Heidelberg, July 2015

Jan-Markus Schwindt

Contents

About the Author

Jan-Markus Schwindt studied physics and mathematics in Heidelberg and Cambridge. He obtained his Ph.D. in 2004 at the Institute of Theoretical Physics in Heidelberg. Subsequently, he spent four years doing research and teaching as a postdoc at the Universities of Mainz and Heidelberg. The focus of his research was on cosmology and quantum gravity. During this time, he often acted as a tutor. Now, he brought his teaching experience into this book.

Part I
Formalism and Interpretation

Chapter 1
Introduction: Nonlocal or Unreal?

Abstract Via Bell's inequality, it is shown that a world described by quantum mechanics must be either nonlocal or unreal, and what that even means.

Quantum mechanics with its counterintuitive claims has created many controversies. In particular, many physicists could not accept the idea that a value measured in an experiment was not determined prior to the measurement but arises spontaneously at the moment of observation. This has lead to speculations about "**hidden variables**" (also called "**hidden parameters**") which render a measurement deterministic.

Only in 1964, 38 years after the formalism of quantum mechanics was formulated, John Bell could show that hidden variables, if they exist at all, must have a certain "ugly" property which makes them quite unattractive to a majority of physicists: they must be "nonlocal", i.e. the change of a variable in one place has instantaneous effects on the rest of the world, without respecting the limit given by the speed of light for the transmission of signals.

In general we assume the **reality** and **locality** of physical phenomena. Here, **reality** means that the properties of an object are valid independent of whether or not we observe them in that moment. In particular, a property is not created by its measurement. The result that a measurement *would* give corresponds to a real property of the object, which exists independent of whether or not the measurement *actually* takes place. **Locality** means that the consequences of events can only propagate through space, and that maximally with the speed of light.

Bell has demonstrated that quantum mechanics cannot be simultaneously real and local, i.e. that at least one of the two assumptions has to be false. For the proof he made up an inequality, **Bell's inequality**, which has to be valid in a real and local world, and then showed that this inequality is violated for quantum phenomena. We want to present his argument for the example of photons passing through polarization filters.

An electromagnetic wave moves in z-direction and is polarized in x-direction. In the (xy)-plane a polarization filter is set up which lets only that part of the wave pass which is polarized in the direction of $\mathbf{r}$, where $\mathbf{r}$ encloses with the x-axis an angle ϕ. The transmitted amplitude $\mathbf{E}'$ is the projection of the original amplitude $\mathbf{E}$

© Springer International Publishing Switzerland 2016

J.-M. Schwindt, *Conceptual Basis of Quantum Mechanics*,
Undergraduate Lecture Notes in Physics, DOI 10.1007/978-3-319-24526-3_1

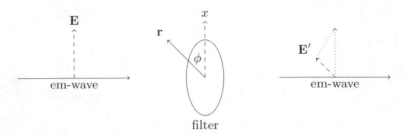

Fig. 1.1 Electromagnetic wave at a polarization filter in **r**-direction

in **r**-direction; its norm is therefore $E' = E \cos \phi$, and the intensity is $I' = I \cos^2 \phi$
(Fig. 1.1).

But taking a closer look, we find that the wave actually consists of photons,
the quanta of the electromagnetic field. This circumstance alone has far-reaching
consequences. For each single photon can pass the filter either completely or not at
all (Fig. 1.2).

If N is the number of photons before and N' the number of photons after the
filter, then $N' = N \cos^2 \phi$ is required for the equation regarding intensity to hold
(the intensity is proportional to the number of photons). Therefore each photon has
the probability $\cos^2 \phi$ to pass the filter. The question arises if this is decided in an
"absolutely random" way for each photon, i.e. not just in a seemingly random way,
due to our missing knowledge about the exact state of the photon. In other words:
whether the photon's destiny is only decided at the moment when it reaches the filter,
or whether there are "hidden parameters" which determine the decision already prior
to that, or which could explain it.

Since the arrival of the photon at the filter can be understood as a measurement
with two possible outcomes, a proponent of the reality principle has to assume that
such hidden parameters exist. For in his view, the ability of a photon to pass the filter
is a property that is not just created at the moment of measurement, but must be given
already before that.

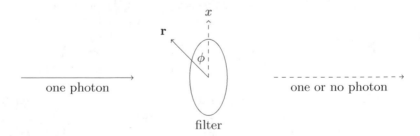

Fig. 1.2 Single photon at a polarization filter in **r**-direction

To decide the question, we have to investigate the combination of measurements of *several* properties, more precisely their **correlation**. Consider a set of objects all of which have three binary properties A, B and C, where binary means that for each object, each of A, B and C can either have a value of true or false. In the example of photons with hidden variables, A could be the ability of a photon to pass a polarization filter in the direction of $\mathbf{r}_A$, and B and C similarly for filters in the directions $\mathbf{r}_B$ and $\mathbf{r}_C$, respectively. It doesn't matter if there is an actual polarization filter filtering in any of these directions. A only means that *if* a filter is set up in direction $\mathbf{r}_A$, the photon will pass it. We denote the negation of A (photon will not pass the filter) by nA, and similarly for B and C.

In the set under consideration, some objects will have the property A, others will not. If an arbitrary object is picked up, there will be a certain probability $p(A)$ that A is true for it, and a certain probability $p(A, B)$ that both A and B are true for it etc. The variant of Bell's inequality we are going to use here reads:

Bell's Inequality

$$p(A, B) \leq p(A, C) + p(B, nC) \tag{1.1}$$

It follows from a simple set theoretical consideration: one has $p(A, B) = p(A, B, C) + p(A, B, nC)$, since for each object with the properties A and B, C is either true or not. Similar relations hold for the other two terms. In this way, the inequality can be rewritten to

$$p(A, B, C) + p(A, B, nC)$$
$$\leq p(A, B, C) + p(A, nB, C) + p(A, B, nC) + p(nA, B, nC).$$

The two terms on the left hand side also appear on the right hand side. Since probabilities are always ≥ 0, this inequality always holds, and (1.1) is thus proven.

The experimental check of Bell's inequalities is however nontrivial. The problem is that one cannot simultaneously measure the polarization of a photon in two different directions. If a filter in $\mathbf{r}_A$-direction is set up, one cannot simultaneously filter in $\mathbf{r}_B$- or $\mathbf{r}_C$-direction. The subsequent application of two filters does not help either, since the measurement in the first filter influences the photon, so that the effect of the second filter is not the same as if the first filter would be missing (Fig. 1.3).

Fig. 1.3 A measurement of polarization in two directions $\mathbf{r}_A$ and $\mathbf{r}_B$ is only possible subsequently. But the first measurement influences the state of the photon

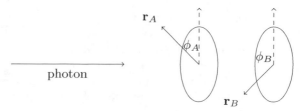

Assume a ray of photons is polarized in x-direction. If the first filter encloses an angle of $45°$ with the x-axis, exactly half of the photons pass. If behind that a second filter is set up in y-direction, again half the photons pass, altogether a quarter of the original ray. However, if already the first filter is set up in y-direction, no photons pass at all. This can be easily understood in the wave picture: at the first filter, the wave is projected into the direction of the filter. Before, it was polarized in x-direction, afterwards in the direction of the angle bisector between x- and y-axis. During the projection at the second filter, the amplitude has now a contribution in y-direction, and so a part of the wave passes the second filter. The intensity decreases at each filter by a factor $\cos^2 45° = \frac{1}{2}$, so altogether it remains $\frac{1}{4}$ of the original intensity. If instead the wave meets the filter in y-direction already in the beginning, it is annihilated by the projection, since it has no contribution in y-direction.

Therefore the filter in y-direction, if placed behind the filter in x-direction, cannot be used to measure the *original* ability of a photon to pass a filter in y-Richtung.

However, Bell's inequalities can be checked using **entangled** photons. Entangled photons are created by certain atomic transitions, for example in calcium. Thereby two photons are emitted in opposite directions such that the following holds: If on opposite sides of the photon source polarization filters are set up, both of them filtering in the same direction **r**, where each of the two photons runs into one of the two filters, then either both photons pass their respective filter, or none of them.

In our terminology for assumed hidden parameters this implies: the second photon has the property A if and only if the first one does. The measurement of such a property on one of the photons is thus equivalent to the same measurement on the other photon.

This observed behavior is remarkable and seems to be an argument *in favor* of hidden variables. This was first pointed out by Einstein in 1935 (together with his colleagues Podolsky and Rosen), by means of a variant of the phenomenon described here. Without hidden variables, photons would make a spontaneous decision at the moment of measurement whether or not to pass their filter. But how should one photon know what the other one has decided? It is as if the first photon shouted to the second one: "Hey, here was a filter in x-direction, and I passed it, so you have to do the same." Einstein spoke of a "**spooky action at a distance**", an impossibility (**Einstein-Podolsky-Rosen paradox**), and concluded there must be hidden parameters.

So, let's consider a source which emits entangled photons in $\pm z$-direction. Assume for the moment that Einstein's conclusion is correct, i.e. that hidden parameters exist. As before, we denote by A, B and C the ability of a photon to pass a polarization filter in some chosen directions $\mathbf{r}_A$, $\mathbf{r}_B$ and $\mathbf{r}_C$, respectively. These properties are assumed to exist independently of whether there is *actually* a filter in any of these directions. Due to entanglement, we can consider A, B and C as properties not of an individual photon, but of a photon pair (since both photons always give the same result at the same kind of filter). Now one can measure two properties simultaneously and therefore determine correlations (Fig. 1.4).

For checking Bell's inequalities (1.1) it is not necessary to measure three properties simultaneously. One only has to measure the combinations (A, B), (A, C) and (B, C) sufficiently many times to infer the probabilities $p(A, B)$, $p(A, C)$ and $p(B, nC)$ from the relative frequencies.

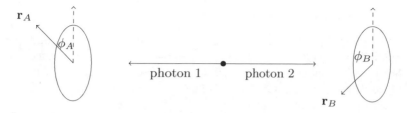

Fig. 1.4 Two entangled photons are emitted in opposite directions. On one photon, the polarization is measured in direction $\mathbf{r}_A$, on the other one in direction $\mathbf{r}_B$

Before that, one can also set up both filters in the same direction $\mathbf{r}_D$ and vary $\mathbf{r}_D$ to verify that the photons are really entangled: either both photons pass their respective filter or none of them. It turns out that independent of $\mathbf{r}_D$ one always has $p(D) = 1/2$, i.e. the ray of photons emitted by the source is completely unpolarized: for any filter direction, half of the pairs pass, the other half does not.

Now the probabilities occurring in Bell's inequalities can be determined. To measure $p(A, B)$, one photon is filtered in direction $\mathbf{r}_A$, the other one in direction $\mathbf{r}_B$. The other probabilities are determined similarly. The measurement yields

$$p(A, B) = \frac{1}{2} \cos^2 \phi_{AB}, \tag{1.2}$$

where ϕ_{AB} is the angle between $\mathbf{r}_A$ and $\mathbf{r}_B$. Equivalently one finds

$$p(A, C) = \frac{1}{2} \cos^2 \phi_{AC}, \tag{1.3}$$

$$p(B, nC) = p(B) - p(B, C) = \frac{1}{2}(1 - \cos^2 \phi_{BC}) = \frac{1}{2} \sin^2 \phi_{BC}. \tag{1.4}$$

Plugged into Bell's inequality (1.1) this gives

$$\cos^2 \phi_{AB} \leq \cos^2 \phi_{AC} + \sin^2 \phi_{BC}. \tag{1.5}$$

If $\mathbf{r}_B$ is chosen as the bisection between $\mathbf{r}_A$ and $\mathbf{r}_C$, e.g. with $\phi_{AB} = \phi_{BC} = 30°$, $\phi_{AC} = 60°$, one gets

$$\frac{3}{4} \leq \frac{1}{4} + \frac{1}{4}, \tag{1.6}$$

which is obviously not true. Bell's inequality is not fulfilled! Our assumption of hidden parameters has led us to a contradiction.

This contradiction can be avoided in two ways:

1. **Giving up reality**: Prior to the measurement, the photons have none of the properties of type A, B, C. Only at the moment of measurement (arrival at the filter) a photon decides spontaneously to pass the filter or not. In the case of an entangled photon pair, the pair decides as a whole, so that both individual measurements are consistent with each other.

2. **Giving up locality**: the measurement on one photon instantaneously influences the other photon, without respecting the limit given by the speed of light. The second photon is thereby "perturbed", and the second filter no longer measures the original property (e.g. B) but a modified one, similar to the subsequent measurements on a single photon.

No matter which variant one chooses, quantum mechanics must be quite a "crazy" theory if it is able to describe such a reality (or a lack of it). Nowadays, a majority of physicists chooses the first variant (giving up reality), but there are also theories with nonlocal hidden variables. In addition, there is an interpretation of quantum mechanics which cannot be assigned to any of these two categories, because it sheds a new light on the entire matter: the Many Worlds Interpretation. Alone the fact that a scientific theory leaves so much room for interpretations is remarkable. We will come back to this issue in Chap. 4.

Chapter 2
Formalism I: Finite-Dimensional Hilbert Spaces

Abstract The weird formalism of QM is introduced, at first in the context of finite-dimensional Hilbert spaces, where the known theorems of Linear Algebra hold. The qubit is used as an example.

2.1 The Postulates of Quantum Mechanics—Overview

At first we want to get a short overview over the postulates of quantum mechanics and their basic meaning, which we will later fill with mathematical and physical life by means of examples. All the terms used in this overview will be later explained in more detail. Our intention is to get a panoramic view on the whole mountain before we dare climbing it.

Yes, we start very axiomatically with the abstract part of QM, because I think it's a good idea to take this hurdle right in the beginning, and it helps to avoid potential misunderstandings. Mathematically the whole thing isn't too complicated for our present purpose: All we need is **Linear Algebra**, and this is not considered to be the most difficult topic in mathematics. You should be familiar with Linear Algebra from previous semesters, but we will also refresh its notions and theorems as far as we need them. The more difficult task is to recognize what kind of physics is hidden inside this formalism, i.e. what the postulates mean physically. It will take some time to figure that out.

In Chap. 3, when we meet wave functions, the physics will become clearer (finally it's about positions and momenta!), but as a tradeoff the math becomes more complicated. For then we need to understand wave functions as vectors in infinite-dimensional vector spaces, for which the somewhat more difficult field of functional analysis is responsible. Everything has a price.

Furthermore it is important to mention that we are dealing with vector spaces over the field of *complex* numbers—and this is already one of the most important differences with respect to classical mechanics, where everything behaves nicely real.

© Springer International Publishing Switzerland 2016
J.-M. Schwindt, *Conceptual Basis of Quantum Mechanics*,
Undergraduate Lecture Notes in Physics, DOI 10.1007/978-3-319-24526-3_2

Here are the Postulates of Quantum Mechanics:

1. **States**: The state of a physical system is represented by a **ray** in a **Hilbert space** $\mathcal{H}$.
   A Hilbert space is a vector space over the field of complex numbers with a **scalar product**. In quantum mechanics, a Hilbert space vector is written in the form $|v\rangle$, the scalar product of two vectors $|u\rangle$ and $|v\rangle$ in the form $\langle u|v\rangle$.
   A ray is a set of vectors $\{\alpha|v\rangle\,|\,\alpha \in \mathbb{C}\}$. In case of a real vector space one would say that a ray is a line through the origin. But since $\mathbb{C}$ rather corresponds to a plane, the notion of a line is misleading here. Anyway, the assignment of a state to a ray implies that vectors differing only by a (complex) factor represent the *same* state.
   Since it is easier to perform calculations with vectors rather than rays, one usually chooses an element $|v\rangle$ of the ray having norm 1, $\langle v|v\rangle = 1$, and denotes this element as the **state vector**. As a language convention, the state vector is often identified with the state itself, i.e. one speaks of the state $|v\rangle$.
   So, a state in quantum mechanics is a quite abstract object, and doesn't seem to match anything we can see in the world around us. The space in which the states "live" is completely different than the three-dimensional space we know. The notion of a distance ("near" and "far") between states needs to be defined and has nothing to do with the distance between points in three-dimensional space. This is already the seed of the nonlocality in quantum mechanics. In the first place it is totally unclear how to obtain our three-dimensional space from the Hilbert space. Only when the state vector takes on the form of a **wave function** (as we will see in Chap. 3), a connection is made between the two spaces, giving the states a more visualizable interpretation.
   The unintuitive character of the notion of a state in quantum mechanics has provoked countless debates about the meaning and the status of reality of the entire description.

2. **Measurement**: The measurement of an **observable**, i.e. of a measurable physical quantity of the observed system, is represented by a **linear, hermitian operator**. The measured value λ is an **eigenvalue** of the operator. After the measurement, the observed system is in the state $|v_\lambda\rangle$, where $|v_\lambda\rangle$ is the projection of the original state vector onto the eigenspace with eigenvalue λ.
   Here, an operator A is a function $\mathcal{H} \to \mathcal{H}$ which maps each Hilbert space vector to a Hilbert space vector, $|v\rangle \to A|v\rangle$. An operator is linear if

$$A\left(\alpha|u\rangle + \beta|v\rangle\right) = \alpha A|u\rangle + \beta A|v\rangle. \tag{2.1}$$

A linear operator A is hermitian if (1) it has only real eigenvalues and (2) the Hilbert space has an orthonormal basis of eigenvectors with respect to A. These two properties are crucial for quantum mechanics. Details about hermiticity will be discussed in Sect. 2.3.

An eigenvalue of A is a number λ for which a vector $|v\rangle$ exists (a so-called **eigenvector**), such that the **eigenvalue equation** $A|v\rangle = \lambda|v\rangle$ holds; i.e., the effect of the operator A on the vector $|v\rangle$ is the multiplication by λ. The eigenspace $\mathcal{H}_\lambda$ for the eigenvalue λ is the subspace of the Hilbert space $\mathcal{H}$ in which this eigenvalue equation is valid. This is indeed a subspace: from the linearity of A follows that if the eigenvalue equation holds for two vectors $|u\rangle$ and $|v\rangle$, it also holds for any linear combination of them.

Finally, projection onto an eigenspace means that the part of the original state vector $|v\rangle$ orthogonal to $\mathcal{H}_\lambda$ is "cut off". If one insists that a state vector should be normalized to 1, one has to multiply the result of the projection by an appropriate number (**normalization factor**). Hence one cannot claim that the state got "smaller" due to the cutoff.

What is the meaning of this abstract definition for a measurement? Well, the process of measurement has a rather strange appearance in quantum mechanics. During a measurement, *two* operators are caught in action:

- The operator A "represents" the observable a being measured. It kind of translates the physical property to be measured into the language of abstract quantum states. This operator A determines the possible measurement values. Since A is hermitian, only real values are possible. In this way, the real world of phenomena is retrieved from the complex quantum world. Now, which observable corresponds to which operator, is a question that will keep us busy throughout the entire book. Remark: In this book we carefully distinguish between the observable a and the corresponding operator A. In many books the operator itself is called an observable.

- The projection operator P_λ projects the original state vector $|v\rangle$ onto the subspace of $\mathcal{H}$ corresponding to the measured value λ. **The measurement therefore inevitably changes the state** (except when $|v\rangle$ was already completely inside the subspace before measurement).

In addition, the measurement postulate explains why certain observables are **quantized**, i.e. appear only in discrete portions: The Hilbert space is continuous, but the eigenvalue spectrum of many observables is not.

The postulate also contains the **Uncertainty Relation**: Two observables can be measured simultaneously only if the corresponding operators A_1 and A_2 have *common* eigenvectors. (In Sect. 2.8 we will see when this is the case.) Otherwise the two projections cannot be performed at the same time. The measurement of one observable creates an "uncertainty" in the other and vice versa. What this exactly means will be explained in Sect. 2.8.

3. **Measurement Probability**: The probability for measuring the value λ is

$$p(\lambda) = \langle v|P_\lambda|v \rangle. \tag{2.2}$$

Here, $|v\rangle$ is the normalized state vector of the observed system *before* the measurement, and P_λ is the projection operator corresponding to the eigenvalue λ. The expression on the right hand side represents the scalar product of the vector $|v\rangle$ with the vector $P_\lambda|v\rangle$. That the sum of all probabilities is 1 is reflected by the fact that the sum of the projection operators for all eigenvalues results in the identity operator (more details in Sect. 2.5).

The probability postulate contains a crucial statement: quantum mechanics is **non-deterministic**. Given a state $|v\rangle$, it is impossible to predict which value will be measured. One can make only a statistical claim. Will a given atom of a radioactive substance have decayed at time t? We can only get the half-life and compute the likelihood from that. Will a given photon pass a certain polarization filter? We can only specify the probability.

4. **Time Evolution**: As long as no measurement takes place, the time evolution of a state is given by the **Schrödinger equation**:

$$i\hbar \frac{d}{dt}|v(t)\rangle = H|v(t)\rangle \tag{2.3}$$

Here, $\hbar = h/(2\pi)$ and h is the **Planck constant**, a fundamental constant of nature. So fundamental that theoretical physicists prefer to use units in which $\hbar$ has the value 1. The **Hamiltonian operator** H represents the energy observable of the system.

So, as long as there is no measurement, the system behaves commendably deterministic: If a state $|v(t)\rangle$ is given at some time t, its time evolution can be unambiguously calculated using the Schrödinger equation, because it is a differential equation of first order in time. Only at the moment of measurement at time t_0, a sudden jump from $|v(t_0)\rangle$ to $P_\lambda|v(t_0)\rangle$ takes place. If a wave function is used to represent the state, this jump is called **collapse of the wave function**.

The Schrödinger equation is valid in the **Schrödinger picture**, where states are time-dependent and operators time-independent. There is another view, the **Heisenberg picture**, in which states do not evolve with time at all, but only the operators representing the observables. Both pictures make exactly the same predictions, so it is a matter of taste which one of them one prefers. This strange feature will be also discussed in more detail later (Sect. 2.9).

We summarize:

Postulates of QM

1. **States**: The state of a physical system is represented by a **ray** in a **Hilbert space** $\mathcal{H}$.
2. **Measurement**: The measurement of an **observable**, that is: a measurable physical quantity of the observed system, is represented by a **linear, hermitian operator**. The measured value λ is an **eigenvalue** of the operator. After the measurement, the observed system is in the state $|v_\lambda\rangle$, where $|v_\lambda\rangle$ is the projektion of the original state vector onto the eigenspace with eigenvalue λ.
3. **Measurement Probability**: The probability for measuring the value λ is

$$p(\lambda) = \langle v|P_\lambda|v\rangle. \tag{2.4}$$

Here, $|v\rangle$ is the normalized state vector of the observed system *before* the measurement, and P_λ is the projection operator corresponding to the eigenvalue λ.
4. **Time Evolution**: As long as no measurement takes place, the time evolution of a state is given by the **Schrödinger equation**

$$i\hbar\frac{d}{dt}|v(t)\rangle = H|v(t)\rangle \tag{2.5}$$

The **Hamiltonian operator** H represents the total energy of the system.

These postulates are strong stuff. For the purpose of lucidity, we have listed them and their basic meaning once completely, but now we have to establish a basis to comprehend them to full extent. That's what we are going to do in the next sections. We will delve into the used terms, extend them, and clarify them by means of a simple example: the two-dimensional Hilbert space. It will however take a while until we are ready to master the two major showcase applications of the quantum formalism: the harmonic oscillator (Sect. 5.3) and the hydrogen atom (Sect. 7.5).

Even when you have mastered this material, there will probably remain a certain discomfort (if you care about these issues at all)—a discomfort that you share with the majority of physicists.

The postulates of Quantum Mechanics, as they are presented above, are written in the spirit of the so-called **Copenhagen Interpretation**, as it was developed by Heisenberg, Bohr and Born. They were reinterpreted in several ways that shed a completely different light on the theory. We will discuss examples of that in Chap. 4.

Different than, for example, the theories of special and general relativity, the postulates of quantum mechanics were not born out of a physical idea, but out of necessity. It turned out that certain heuristics are qualified to predict the outcomes of certain experiments and observations (for example the spectral lines of hydrogen). These heuristics (Heisenberg's matrix mechanics) were further developed into the form of the postulates presented above, which turned out to be extremely successful. They form a conceptual frame for the explanation of many phenomena which cannot be understood in a classical way, e.g. the photo effect, the stability of atoms, chemical bindings etc.

The predictions of quantum mechanics were confirmed again and again, despite all their strangeness.

2.2 States in Hilbert Space

QM is expressed in terms of vectors in Hilbert space. Therefore we first have to clarify what a Hilbert space is.

A Hilbert space $\mathcal{H}$ is a complete vector space over $\mathbb{C}$ with a scalar product. The first property, **completeness**, means that each Cauchy sequence converges inside the vector space, which is not so relevant to us. A **scalar product** is a map $\mathcal{H} \times \mathcal{H} \to \mathbb{C}$ with the following properties: It is

- **anti-linear** in the first argument:

$$\langle \alpha u + \beta v | w \rangle = \alpha^* \langle u | w \rangle + \beta^* \langle v | w \rangle \tag{2.6}$$

Here, α and β are complex numbers, and the "anti" in anti-linear means that α and β need to be complex conjugated when extracted from the left part of the scalar product.
- **linear** in the second argument:

$$\langle u | \alpha v + \beta w \rangle = \alpha \langle u | v \rangle + \beta \langle u | w \rangle \tag{2.7}$$

- **hermitian**:
$$\langle u | v \rangle = \langle v | u \rangle^* \tag{2.8}$$

As a consequence, $\langle v | v \rangle$ is always real.
- **positive definite**:
$$\langle v | v \rangle \geq 0, \tag{2.9}$$

where equality holds if and only if $|v\rangle = 0$.

A Hilbert space is finite-dimensional if a finite number of **basis vectors** $|e_i\rangle$, $i = 1, 2, \ldots, n$ exist, such that any vector can be written as a **linear combination** of the $|e_i\rangle$,

$$|v\rangle = \sum_{i=1}^{n} \alpha_i |e_i\rangle. \tag{2.10}$$

One can then write $|v\rangle$ as a columnar vector $|v^{(e)}\rangle$ with respect to the chosen basis, with complex components $|v^{(e)}\rangle_i = \alpha_i$,

$$|v^{(e)}\rangle = \begin{pmatrix} \alpha_1 \\ \vdots \\ \alpha_n \end{pmatrix}. \tag{2.11}$$

A theorem in Linear Algebra says that the $|e_i\rangle$ can be chosen as an **orthonormal basis** such that

$$\langle e_i | e_j \rangle = \delta_{ij}. \tag{2.12}$$

In this case the scalar product can be expressed in terms of the components: If $|u\rangle$ has the components α_i and $|v\rangle$ the components β_i, then

$$\langle u|v \rangle = \langle \sum_i \alpha_i e_i | \sum_j \beta_j e_j \rangle = \sum_{i,j} \alpha_i^* \beta_j \langle e_i | e_j \rangle = \sum_{i,j} \alpha_i^* \beta_j \delta_{ij} = \sum_i \alpha_i^* \beta_i. \tag{2.13}$$

Please note the complex conjugation of the α_i, which follows from the anti-linearity of the scalar product in the first argument. The orthonormality in the form (2.12) and the resulting equation (2.13) automatically confirms that the scalar product is positive definite:

$$\langle v|v \rangle = \sum_{i=1}^{n} \beta_i^* \beta_i = \sum_{i=1}^{n} |\beta_i|^2 \geq 0 \tag{2.14}$$

where equality holds if and only if all components vanish. We can also argue in the other direction: only because the scalar product is positive definite, we can choose an orthonormal basis of the form (2.12).

The value $||v|| = \sqrt{\langle v|v \rangle}$ is called the **norm**, or **length** of the vector $|v\rangle$. We can always take the square root, since the scalar product is positive definite.

There are two important inequalities for the norms of vectors, the **Schwarz inequality** and the **triangle inequality**. Die Schwarz inequality reads

$$\langle u|v \rangle \langle v|u \rangle \leq \langle u|u \rangle \langle v|v \rangle \tag{2.15}$$

or, after taking the square root on both sides,

$$|\langle u|v \rangle| \leq ||u|| \, ||v||, \tag{2.16}$$

since $\langle u|v \rangle \langle v|u \rangle = \langle u|v \rangle \langle u|v \rangle^* = |\langle u|v \rangle|^2$. The inequality means that the scalar product of two vectors cannot be larger than the product of their norms. To prove that, we use the fact that the norm of the vector

$$|w\rangle = |u\rangle - \frac{\langle v|u \rangle}{\langle v|v \rangle} |v\rangle \tag{2.17}$$

is non-negative:

$$0 \leq \langle u - \frac{\langle v|u\rangle}{\langle v|v\rangle} v | u - \frac{\langle v|u\rangle}{\langle v|v\rangle} v \rangle$$

$$= \langle u|u\rangle - \frac{\langle v|u\rangle \langle u|v\rangle}{\langle v|v\rangle} - \frac{\langle v|u\rangle^* \langle v|u\rangle}{\langle v|v\rangle} + \frac{\langle v|u\rangle^* \langle v|u\rangle \langle v|v\rangle}{\langle v|v\rangle^2}$$

$$= \langle u|u\rangle - \frac{\langle v|u\rangle \langle u|v\rangle}{\langle v|v\rangle}$$

Bringing the second term to the left hand side and multiplying both sides by $\langle v|v\rangle$, one gets (2.15). In the case that $\langle v|v\rangle = 0$ and therefore (2.17) is not well-defined, one must have $|v\rangle = 0$, and now (2.15) holds because both sides vanish.

The triangle inequality reads

$$||u + v|| \leq ||u|| + ||v||. \tag{2.18}$$

It means that the norm of the sum of two vectors cannot exceed the sum of the norms of the individual vectors. The name "triangle inequality" reflects that real (as opposed to complex) vectors can be visualized as "arrows". If one draws the "arrows" for $|u\rangle$ and $|u + v\rangle$ beginning at the same point, but the "arrow" for $|v\rangle$ beginning at the "arrowhead" of $|u\rangle$, then these three "arrows" form a triangle.

The triangle inequality can be proven in the following way:

$$||u + v||^2 = \langle u + v|u + v\rangle$$

$$= \langle u|u\rangle + \langle u|v\rangle + \langle v|u\rangle + \langle v|v\rangle$$

$$= \langle u|u\rangle + \langle u|v\rangle + \langle u|v\rangle^* + \langle v|v\rangle$$

$$= \langle u|u\rangle + 2\mathrm{Re}(\langle u|v\rangle) + \langle v|v\rangle$$

$$\leq \langle u|u\rangle + 2|\langle u|v\rangle| + \langle v|v\rangle$$

$$\leq \langle u|u\rangle + 2||u|| \, ||v|| + \langle v|v\rangle$$

$$= (||u|| + ||v||)^2$$

In the fifth line we used that the real part of a complex number is smaller or equal to its absolute value. $\mathrm{Re}(z) \leq |z|$. In the sixth line the Schwarz inequality was applied.

To any object of a quantum mechanical measurement corresponds a specific Hilbert space. The state of such an object—or our knowledge about it (there are different opinions about this subtle difference, see Chap. 4)—is represented by a state in the Hilbert space. A state in Hilbert space is, as already mentioned, a **ray**, i.e. a set of vectors of the form $\{\alpha|v\rangle | \alpha \in \mathbb{C}\}$ (one can also say: a one-dimensional subspace of $\mathcal{H}$). In other words: all vectors differing only by a complex factor α represent the same state. The factor α can be written as a product of the norm r and the phase $\exp(i\varphi)$. Hence two vectors $|v_1\rangle$ und $|v_2\rangle$ correspond to the same state if

$$|v_2\rangle = \alpha|v_1\rangle = re^{i\varphi}|v_1\rangle, \qquad r, \varphi \in \mathbb{R}. \tag{2.19}$$

Usually one chooses a **normalized** vector to represent the state and denotes this vector itself as the state, or the state vector. "Normalized" hereby means that the vector has norm 1, $\langle v|v \rangle = 1$.

A not yet normalized vector $|v\rangle$ can be normalized by multiplying it with a **normalization factor** N, $N = 1/||v||$. (One divides by the length, such that the length is 1 afterwards.) For then

$$\langle Nv|Nv \rangle = N^2 \langle v|v \rangle = \frac{1}{\langle v|v \rangle} \langle v|v \rangle = 1. \tag{2.20}$$

Please note that the vector is not unambiguously determined by the normalization. The normalization only fixes the length, but not the phase. Vectors differing only by a phase have the same length. This is rather obvious, but mathematically it is a consequence of the anti-linearity of the scalar product in its first argument:

$$\langle e^{i\varphi}v|e^{i\varphi}v \rangle = e^{-i\varphi}e^{i\varphi}\langle v|v \rangle = \langle v|v \rangle \tag{2.21}$$

The simplest of all Hilbert spaces is one-dimensional: the set $\mathbb{C}$ of complex numbers itself. In this Hilbert space there is only one single state, since all one-dimensional vectors (i.e. complex numbers) differ only by a complex factor. This Hilbert space is pretty boring.

The simplest non-trivial Hilbert space is two-dimensional, $\mathbb{C}^2$. We will clarify all aspects of our new terminology by means of the example $\mathbb{C}^2$.

For $\mathcal{H} = \mathbb{C}^2$, an orthonormal basis consists of two vectors $|e_1\rangle$ and $|e_2\rangle$. A vector $|v\rangle$ is then determined by two complex components α_1 and α_2, or by four real numbers, writing $\alpha_1 = x_1 + iy_1$ and $\alpha_2 = x_2 + iy_2$.

The set of states is obtained by identifying vectors which differ only by a complex factor. One can easily see that a state is determined by the ratio α_1/α_2. Vectors differing only in norm and phase,

$$\begin{pmatrix} \alpha_1' \\ \alpha_2' \end{pmatrix} = re^{i\varphi} \begin{pmatrix} \alpha_1 \\ \alpha_2 \end{pmatrix}, \tag{2.22}$$

have the same α_1/α_2, because norm and phase are canceled in the division. On the other hand, to each α_1/α_2 belongs exactly one state. (There is one additional state for $\alpha_2 = 0$, that is the state with state vector $|e_1\rangle$). A state is therefore determined by one complex number $\lambda = \alpha_1/\alpha_2$, or by two real numbers, if one decomposes λ into real and imaginary parts.

In practice one uses normalized vectors to represent states. If one for example speaks about the state

$$|v\rangle = \frac{1}{\sqrt{2}}(|e_1\rangle + i|e_2\rangle) \tag{2.23}$$

then the state is meant for which the vector $|v\rangle$ is a normalized representative. One should keep in mind that one could have also chosen

$$|v\rangle = \frac{1}{\sqrt{2}}(-i|e_1\rangle + |e_2\rangle), \tag{2.24}$$

since this is also a normalized representative of the same state, differing only by a factor of i from the original choice (or $-i$, depending on which vector one starts with).

So far we have seen the "reverted" vectors $\langle v|$ only as a part of our notation for the scalar product $\langle u|v\rangle$. However, one can also consider them independently of that, as elements of the **dual space** $\mathcal{H}^*$ of $\mathcal{H}$. The dual space V^* of a complex vector space V is the set of all **linear forms** on V, i.e. all linear maps $V \to \mathbb{C}$. A theorem in Linear Algebra says that V^* is itself a vector space which is even isomorphic to V in the finite-dimensional case. Each linear form can be written as the scalar product with a specific vector, i.e. for each linear form λ there is a vector $|u\rangle$, such that for any vector $|v\rangle \in V$ one has:

$$\lambda(|v\rangle) = \langle u|v\rangle \tag{2.25}$$

It is therefore convenient to write $\langle u|$ instead of λ for an element of the dual space. The notions **"bra vector"** for $\langle u|$ and **"ket vector"** for $|v\rangle$ have become a popular convention; together they result in a "bracket".

Due to the anti-linearity of the scalar product in its first argument, the components of a bra vector with respect to a basis $\{|e_i\rangle\}$ are complex conjugate compared to the ket vector:

$$|v\rangle = \sum_i \alpha_i |e_i\rangle \iff \langle v| = \sum_i \langle e_i| \alpha_i^* \tag{2.26}$$

We can write the ket vector $|v\rangle$ as a columnar vector $|v^{(e)}\rangle$ and the bra vector $\langle u|$ as a row vector $\langle u^{(e)}|$ with respect to a basis $\{|e_i\rangle\}$. If it is an orthonormal basis, the scalar product can by means of (2.13) be expressed as a matrix multiplication: With $|u\rangle = \sum_i \alpha_i |e_i\rangle$ and $|v\rangle = \sum_i \beta_i |e_i\rangle$ one has

$$\langle u|v\rangle = \begin{pmatrix} \alpha_1^* & \alpha_2^* & \cdots & \alpha_n^* \end{pmatrix} \cdot \begin{pmatrix} \beta_1 \\ \beta_2 \\ \vdots \\ \beta_n \end{pmatrix} = \alpha_1^* \beta_1 + \alpha_2^* \beta_2 + \cdots + \alpha_n^* \beta_n. \tag{2.27}$$

Hereby the row vector $\langle u^{(e)}|$ (with components $\langle u^{(e)}|_i = \alpha_i^*$) was interpreted as a $1 \times n$ matrix, the column vector $|v^{(e)}\rangle$ (with components $|v^{(e)}\rangle_i = \beta_i$) as an $n \times 1$ matrix.

Self-check questions:

1. What are the properties of a scalar product?
2. How many state vectors represent a given state?
3. What is a bra vector?

2.3 Linear Hermitian Operators

In QM, measurements are described by linear hermitian operators. A linear operator A is a linear map

$$\mathcal{H} \to \mathcal{H}, \quad |v\rangle \to A|v\rangle. \tag{2.28}$$

With respect to a basis $\{|e_i\rangle\}$, A can be written as a matrix $A^{(e)}$ with components $A^{(e)}_{ij}$, i.e. for a vector $|v\rangle = \sum_i \alpha_i |e_i\rangle$ one has

$$A|v\rangle = \sum_i (\sum_j A^{(e)}_{ij} \alpha_j)|e_i\rangle, \tag{2.29}$$

i.e. $\sum_j A^{(e)}_{ij} \alpha_j$ is the i-th component of the vector $A|v\rangle$,

$$(A^{(e)}|v^{(e)}\rangle)_i = \sum_j A^{(e)}_{ij}|v^{(e)}\rangle_j. \tag{2.30}$$

The set of linear operators on $\mathcal{H}$ forms a vector space. That is, the sum of two linear operators $A + B$ is again a linear operator, and the product of a complex number with a linear operator λA is also a linear operator.

The successive application of several linear maps is again a linear map. This is true for the successive application of linear operators A and B, as well as for the successive application of a linear operator A and a linear form $\langle u|$. In the first case the result is a vector, in the second case a complex number.

Let's consider the second case: The combination $(\langle u|A)$ of $\langle u|$ and A is, according to what we just said, again a linear form: It maps a vector $|v\rangle$ to the complex number $\langle u|(A|v\rangle)$:

$$(\langle u|A)|v\rangle) = \langle u|(A|v\rangle) \tag{2.31}$$

The position of the brackets is therefore arbitrary:

- One can imagine that A acts first towards the *right*, on the ket vector $|v\rangle$, resulting in a new ket vector $|Av\rangle$, and that afterwards the scalar product with $\langle u|$ is performed, $\langle u|Av\rangle$.
- Similarly one can also imagine that A acts first towards the *left*, on the bra vector $\langle u|$, resulting in a new bra vector (new linear form) $\langle uA|$, and that afterwards the scalar product with $|v\rangle$ is performed, $\langle uA|v\rangle$.

Due to this ambiguity, one simply writes $\langle u|A|v\rangle$.

Expressed in components w.r.t. an orthonormal basis $\{|e_i\rangle\}$, the same considerations apply. According to (2.13),

$$\langle u^{(e)}|A^{(e)}v^{(e)}\rangle = \sum_i \langle u^{(e)}|_i \, |A^{(e)}v^{(e)}\rangle_i = \sum_{i,j} \langle u^{(e)}|_i \, A^{(e)}_{ij}|v^{(e)}\rangle_j. \tag{2.32}$$

The right hand side can again be read in two ways:

- The column vector $|v^{(e)}\rangle$ is multiplied from the left by the matrix $A^{(e)}$. Then one performs the scalar product of $|u^{(e)}\rangle$ with the result of this multiplication.
- The row vector $\langle u^{(e)}|$ is multiplied from the right by the matrix $A^{(e)}$. Then one performs the scalar product of $|v^{(e)}\rangle$ with the result of this multiplication.

With an orthonormal basis, one can therefore leave out the brackets also when one uses components, $\langle u^{(e)}|A^{(e)}|v^{(e)}\rangle$. However, this works only in an orthonormal basis!

Exercise 2.1

(a) Let $\mathcal{H} = \mathbb{C}^2$ with orthonormal basis $|e_1\rangle$, $|e_2\rangle$. An operator A is in this basis given by the matrix $A^{(e)} = \begin{pmatrix} 0 & 1 \\ 1 & 0 \end{pmatrix}$. Given are also two states $|u\rangle = |e_1\rangle$, $|v\rangle = |e_2\rangle$. Compute $\langle u^{(e)}|A^{(e)}v^{(e)}\rangle$ and $\langle u^{(e)}A^{(e)}|v^{(e)}\rangle$. The results are the same.

(b) Now we define a second basis, $|f_1\rangle = |e_1\rangle$, $|f_2\rangle = 2|e_2\rangle$, which is orthogonal, but not orthonormal, since $\langle f_2|f_2\rangle = 4$. What are the components of $\langle u^{(f)}|$, $|v^{(f)}\rangle$ and $A^{(f)}$? How is the scalar product to be performed in this basis? Show that $\langle u^{(f)}|A^{(f)}v^{(f)}\rangle \neq \langle u^{(f)}A^{(f)}|v^{(f)}\rangle$.

For the sake of simplicity, in the following we will always assume that $\{|e_i\rangle\}$ is an orthonormal basis.

For each linear operator A there is an **adjoint** or **hermitian conjugate** operator $A^\dagger$, defined by the requirement that its effect on a bra vector ist the same as the effect of A on a ket vector:

$$A|u\rangle = |v\rangle \iff \langle u|A^\dagger = \langle v| \tag{2.33}$$

What does the corresponding matrix $A^{\dagger(e)}$ look like? The row vector $\langle u|$ is, compared the column vector $|u\rangle$, transposed and complex conjugate. The same is true for the matrix $A^{\dagger(e)}$: It is obtained from $A^{(e)}$ by transposition and complex conjugation:

$$A^{\dagger(e)} = A_t^{(e)*} \tag{2.34}$$

For $|v^{(e)}\rangle_i = \sum_j A_{ij}^{(e)}|u^{(e)}\rangle_j$ implies

$$\langle v^{(e)}|_i = |v^{(e)}\rangle_i^* = \sum_j A_{ij}^{(e)*}|u^{(e)}\rangle_j^* = \sum_j \langle u^{(e)}|_j (A_t^{(e)})_{ji}^*, \tag{2.35}$$

and hence

$$A_{ji}^{(e)\dagger} = (A_t^{(e)})_{ji}^* = A_{ij}^{(e)*}. \tag{2.36}$$

(Since $\langle u^{(e)}|$ is a row vector, the matrix always needs to be placed to its right, otherwise the matrix multiplication makes no sense.)

The product AB of two linear operators A and B is defined by successive application:

$$(AB)|v\rangle = A(B|v\rangle), \qquad \langle u|(AB) = (\langle u|A)B. \tag{2.37}$$

As already mentioned, AB is again a linear operator. The convention to define a product as successive application is motivated by the corresponding matrices: The matrix of AB is the product of the matrices of A and B.

Exercise 2.2
Show that $(AB)^{(e)} = A^{(e)}B^{(e)}$.

The adjoint operator fulfills a number of important relations:

$$(A^\dagger)^\dagger = A, \tag{2.38}$$

$$(\lambda A)^\dagger = \lambda^* A^\dagger, \quad \lambda \in \mathbb{C}, \tag{2.39}$$

$$(A + B)^\dagger = A^\dagger + B^\dagger, \tag{2.40}$$

$$(AB)^\dagger = B^\dagger A^\dagger. \tag{2.41}$$

Similar relations hold for the corresponding matrices. Please note the reverse order in the last relation. It is due to the fact that in (2.37), the operation on a ket vector is done by the right operator first, but when operating on a bra vector, the left operator comes first. From (2.33) then follows

$$A(B|u\rangle) = |v\rangle \iff (\langle u|B^\dagger)A^\dagger = \langle v|. \tag{2.42}$$

For matries, the reverse order is also easy to show: If $C = AB$, i.e. $C_{ik} = \sum_j A_{ij}B_{jk}$, then

$$C_{ik}^\dagger = C_{ki}^* = \sum_j A_{kj}^* B_{ji}^* = \sum_j B_{ji}^* A_{kj}^* = \sum_j B_{ij}^\dagger A_{jk}^\dagger = (B^\dagger A^\dagger)_{ik}. \tag{2.43}$$

Here we dropped the superscript (e) for the sake of readability. We will keep this convention for the future: If there is no danger of confusion, we will identify an operator with its matrix w.r.t. a given orthonormal basis.

A linear operator is **hermitian** if $A^\dagger = A$. Previously we have defined "hermitian" in a different way: we said that an operator is hermitian if it has only real eigenvalues and the Hilbert space has an orthonormal basis consisting of eigenvectors. The first of these properties is necessary for using these operators as observables, because results of measurements happen to be real. Further below we will see that both definitions are equivalent.

The sum of two hermitian operators is again hermitian. The product λA of a complex number λ with a hermitian operator A however is, due to (2.39), only hermitian if $\lambda \in \mathbb{R}$. This means that the hermitian operators form only a real, not a complex vector space.

The product of two hermitian operators A, B is, due to (2.41), only hermitian if A and B **commute**, i.e. if $AB = BA$.

According to (2.36), the components of the matrix $A^{(e)}$ (w.r.t. an orthonormal basis $\{|e_i\rangle\}$) of an hermitian operator A obey

$$A_{ij}^{(e)} = A_{ji}^{(e)*}. \tag{2.44}$$

A matrix with this property is also called hermitian. In particular, the diagonal entries of $A^{(e)}$ are real. Since hermiticity was defined on the level of operators, the property (2.44) is independent of the concrete choice of an orthonormal basis. For the change to a different orthonormal basis $\{|f_i\rangle\}$ this means that the new matrix $A^{(f)}$ is again hermitian. However, this works only for orthonormal bases!

Exercise 2.3
Have another look at the transformation in Exercise 2.1. There a hermitian matrix $A^{(e)}$ is transformed into the non-hermitian matrix $A^{(f)}$!

This is again related to the difference between a matrix acting to the right on a ket vector and a matrix acting to the left on a bra vector, which (the difference) exists only in a non-orthonormal basis, cf. Exercise 2.1.

A linearer operator is called **antihermitian** if $A^\dagger = -A$. If A is hermitian, then iA is antihermitian. Any linear operator A can be written as the sum of a hermitian operator A_h and an antihermitian operator A_a:

$$A = A_h + A_a, \quad A_h = \frac{A + A^\dagger}{2}, \quad A_a = \frac{A - A^\dagger}{2}. \tag{2.45}$$

In the example $\mathcal{H} = \mathbb{C}^2$, a hermitian matrix is determined by four independent real components: It is of the form

$$A = \begin{pmatrix} a & b + ci \\ b - ci & d \end{pmatrix}, \quad a, b, c, d \in \mathbb{R}. \tag{2.46}$$

Thus, A can be written as a real linear combination of the unit matrix

$$\mathbf{1} = \begin{pmatrix} 1 & 0 \\ 0 & 1 \end{pmatrix} \tag{2.47}$$

and the three **Pauli matrices**

$$\sigma_x = \begin{pmatrix} 0 & 1 \\ 1 & 0 \end{pmatrix}, \quad \sigma_y = \begin{pmatrix} 0 & -i \\ i & 0 \end{pmatrix}, \quad \sigma_z = \begin{pmatrix} 1 & 0 \\ 0 & -1 \end{pmatrix}, \tag{2.48}$$

$$A = \frac{a+d}{2}\mathbf{1} + b\sigma_x - c\sigma_y + \frac{a-d}{2}\sigma_z. \tag{2.49}$$

Self-check questions:

1. What is the meaning of the expression $\langle u|A|v\rangle$? Why does it make sense without brackets?
2. What does it mean to say that $A^\dagger$ is the adjoint operator of A?
3. What is a hermitian operator? What is a hermitian matrix?

2.4 Eigenvalues and Eigenvectors

The possible outcomes of a quantum mechanical measurement are the **eigenvalues** of a hermitian operator A, i.e. the measured value λ obeys

$$A|v\rangle = \lambda|v\rangle \tag{2.50}$$

for an appropriate **eigenvector** $|v\rangle$. The eigenvectors of a given eigenvalue λ form a vector space, a subspace of $\mathcal{H}$, the **eigenspace** $\mathcal{H}_\lambda$ of the eigenvalue λ.

Maybe you remember from your Linear Algebra course how the eigenvalues are determined: one calculates the zeroes of the **characteristic polynomial**. The right hand side of equation (2.50) can be written as $\lambda\mathbf{1}|v\rangle$, where $\mathbf{1}$ is the identity or unit operator (or the unit matrix, if we identify the operator with its matrix). Bringing everything on one side, one gets

$$(\lambda\mathbf{1} - A)|v\rangle = 0. \tag{2.51}$$

Now we have a matrix $(\lambda\mathbf{1} - A)$ and a column vector $|v\rangle$, and thereby obtain an n-dimensional system of linear equations (where n is the dimension of $\mathcal{H}$): Each row of the matrix $(\lambda\mathbf{1} - A)$, applied to the column vector $|v\rangle$, produces a linear equation. As you know, such a system of equations has non-vanishing solutions for $|v\rangle$ only if the determinant vanishes,

$$\det(\lambda\mathbf{1} - A) = 0. \tag{2.52}$$

The left hand side of (2.52) is a polynomial in λ, the **characteristic polynomial** of the matrix A. The solutions of (2.52) are the eigenvalues of A. These solutions are then plugged into (2.51), in order to determine the corresponding eigenvectors. By the way: Since the determinant is independent of the chosen orthonormal basis, the characteristic polynomial doesn't change under a basis transformation. This is necessary, because the eigenvalues of an operator don't depend on the choice of a basis, of course.

We want to apply the procedure to a simple example, the matrix σ_x which we met when we discussed hermitian operators in our example Hilbert space $\mathcal{H} = \mathbb{C}^2$. The characteristic polynomial of σ_x is:

$$\det(\lambda\mathbf{1} - \sigma_x) = \det \begin{pmatrix} \lambda & -1 \\ -1 & \lambda \end{pmatrix} = \lambda^2 - 1 \tag{2.53}$$

The zeroes of this polynomial are 1 and -1. Hence, these are our eigenvalues. Now we write $|v\rangle = \binom{\alpha}{\beta}$ and set $\lambda = 1$. The corresponding eigenvectors are the solutions of

$$\begin{pmatrix} 1 & -1 \\ -1 & 1 \end{pmatrix} \begin{pmatrix} \alpha \\ \beta \end{pmatrix} = 0, \tag{2.54}$$

i.e. $\alpha = \beta$. The eigenspace $\mathcal{H}_1$ therefore consists of all vectors of the form $\binom{\alpha}{\alpha}$.
The eigenvectors for $\lambda = -1$ are the solutions of

$$\begin{pmatrix} -1 & -1 \\ -1 & -1 \end{pmatrix} \begin{pmatrix} \alpha \\ \beta \end{pmatrix} = 0, \tag{2.55}$$

i.e. $\alpha = -\beta$.
The eigenspace $\mathcal{H}_{-1}$ therefore consists of the vectors of the form $\binom{\alpha}{-\alpha}$.

Exercise 2.4
Determine the eigenvalues and eigenspaces of σ_y.

The matrix σ_z is already diagonal, so the eigenvalues can be read off directly, they are $\lambda = 1$ und -1. The eigenvectors have the form $\binom{\alpha}{0}$ and $\binom{0}{\alpha}$, respectively.
It still remains to be proven that our two definitions of hermiticity are equivalent. This is what we are going to do now:

1. A linear operator A is called hermitian if $A^\dagger = A$.
2. A linear operator A is called hermitian if it has only real eigenvalues and $\mathcal{H}$ has an orthonormal basis of eigenvectors w.r.t. A.

$(1) \Rightarrow (2)$:

- At first we prove that all eigenvalues are real. In (2.8) we set $|u\rangle = A|v\rangle$ and therefore $\langle u| = \langle v|A^\dagger = \langle v|A$:

$$\langle v|A|v\rangle = \langle v|A|v\rangle^* \tag{2.56}$$

If $A|v\rangle = \lambda|v\rangle$, it follows

$$\lambda\langle v|v\rangle = (\lambda\langle v|v\rangle)^*. \tag{2.57}$$

Since $\langle v|v\rangle$ is real, λ has to be real too.

- As a next step we show that eigenvectors of different eigenvalues are orthogonal. Let

$$A|u\rangle = \lambda_1|u\rangle, \quad A|v\rangle = \lambda_2|v\rangle, \quad \lambda_1 \neq \lambda_2. \tag{2.58}$$

From the second equation follows $\langle v|A = \lambda_2\langle v|$. If one takes the scalar product with $|u\rangle$ in this equation and the scalar product with $\langle v|$ in the first equation, the difference between the two equation gives

$$\langle v|A|u\rangle - \langle v|A|u\rangle = \lambda_1\langle v|u\rangle - \lambda_2\langle v|u\rangle, \tag{2.59}$$

so

$$(\lambda_1 - \lambda_2)\langle v|u\rangle = 0, \tag{2.60}$$

and therefore $\langle v|u\rangle = 0$.

- As a last step it remains to show that the eigenvectors span the complete space $\mathcal{H}$. For that purpose we make use of the matrix $A^{(e)}$ corresponding to A w.r.t. a basis $\{|e_i\rangle\}$. Since $\mathbb{C}$ is algebraically closed, the characteristic polynomial of $A^{(e)}$ has at least one zero, hence A has at least one eigenvalue λ. In the corresponding eigenspace $\mathcal{H}_\lambda$ we choose a normalized eigenvector $|v\rangle$. Then we choose a new orthonormal basis $\{|f_i\rangle\}$ of $\mathcal{H}$, in which $|f_1\rangle = |v\rangle$. In this basis, $A^{(f)}$ has the form

$$A^{(f)} = \begin{pmatrix} \lambda & A_{12} & \cdots & A_{1n} \\ 0 & A_{22} & \cdots & A_{2n} \\ \vdots & \vdots & \ddots & \vdots \\ 0 & A_{n2} & \cdots & A_{nn} \end{pmatrix}. \tag{2.61}$$

In particular, the components of the first column, A_{i1}, all vanish for $i > 1$, because otherwise we wouldn't have $A|f_1\rangle = \lambda|f_1\rangle$. Since $A^{(f)}$ is hermitian, all the components A_{1i} with $i > 1$ have to vanish too. So, $A^{(f)}$ has the block diagonal form

$$A^{(f)} = \begin{pmatrix} \lambda & 0 \\ 0 & A'^{(f)} \end{pmatrix}, \quad A'^{(f)} = \begin{pmatrix} A_{22} & \cdots & A_{2n} \\ \vdots & \ddots & \vdots \\ A_{n2} & \cdots & A_{nn} \end{pmatrix}. \tag{2.62}$$

It follows that the effect of $A^{(f)}$ (and so that of A) on the two parts of the decomposition $\mathcal{H} = \mathcal{H}_v \oplus \mathcal{H}'$ is independent, i.e. $|u\rangle \in \mathcal{H}_v \Rightarrow A|u\rangle \in \mathcal{H}_v$ and $|u\rangle \in \mathcal{H}' \Rightarrow A|u\rangle \in \mathcal{H}'$. Here, $\mathcal{H}_v$ is the one-dimensional subspace spanned by $|v\rangle$ (which is not necessarily the entire eigenspace $\mathcal{H}_\lambda$) and $\mathcal{H}'$ the subspace spanned by all other basis vectors. In particular, each vector in $\mathcal{H}'$ is orthogonal to $|v\rangle$.

In $\mathcal{H}'$, A is represented by the matrix $A'^{(f)}$, and we can proceed with it just as with $A^{(e)}$. We find an eigenvalue, a corresponding normalized eigenvector $|w\rangle$, construct a new orthonormal basis $\{|g_i\rangle\}$ $(i = 2, \dots n)$ of $\mathcal{H}'$ with $|g_2\rangle = |w\rangle$, thereby separating the subspace spanned by $|w\rangle$, $\mathcal{H}' = \mathcal{H}'_w \oplus \mathcal{H}''$. We proceed like that until A is completely diagonalized and $\mathcal{H}$ completely decomposed. In the resulting orthonormal basis, each element is an eigenvector.

(2) $\Rightarrow$ (1):
Given an orthonormal basis $\{|e_i\rangle\}$ of eigenvectors with real eigenvalues, $A^{(e)}$ is diagonal with the real eigenvalues on the diagonal. In particular, $A^{(e)}$ is hermitian, and therefore A is.

Exercise 2.5
Let $\mathcal{H} = \mathbb{C}^2$ and $\{|e_i\rangle\}$ an orthonormal basis.

(a) Show that the nonhermitian operator A with $A^{(e)} = \begin{pmatrix} 1 & 1 \\ 0 & 1 \end{pmatrix}$ has only real eigenvalues, but that $\mathcal{H}$ is not spanned by the corresponding eigenvectors.

(b) Show that the nonhermitian operator B with $B^{(e)} = \begin{pmatrix} 2 & 1 \\ 0 & 1 \end{pmatrix}$ has only real eigenvalues, and that $\mathcal{H}$ is spanned by the corresponding eigenvectors. But there is no orthonormal basis of eigenvectors.

Self-check questions:

1. How do you determine the eigenvalues of a hermitian operator?
2. What ensures that the result does not depend on a choice of basis?
3. What characterizes eigenvalues, eigenvectors and eigenspaces of hermitian operators?

2.5 Projection and Measurement

So far we have only repeated Linear Algebra of vector spaces $\mathbb{C}$, and that with the notation conventional in quantum mechanics. Now it's getting time to see what this has to do with physics. It is very convenient to use the electron's spin as a first example, for this is a property of the electron described in terms of the two-dimensional Hilbert space and the Pauli matrices we became familiar with in the previous sections. Spin is a property which is related to angular momentum in a subtle way, as we will see in Chap. 9. It is therefore often described as "a kind of intrinsic rotation" of the electron. But this is misleading; electrons are not little globules that can be imagined as rotating. Spin is something quite abstract. It is nevertheless—or even just because of that—perfectly suitable to explain the principles of a quantum mechanical measurement.

The **spin state** of an electron is represented by a vector in $\mathcal{H} = \mathbb{C}^2$. (We remember: actually by a ray, but for simplicity one chooses a normalized vector as a representative of the ray.) Spin itself is a vectorial property, namely vectorial with respect to our three-dimensional space, i.e. it has components in x-, y- and z-direction. That is quite remarkable, for it means that the information about a three-dimensional real vector is contained in a two-dimensional complex state vector.

The three spin components s_x, s_y, s_z are as observables associated with the operators S_x, S_y, S_z, and these are, in an appropriate orthonormal basis $|e_1\rangle$, $|e_2\rangle$, given by the matrices $\frac{\hbar}{2}\sigma_x$, $\frac{\hbar}{2}\sigma_y$ and $\frac{\hbar}{2}\sigma_z$,

$$S_x = \frac{\hbar}{2}\sigma_x, \quad S_y = \frac{\hbar}{2}\sigma_y, \quad S_z = \frac{\hbar}{2}\sigma_z. \tag{2.63}$$

In a conventional notation, the spin operator is often written as a vector: $\mathbf{S} = \frac{\hbar}{2}\boldsymbol{\sigma}$, such as if the three Pauli matrices formed a vector in three dimensions, whose components are 2×2-matrices.

Now consider a measurement of the z-component of an electron's spin whose spin state vector $|v\rangle$ has components (α, β) in the mentioned basis. By the second postulate of quantum mechanics, the possible measurement values are the eigenvalues of $\frac{\hbar}{2}\sigma_z$, i.e. $\pm\frac{\hbar}{2}$. The eigenspace of the eigenvalue $+\frac{\hbar}{2}$ is spanned by $|e_1\rangle$, the eigenspace of the eigenvalue $-\frac{\hbar}{2}$ by $|e_2\rangle$. After the measurement, the electron is in the state $|e_1\rangle$ if $+\frac{\hbar}{2}$ was measured, or in the state $|e_2\rangle$ if $-\frac{\hbar}{2}$ was measured. The state $|e_1\rangle$ is therefore denoted as "spin up" (spin points to positive z-direction), $|e_2\rangle$ as "spin down" (spin points to negative z-direction). Usually these two states are written $|\uparrow\rangle$ and $|\downarrow\rangle$. The bra/ket notation is often used in this way: Instead of identifiers like e_i or v, significant symbols are placed between the brackets. Enumerated basis vectors are written $|1\rangle$, $|2\rangle$, ..., $|n\rangle$ instead of $|e_1\rangle$, $|e_2\rangle$, ..., $|e_n\rangle$. Often the eigenvalues themselves are used to denote the states; in our case that would be $|\frac{\hbar}{2}\rangle$ and $|-\frac{\hbar}{2}\rangle$. This obviously works only if the eigenspaces are one-dimensional, with a unique state corresponding to an eigenvalue.

We want to follow our own convention here, using the notation $|z+\rangle$ for $|e_1\rangle$ and $|z-\rangle$ for $|e_2\rangle$. The reason is that we will often compare spins in different spatial directions, and for that arrows and eigenvalues alone are useless.

Projecting a vector $|v\rangle$ onto a vector $|u\rangle$ means taking the part of $|v\rangle$ parallel to $|u\rangle$. The projection operators onto $|z+\rangle$ and $|z-\rangle$ are, written a matrices,

$$P_{z+} = \begin{pmatrix} 1 & 0 \\ 0 & 0 \end{pmatrix}, \quad P_{z-} = \begin{pmatrix} 0 & 0 \\ 0 & 1 \end{pmatrix}, \tag{2.64}$$

since

$$P_{z+}|v\rangle = \begin{pmatrix} 1 & 0 \\ 0 & 0 \end{pmatrix}\begin{pmatrix} \alpha \\ \beta \end{pmatrix} = \begin{pmatrix} \alpha \\ 0 \end{pmatrix}, \tag{2.65}$$

$$P_{z-}|v\rangle = \begin{pmatrix} 0 & 0 \\ 0 & 1 \end{pmatrix}\begin{pmatrix} \alpha \\ \beta \end{pmatrix} = \begin{pmatrix} 0 \\ \beta \end{pmatrix}. \tag{2.66}$$

If $|v\rangle$ is normalized, i.e. $\alpha^*\alpha + \beta^*\beta = 1$, the third postulate of quantum mechanics claims that the probabilities for the two possible measurement results are given by

$$p(z+) = \langle v|P_{z+}|v\rangle = \begin{pmatrix} \alpha^* & \beta^* \end{pmatrix} \begin{pmatrix} 1 & 0 \\ 0 & 0 \end{pmatrix} \begin{pmatrix} \alpha \\ \beta \end{pmatrix} = \begin{pmatrix} \alpha^* & \beta^* \end{pmatrix} \begin{pmatrix} \alpha \\ 0 \end{pmatrix} = \alpha^*\alpha, \qquad (2.67)$$

$$p(z-) = \langle v|P_{z-}|v\rangle = \begin{pmatrix} \alpha^* & \beta^* \end{pmatrix} \begin{pmatrix} 0 & 0 \\ 0 & 1 \end{pmatrix} \begin{pmatrix} \alpha \\ \beta \end{pmatrix} = \begin{pmatrix} \alpha^* & \beta^* \end{pmatrix} \begin{pmatrix} 0 \\ \beta \end{pmatrix} = \beta^*\beta. \qquad (2.68)$$

This result can be generalized easily: The probability to measure an eigenvalue λ is equal to the squared component of the normalized state vector parallel to the eigenspace $\mathcal{H}_\lambda$. This is what the third postulate boils down to.

The sum of the two probabilities is

$$p(z+) + p(z-) = \alpha^*\alpha + \beta^*\beta = 1, \qquad (2.69)$$

as it has to be (the measurement has exactly one result, in any case). The normalized state vector is not unique, it can be multiplied by a phase $\exp(i\varphi)$, which is however canceled in the probabilities $\alpha^*\alpha$ and $\beta^*\beta$ (the norm squared). Probabilities cannot depend on a choice of phase.

If one wants to use state vectors which are not normalized, one has to adapt the rule for probabilities, dividing by the norm squared of the state vector:

$$p(\lambda) = \frac{\langle v|P_\lambda|v\rangle}{\langle v|v\rangle} \qquad (2.70)$$

At this point we want to clarify what it means to have only hermitian operators as observables. Hermitian operators have three important properties:

- They have only real eigenvalues. That's important, since measured values are always real.
- Eigenvectors of different eigenvalues are orthogonal. That's important for the measurement to be consistent: If "spin up" was measured, the electron is in the state $|z+\rangle$ afterwards. If the same measurement is performed again immediately, the result has to be again "spin up" (with probability 1) for the state is now orthogonal to $|z-\rangle$. If $|z+\rangle$ had a non-vanishing component in the direction of $|z-\rangle$, there would be a positive probability for the second measurement to result in "spin down", and hence to disagree with the first one.
- The eigenvectors of all eigenvalues span the entire Hilbert space. That's important, because otherwise there would be states which are orthogonal to all eigenspaces. A measurement on such a state could not have any result.

The question comes up why nature plays this operators game at all, and in what way the operators actually participate in a measurement. In fact, in all real experiments electrons are tackled with measurement devices, not with matrices. The measurement device must have something about it which acts like a hermitian matrix (a hermitian operator) on the electron. But what, and how? This question is not so easy to answer, and again there are various opinions. We will come back to this in the interpretation Chap. 4.

Projection operators apparently play a big role in quantum mechanics. We are therefore going to discuss them in more detail. Let $\{|i\rangle\}$ with $i = 1, \ldots, n$ be an orthonormal basis of $\mathcal{H}$ and $|v\rangle$ a vector. The representation of $|v\rangle$ in this basis,

$$|v\rangle = \sum_i \alpha_i |i\rangle, \tag{2.71}$$

consists of components $\alpha_i |i\rangle$, the projections of $|v\rangle$ onto the corresponding basis vector $|i\rangle$. The coefficients α_i are obtained via the scalar products

$$\alpha_i = \langle i|v\rangle. \tag{2.72}$$

Hence one can write $|v\rangle$ as

$$|v\rangle = \sum_i |i\rangle\langle i|v\rangle. \tag{2.73}$$

In this way we get an interesting expression for the identity operator:

$$\mathbf{1} = \sum_i |i\rangle\langle i|. \tag{2.74}$$

The left hand side of the previous equation (2.73) is equal to $\mathbf{1}|v\rangle$, and since it is valid for all vectors $|v\rangle$, it implies (2.74). The identity operator is decomposed into a set of projection operators: The projection onto a basis vector $|i\rangle$ is produced via the operator

$$P_i = |i\rangle\langle i|, \tag{2.75}$$

because, as we saw,

$$P_i|v\rangle = |i\rangle\langle i|v\rangle = \alpha_i |i\rangle. \tag{2.76}$$

With the help of the identity operator we can deduce the so-called completeness relation:

$$\langle u|v\rangle = \langle u|\mathbf{1}|v\rangle = \sum_i \langle u|i\rangle\langle i|v\rangle \tag{2.77}$$

Here we applied a trick we will meet again many times: We inserted a unit operator in the form of (2.74). From (2.77) follows the **completeness relation**

$$\langle v|v\rangle = \sum_i |\langle i|v\rangle|^2. \tag{2.78}$$

One can use it to check whether a supposed basis $\{|i\rangle\}$ really spans the entire Hilbert space. In finite-dimensional spaces this can be checked much more easily by counting the basis vectors (assuming one knows that the $\{|i\rangle\}$ are linearly independent). The completeness relation therefore becomes really interesting only when we consider infinite dimensional Hilbert spaces. Since (2.78) is equivalent to (2.74), the equation (2.74) itself is often called completeness relation.

From the form (2.74) of the unit operator we can obtain an interesting representation of an operator A. Let $|u\rangle = \sum_i \alpha_i |i\rangle$ and $|v\rangle = \sum_i \beta_i |i\rangle$. One has

$$A = 1A1 = \sum_{i,j} |i\rangle\langle i|A|j\rangle\langle j| \tag{2.79}$$

and so

$$\langle u|A|v\rangle = \sum_{i,j} \langle u|i\rangle\langle i|A|j\rangle\langle j|v\rangle = \sum_{i,j} \alpha_i^* \langle i|A|j\rangle \beta_j. \tag{2.80}$$

Comparing with $\langle u|A|v\rangle = \sum_{i,j} \alpha_i^* A_{ij} \beta_j$ we conclude

$$\langle i|A|j\rangle = A_{ij} \tag{2.81}$$

and hence

$$A = \sum_{i,j} |i\rangle A_{ij} \langle j|. \tag{2.82}$$

The matrix components A_{ij} use $|i\rangle$ and $\langle j|$ to "grab" the right components from $\langle u|$ and $|v\rangle$. One should keep this representation of an operator in mind, it will be used from time to time.

The right hand side of equation (2.75) can be understood as a matrix multiplication of the column vector $|i\rangle$ with the row vector $\langle i|$. Indeed, for our spin example this yields

$$P_{z+} = |z+\rangle\langle z+| = \begin{pmatrix} 1 \\ 0 \end{pmatrix} (1\ 0) = \begin{pmatrix} 1 & 0 \\ 0 & 0 \end{pmatrix}, \tag{2.83}$$

and similarly for P_{z-}.

If we want to measure the x-component of the spin, we have to work with the eigenspaces of $\frac{\hbar}{2}\sigma_1$. The normalized eigenvectors for the eigenvalues $+\frac{\hbar}{2}$ and $-\frac{\hbar}{2}$ are

$$|x+\rangle = \frac{1}{\sqrt{2}} \begin{pmatrix} 1 \\ 1 \end{pmatrix}, \qquad |x-\rangle = \frac{1}{\sqrt{2}} \begin{pmatrix} 1 \\ -1 \end{pmatrix}, \tag{2.84}$$

respectively. The associated projection operators are

$$P_{x+} = |x+\rangle\langle x+| = \frac{1}{2} \begin{pmatrix} 1 \\ 1 \end{pmatrix} (1\ 1) = \frac{1}{2} \begin{pmatrix} 1 & 1 \\ 1 & 1 \end{pmatrix}, \tag{2.85}$$

$$P_{x-} = |x-\rangle\langle x-| = \frac{1}{2}\begin{pmatrix} 1 \\ -1 \end{pmatrix}(1 \ -1) = \frac{1}{2}\begin{pmatrix} 1 & -1 \\ -1 & 1 \end{pmatrix}. \tag{2.86}$$

The sum of them yields the unit operator, as it has to be:

$$P_{x+} + P_{x-} = \frac{1}{2}\begin{pmatrix} 1 & 1 \\ 1 & 1 \end{pmatrix} + \frac{1}{2}\begin{pmatrix} 1 & -1 \\ -1 & 1 \end{pmatrix} = \begin{pmatrix} 1 & 0 \\ 0 & 1 \end{pmatrix}. \tag{2.87}$$

Given a state $|v\rangle = \begin{pmatrix} \alpha \\ \beta \end{pmatrix}$, the probabilities for the two possible measurement results are

$$p(x+) = \langle v|P_{x+}|v\rangle = \frac{1}{2}(\alpha^* \ \beta^*)\begin{pmatrix} 1 & 1 \\ 1 & 1 \end{pmatrix}\begin{pmatrix} \alpha \\ \beta \end{pmatrix} \tag{2.88}$$

$$= \frac{1}{2}(\alpha^* \ \beta^*)\begin{pmatrix} \alpha + \beta \\ \alpha + \beta \end{pmatrix} = \frac{1}{2}(\alpha^* + \beta^*)(\alpha + \beta), \tag{2.89}$$

$$p(x-) = \langle v|P_{x-}|v\rangle = \frac{1}{2}(\alpha^* \ \beta^*)\begin{pmatrix} 1 & -1 \\ -1 & 1 \end{pmatrix}\begin{pmatrix} \alpha \\ \beta \end{pmatrix} \tag{2.90}$$

$$= \frac{1}{2}(\alpha^* \ \beta^*)\begin{pmatrix} \alpha - \beta \\ -\alpha + \beta \end{pmatrix} = \frac{1}{2}(\alpha^* - \beta^*)(\alpha - \beta). \tag{2.91}$$

The sum of these probabilities is

$$p(x+) + p(x-) = \frac{1}{2}\left[(\alpha^* + \beta^*)(\alpha + \beta) + (\alpha^* - \beta^*)(\alpha - \beta)\right] \tag{2.92}$$

$$= \alpha^*\alpha + \beta^*\beta = 1. \tag{2.93}$$

Exercise 2.6
Perform the same calculation for a measurement of the spin in y-direction.

Exercise 2.7
Verify the following: if the electron is in an eigenstate of one of the spin operators S_x, S_y, S_z, then the probabilities for the results regarding any of the other two operators equals 1/2. For example, if the electron is in the state $|z+\rangle$, then $p(x+) = p(x-) = \frac{1}{2}$. So, if any of the three spin variables is **certain** (i.e. we have an eigenstate; the probability for one outcome is 1, for the other outcome 0), then the two other spin variables are **maximally uncertain**, i.e. the probabilities for the spin values in the other directions are equally distributed among the possible outcomes. This is an example for an **Uncertainty Relation**.

The eigenspace for an eigenvalue λ of an operator can be more than one-dimensional. In this case, given an appropriate orthonormal basis $\{|i\rangle\}, i = 1, \ldots, n$, there will be several basis vectors $|i_1\rangle, \ldots, |i_k\rangle$ spanning the eigenspace $\mathcal{H}_\lambda$. The associated projection operator is then

$$P_\lambda = |i_1\rangle\langle i_1| + \cdots + |i_k\rangle\langle i_k|. \tag{2.94}$$

Exercise 2.8
Projection operators always have the eigenvalues 0 and 1. The components of a vector orthogonal to the space on which is projected are annihilated (eigenvalue 0), the components parallel to the space on which is projected are conserved (eigenvalue 1). Verify this for P_{x+}, P_{x-}, using the characteristic polynomial.

Exercise 2.9
Since a projection operator P_λ always has eigenvalues 0 and 1 and is hermitian (how can that be inferred from the representation 2.94?), there is an orthonormal basis in which P_λ is diagonal and has only ones and zeroes on the diagonal. It follows that $P_\lambda^2 = P_\lambda$. Verify this equation for P_{x+}, P_{x-}.

Formally, a projection operator is even defined in this way: A projection operator is a linear operator P for which $P^2 = P$.

In our examples, we started with the state vector $|v\rangle$ and calculated the measurement probabilities based on that. In real experiments, however, one often has to deduce the state from the measurement results. But how can we do that? The problem is that after an individual measurement we only know the state **after** the measurement. In the spin example these are only two possible results. **There are infinitely many different states in a two-dimensional Hilbert space, but through a measurement, we always obtain only 1 bit of information.** No matter in which direction we measure the spin, there are always only two possible results. And not only that: the components of the state orthogonal to the obtained outcome are irrecoverably lost. If the electron is in the state $|x+\rangle$, the result of an s_x-measurement will be $s_x = +\frac{\hbar}{2}$ with probability 1. But if the experimenter decides to measure the spin in z-direction, and obtains "spin up", the electron is in the state $|z+\rangle$ after the measurement. The information about the x-component of the spin is unrecoverably lost. If the experimenter now decides to measure s_x after all, he will obtain $s_x = +\frac{\hbar}{2}$ and $s_x = -\frac{\hbar}{2}$ with equal probabilities.

Since a measurement on a two-dimensional Hilbert space always gives you only 1 bit of information, such a system is also called a **Qubit** (short for quantum bit). The measurement device can be gauged such that it shows a zero in case of "spin up" and a one in case of "spin down". The associated operator is then $\begin{pmatrix} 0 & 0 \\ 0 & 1 \end{pmatrix}$. In a series of measurements the device will write a sequence of ones and zeroes—that is, of bits. However, somehow qubits are more than just bits, for behind them there is a two-dimensional state space that can be "queried" in different ways—with any of the three Pauli matrices, or a linear combination of them. With each query the state is projected on a specific subspace and gives 1 bit of information, representing the choice of this subspace rather than the orthogonal one. But the most important difference to a normal bit shows up when several qubits are combined. Then there is the possibility of **entanglement**, which we will discuss in Sect. 2.10 on tensor products. Qubits offer new possibilities of information processing, which are studied in the relatively new field of **Quantum Information**.

In order to get more than 1 bit of information about a spin state, one needs to have a whole bunch of electrons demonstrably being in the same state. Then one can perform a measurement on each electron—sometimes in x-, sometimes in y-, sometimes in z-direction. Each time one gets 1 bit of information about the same state, information that sums up. One gets statistical distributions for each of the three spin operators, and from than one can more and more single out a state.

But how do you know that all electrons are in the same state, and not a statistical mixture of different states? The experimenter may claim that he has prepared all electrons in the same way, and so they have to be in the same state, but how can we verify that? The answer is: by looking at the statistical distributions.

Example: In a series of measurements we found that

$$p(z+) = p(z-) = p(x+) = p(x-) = \frac{1}{2}. \tag{2.95}$$

Let's assume this is the probability distribution for a single quantum state $|v\rangle$. Then we can conclude:

$$|\langle v|z+\rangle| = |\langle v|z-\rangle| = |\langle v|x+\rangle| = |\langle v|x-\rangle| \tag{2.96}$$

If $|v\rangle = \alpha|z+\rangle + \beta|z-\rangle$, we have

$$|\alpha| = |\beta| = \frac{1}{\sqrt{2}}|\alpha + \beta| = \frac{1}{\sqrt{2}}|\alpha - \beta|. \tag{2.97}$$

This is only possible if $\alpha = \pm i\beta$, that is to say if $|v\rangle$ is an eigenstate of S_y. So, if the measurements show that $p(y+) = 1$ or $p(y-) = 1$, then all electrons are in the same spin state, otherwise not. We will come back to the difference between pure states and statistical mixtures in Sect. 12.3.

The **expectation value** $\langle A\rangle_v$ for the measurement of an observable with operator A on a state $|v\rangle$ is defined as the average value one will get if one performs the same measurement many times on the same state (that is, on many quantum objects being in the same state). Mathematically, it is the sum of possible outcomes, multiplied by the corresponding probabilities,

$$\langle A \rangle_v = \sum_i p(\lambda_i)\lambda_i = \sum_i \lambda_i \langle v | P_{\lambda_i} | v \rangle. \tag{2.98}$$

With

$$A P_{\lambda_i} | v \rangle = \lambda_i P_{\lambda_i} | v \rangle \tag{2.99}$$

(since P_{λ_i} projects $|v\rangle$ onto the eigenspace of eigenvalue λ_i) and $\sum_i P_{\lambda_i} = \mathbf{1}$ follows

$$\langle A \rangle_v = \langle v | A | v \rangle. \tag{2.100}$$

The **standard deviation** or **uncertainty** $(\Delta A)_v$ of an observable with operator A in a state $|v\rangle$ is defined as the square root of the average quadratic deviation from the expectation value,

$$(\Delta A)_v = \sqrt{\langle (A - \langle A \rangle_v)^2 \rangle_v} = \sqrt{\langle A^2 \rangle_v - \langle A \rangle_v^2}. \tag{2.101}$$

To understand the second part of the equation, consider that $\langle A\langle A \rangle_v \rangle_v = \langle A \rangle_v^2$. If $|v\rangle$ is an eigenstate of A with eigenvalue λ, then λ is the only possible measurement value and therefore equals the expectation value, and so $(\Delta A)_v = 0$.

Self-check questions:

1. Why is it important that the operator associated with a measurement is hermitian?
2. What does the completeness relation say, and why does it have this name?
3. How do you calculate the expectation value and the standard deviation for a given operator and a given state?

2.6 Unitary Operators

With the hermitian operators and the projection operators (which are a subset of the former), we have met two specific types of operators playing an important role in QM. In this section we are going to deal with another type: the unitary operators. These are operators which do not change the scalar product between vectors. Unitary operators are relevant for the transformation between two orthonormal bases, as well as for the time evolution of quantum states according to the Schrödinger equation.

We will need the exponential of operators as a tool: The exponential e^A of an operator A is defined as the power series

$$e^A = \sum_{n=0}^{\infty} \frac{1}{n!} A^n. \tag{2.102}$$

As an example we calculate $U_y(\alpha) := e^{-i\alpha\sigma_y}$ with $\alpha \in \mathbb{R}$. The powers of σ_y take on only two different values, depending on whether the exponent is even or odd:

$$\sigma_y^{2n+1} = \sigma_y = \begin{pmatrix} 0 & -i \\ i & 0 \end{pmatrix}, \quad \sigma_y^{2n} = 1 = \begin{pmatrix} 1 & 0 \\ 0 & 1 \end{pmatrix} \tag{2.103}$$

From there we get to:

$$U_y(\alpha) = e^{-i\alpha\sigma_y} = \sum_{n=0}^{\infty} \frac{(-i)^n \alpha^n}{n!} \sigma_y^n \tag{2.104}$$

$$= \begin{pmatrix} \sum_{k=0}^{\infty} (-1)^k \frac{\alpha^{2k}}{(2k)!} & -\sum_{k=0}^{\infty} (-1)^k \frac{\alpha^{2k+1}}{(2k+1)!} \\ \sum_{k=0}^{\infty} (-1)^k \frac{\alpha^{2k+1}}{(2k+1)!} & \sum_{k=0}^{\infty} (-1)^k \frac{\alpha^{2k}}{(2k)!} \end{pmatrix} \tag{2.105}$$

$$= \begin{pmatrix} \cos\alpha & -\sin\alpha \\ \sin\alpha & \cos\alpha \end{pmatrix} \tag{2.106}$$

Exercise 2.10
Verify:

$$U_x(\alpha) = e^{-i\alpha\sigma_x} = \begin{pmatrix} \cos\alpha & -i\sin\alpha \\ -i\sin\alpha & \cos\alpha \end{pmatrix} \tag{2.107}$$

$$U_z(\alpha) = e^{-i\alpha\sigma_z} = \begin{pmatrix} e^{-i\alpha} & 0 \\ 0 & e^{i\alpha} \end{pmatrix} \tag{2.108}$$

The three matrices $U_x(\alpha)$, $U_y(\alpha)$, $U_z(\alpha)$ are **unitary**. A matrix/an operator U is called unitary if

$$U U^\dagger = U^\dagger U = 1. \tag{2.109}$$

Exercise 2.11
Verify that $U_x(\alpha)$, $U_y(\alpha)$, $U_z(\alpha)$ are unitary.

Unitary operators conserve the scalar product: Let $|u'\rangle = U|u\rangle$ and $|v'\rangle = U|v\rangle$. Then

$$\langle u'|v'\rangle = \langle u|U^\dagger U|v\rangle = \langle u|1|v\rangle = \langle u|v\rangle. \tag{2.110}$$

In particular, U maps an orthonormal basis $\{|e_i\rangle\}$ to a new orthonormal basis $\{|f_i\rangle\}$. On the other hand, for two given orthonormal bases $\{|e_i\rangle\}$ and $\{|f_i\rangle\}$, there is always a unitary operator U which maps $\{|e_i\rangle\}$ to $\{|f_i\rangle\}$. $U^\dagger$ then automatically maps $\{|f_i\rangle\}$ to $\{|e_i\rangle\}$:

$$|f_i\rangle = U|e_i\rangle \quad \Rightarrow \quad U^\dagger|f_i\rangle = U^\dagger U|e_i\rangle = |e_i\rangle \tag{2.111}$$

Unitary matrices are the complex generalizations of orthogonal matrices (rotation matrices). You may remember from Linear Algebra, that a matrix O is called orthogonal if

$$OO_t = O_tO = 1 \tag{2.112}$$

(where t stands for transposed), and that these are just the matrices inducing rotations and reflections in a real vector space. Among our three examples, $U_y(\alpha)$ is the only one with real entries. Indeed, we identify in it the rotation matrix in two dimensions, which rotates a cartesian coordinate system by the angle α (or, equivalently, the basis vectors $\mathbf{e}_1$, $\mathbf{e}_2$).

In our three examples, we obtained the unitary matrices as exponentials of hermitian matrices. This can be generalized: e^{iH} is unitary if H is hermitian, since

$$(e^{iH})^{\dagger}e^{iH} = e^{-iH^{\dagger}}e^{iH} = e^{-iH}e^{iH} = e^{-iH+iH} = e^{0} = 1. \tag{2.113}$$

Here we used that $e^{A+B} = e^{A}e^{B}$ if A and B commute, that is, if $AB = BA$. For then we can move the powers of A and B forth and back inside a product, just as for numbers. If A and B don't commute, however, one has in general $e^{A+B} \neq e^{A}e^{B}$!

We can apply unitary operators in two different ways:

- as an "active" operation on the vectors/states, $|u\rangle \to U|u\rangle$;
- as a "passive" coordinate transformation. This means one lets U act only on the basis vectors and creates a new basis in this way:

$$|e_i\rangle \to |f_i\rangle = U|e_i\rangle \tag{2.114}$$

The unchanged states $|u\rangle$ can now be expressed in terms of the new basis vectors. In components, the basis vector $|e_i\rangle$ has a 1 on the i-th position and otherwise only zeroes, $|e_i^{(e)}\rangle_j = \delta_{ij}$. From the definition (2.114) of the f-basis follows

$$|f_i^{(e)}\rangle_j = \sum_k U_{jk}|e_i^{(e)}\rangle_k = \sum_k U_{jk}\delta_{ik} = U_{ji}. \tag{2.115}$$

This implies a practical rule about the connection between the new basis vectors $\{|f_i\rangle\}$ and the transformation matrix U: If the $\{|f_i\rangle\}$ are given, one obtains U by taking the components of $|f_i\rangle$ as the i-th column of U. On the other hand, if U is given, one can read off the basis vector $|f_i\rangle$ from the i-th column of U.

One furthermore gets

$$|f_i\rangle = \sum_j |f_i^{(e)}\rangle_j |e_j\rangle = \sum_j U_{ji}|e_j\rangle \tag{2.116}$$

$$|u\rangle = \sum_i |u^{(e)}\rangle_i |e_i\rangle = \sum_i |u^{(f)}\rangle_j |f_j\rangle = \sum_{i,j} |u^{(f)}\rangle_j U_{ij}|e_i\rangle \tag{2.117}$$

and comparison of coefficients yields

$$|u^{(e)}\rangle_i = \sum_j U_{ij}|u^{(f)}\rangle_j, \qquad (2.118)$$

thus

$$|u^{(e)}\rangle = U|u^{(f)}\rangle, \quad |u^{(f)}\rangle = U^\dagger|u^{(e)}\rangle. \qquad (2.119)$$

Note that U acts here only as a matrix on the components of $|u\rangle$, in order to transform them into new coordinates. U does *not* act as an operator on the vector $|u\rangle$ itself. The norm of $|u\rangle$ cannot depend on the used basis. Therefore one must have

$$\langle u^{(e)}| = \langle u^{(f)}|U^\dagger, \quad \langle u^{(f)}| = \langle u^{(e)}|U \qquad (2.120)$$

so that

$$\langle u^{(f)}|u^{(f)}\rangle = \langle u^{(e)}|UU^\dagger|u^{(e)}\rangle = \langle u^{(e)}|u^{(e)}\rangle \qquad (2.121)$$

holds for any $|u\rangle$. The matrix representation of an operator A is also transformed when changing to a new basis. It must obey

$$A^{(f)} = U^\dagger A^{(e)}U, \qquad (2.122)$$

so that any scalar quantity $\langle u|A|v\rangle$ is independent of the chosen basis:

$$\langle u^{(f)}|A^{(f)}|u^{(f)}\rangle = \langle u^{(e)}|UU^\dagger A^{(e)}UU^\dagger|u^{(e)}\rangle = \langle u^{(e)}|A^{(e)}|u^{(e)}\rangle. \qquad (2.123)$$

U itself is not affected by the transformation:

$$U^{(f)} = U^{(e)\dagger}U^{(e)}U^{(e)} = U^{(e)}, \qquad (2.124)$$

hence, it is justified that we didn't use indicators (e) or (f) for U.

Let's have a look how $U_y(\alpha)$ acts on our spin vectors. For that, we summarize the results about the eigenvectors from the previous section:

$$|x+\rangle = \frac{1}{\sqrt{2}}\begin{pmatrix} 1 \\ 1 \end{pmatrix}, \quad |y+\rangle = \frac{1}{\sqrt{2}}\begin{pmatrix} 1 \\ i \end{pmatrix}, \quad |z+\rangle = \begin{pmatrix} 1 \\ 0 \end{pmatrix} \qquad (2.125)$$

$$|x-\rangle = \frac{1}{\sqrt{2}}\begin{pmatrix} 1 \\ -1 \end{pmatrix}, \quad |y-\rangle = \frac{1}{\sqrt{2}}\begin{pmatrix} 1 \\ -i \end{pmatrix}, \quad |z-\rangle = \begin{pmatrix} 0 \\ 1 \end{pmatrix} \qquad (2.126)$$

We start with the basis vectors

$$|e_1\rangle = |z+\rangle, \quad |e_2\rangle = |z-\rangle \qquad (2.127)$$

and investigate the effect of

$$U_y(\frac{\pi}{4}) = \frac{1}{\sqrt{2}} \begin{pmatrix} 1 & -1 \\ 1 & 1 \end{pmatrix}. \tag{2.128}$$

From the columns of this matrix, we can read off the new basis vectors

$$|f_1\rangle = U_y(\frac{\pi}{4})|z+\rangle = |x+\rangle, \quad |f_2\rangle = U_y(\frac{\pi}{4})|z-\rangle = -|x-\rangle. \tag{2.129}$$

Similarly we find

$$U_y(\frac{\pi}{4})|x+\rangle = |z-\rangle, \quad U_y(\frac{\pi}{4})|x-\rangle = |z+\rangle \tag{2.130}$$

and in general

$$U_y(\alpha)|y+\rangle = \frac{1}{\sqrt{2}} \begin{pmatrix} \cos\alpha - i\sin\alpha \\ \sin\alpha + i\cos\alpha \end{pmatrix} = e^{-i\alpha}|y+\rangle \tag{2.131}$$

$$U_y(\alpha)|y-\rangle = \frac{1}{\sqrt{2}} \begin{pmatrix} \cos\alpha + i\sin\alpha \\ \sin\alpha - i\cos\alpha \end{pmatrix} = e^{i\alpha}|y-\rangle. \tag{2.132}$$

Exercise 2.12
Verify all that!

We can see that $U_y(\frac{\pi}{4})$ rotates $|z+\rangle$ to $|x+\rangle$, and $|x+\rangle$ to $|z-\rangle$, while $|y+\rangle$ remains unchanged up to a phase. Our statement that the two-dimensional complex state vector encodes a three-dimensional real spin vector is thus confirmed. The two-dimensional U_y-rotation by the angle $\pi/4$ seems to correspond to a three-dimensional rotation about the y-axis by the angle $\pi/2$. The factor of 2 between the two rotation angles is noteworthy.

Exercise 2.13
Show that similar results hold for $U_x(\frac{\pi}{4})$ and $U_z(\frac{\pi}{4})$.

Increasing α up to $\alpha = 2\pi$, the $(|x+\rangle, |z+\rangle, |x-\rangle, |z-\rangle)$-"plane" is rotated twice in a complete circle

$$U_y(\frac{\pi}{2})|z+\rangle = |z-\rangle, \quad U_y(\frac{3\pi}{4})|z+\rangle = -|x-\rangle, \tag{2.133}$$

$$U_y(\pi)|z+\rangle = -|z+\rangle, \quad U_y(\frac{5\pi}{4})|z+\rangle = -|x+\rangle, \tag{2.134}$$

$$U_y(\frac{3\pi}{2})|z+\rangle = -|z-\rangle, \quad U_y(\frac{7\pi}{4})|z+\rangle = |x-\rangle. \tag{2.135}$$

Here, the signs in front of the ket vectors are irrelevant phases.

The transformation of the Pauli matrices confirms this relation between $\mathbb{C}^2$ and $\mathbb{R}^3$, more precisely between the unitary transformations in $\mathbb{C}^2$ and the rotations in $\mathbb{R}^3$. For example, one has

$$U_y^\dagger(\alpha)\sigma_z U_y(\alpha) = \cos(2\alpha)\,\sigma_z - \sin(2\alpha)\,\sigma_x. \tag{2.136}$$

Exercise 2.14
Verify this.

The set of 2×2-matrices resulting from all linear combinations

$$B = \beta_x\sigma_x + \beta_y\sigma_y + \beta_z\sigma_z \tag{2.137}$$

(with $\beta := (\beta_x, \beta_y, \beta_z) \in \mathbb{R}^3$) form a three-dimensional real vector space (meaning that the coefficients are real; the matrices don't have to be). That is, we construct an isomorphism (or equivalence) between the matrix B and the vector β and write $B \cong \beta$. For example,

$$\sigma_z \cong \begin{pmatrix} 0 \\ 0 \\ 1 \end{pmatrix}, \quad \cos(2\alpha)\,\sigma_z - \sin(2\alpha)\,\sigma_x \cong \begin{pmatrix} -\sin 2\alpha \\ 0 \\ \cos 2\alpha \end{pmatrix}. \tag{2.138}$$

With some calculation effort one can generalize (2.136) to:

$$\begin{aligned} U_x^\dagger(\alpha)BU_x(\alpha) &\cong R_x(2\alpha)\beta \\ U_y^\dagger(\alpha)BU_y(\alpha) &\cong R_y(2\alpha)\beta \\ U_z^\dagger(\alpha)BU_z(\alpha) &\cong R_z(2\alpha)\beta \end{aligned} \tag{2.139}$$

Here, $R_{x,y,z}$ are the rotation matrices in three dimensions:

$$R_x(\phi) = \begin{pmatrix} 1 & 0 & 0 \\ 0 & \cos\phi & \sin\phi \\ 0 & -\sin\phi & \cos\phi \end{pmatrix}, \quad R_y(\phi) = \begin{pmatrix} \cos\phi & 0 & -\sin\phi \\ 0 & 1 & 0 \\ \sin\phi & 0 & \cos\phi \end{pmatrix},$$

$$R_z(\phi) = \begin{pmatrix} \cos\phi & \sin\phi & 0 \\ -\sin\phi & \cos\phi & 0 \\ 0 & 0 & 1 \end{pmatrix} \tag{2.140}$$

The unitary transformation of the matrix B thus corresponds to a rotation of the vector β by the doubled angle. The Pauli matrices correspond to the basis vectors in $\mathbb{R}^3$. The interpretation of spin as a vector in $\mathbb{R}^3$ is based on this analogy. Now we are able to define the spin in an arbitrary direction. For example, to describe the measurement of a spin in the direction $\frac{1}{\sqrt{2}}(\mathbf{e}_x + \mathbf{e}_z)$ one uses the operator

$$\frac{\hbar}{2}\frac{1}{\sqrt{2}}(\sigma_x + \sigma_z) = \frac{\hbar}{2}\frac{1}{\sqrt{2}}\begin{pmatrix} 1 & 1 \\ 1 & -1 \end{pmatrix}. \tag{2.141}$$

In Sect. 9.3 we will discuss the deeper mathematical background of the relation (2.139). At that point we will also understand how spin is related to angular momentum.

Exercise 2.15

Any hermitian 2×2-matrix M can be written as

$$M = t\mathbf{1} + x\sigma_x + y\sigma_y + z\sigma_z \tag{2.142}$$

(cf. (2.49)). Show that

$$\det M = t^2 - x^2 - y^2 - z^2 \tag{2.143}$$

Self-check questions:

1. Why do unitary operators map orthonormal bases to orthonormal bases?
2. How are hermitian and unitary operators connected to each other?
3. How are unitary operators in two complex dimensions related to rotations in three real dimensions?

2.7 Time Evolution and Schrödinger Equation

So far we have developed the mathematical apparatus to analyze the state of a quantum system at a given moment in time. Now we want to concern ourselves with the time evolution of the system, which is given by the Schrödinger equation:

$$i\hbar\frac{d}{dt}|v(t)\rangle = H|v(t)\rangle \tag{2.144}$$

Here, H is the **Hamiltonian operator**. The Hamiltonian operator describes the energy of a system, i.e. it is the operator associated with the energy observable. In most cases it is the classical description of a system which is known at first, with a classical expression for the energy, E_{cl}. This classical description is then **quantized**, by substituting all classical variables appearing in E_{cl} with the associated operators. The resulting operator then is the Hamiltonian operator H. In the chapter on infinite-dimensional Hilbert spaces, we will start with the classical description, where E_{cl} is given as a function of spatial positions x_i and momentum variables p_i, the Hamiltonian function. The x_i and p_i are then replaced by operators. This is where the name Hamiltonian operator comes from.

> **Exercise 2.16**
> Let A be an operator for the classical variable (observable) a. Using an ortho-normal basis of eigenvectors, show that an observable b, given by the power series $\sum_n \alpha_n a^n$, is associated with the operator $B = \sum_n \alpha_n A^n$.

However, the replacement of classical variables with operators is not always unique. For example, let $E_{kl} = ab$ with two classical variables a and b, with corresponding hermitian operators A and B which do not commute, $AB \neq BA$. If we simply replace ab with AB, then H is not hermitian:

$$H^\dagger = (AB)^\dagger = B^\dagger A^\dagger = BA \neq AB \qquad (2.145)$$

For products of non-commuting observables, H must be symmetrized,

$$H = \frac{1}{2}(AB + BA), \qquad (2.146)$$

in order to get a hermtian result. But if the product contains higher powers, e.g. $E_{cl} = ab^2$, there will be several possibilities how to symmetrize, for example

$$H = \frac{1}{2}(AB^2 + B^2 A) \quad \text{or} \quad H = \frac{1}{3}(AB^2 + BAB + B^2 A). \qquad (2.147)$$

For higher powers appearing in the products, there will be more ambiguities. The classical description of a system contains less information than the quantum mechanical one. Therefore, the reconstruction of the latter from the former is not always unique. Fortunately, all the systems we are going to study in this book are free from such problems.

The example we are going to work with in this section is the time evolution of a spin in an external magnetic field. In this case, the classical energy is given by

$$E_{kl} = \alpha\, \mathbf{s} \cdot \mathbf{B} \qquad (2.148)$$

with a real coupling parameter α. This means that the energetically most favored orientation of the spin is antiparallel to the magnetic field (if $\alpha > 0$), the least favored one parallel to $\mathbf{B}$. From a classical perspective, there is a force which tries to orient the spin in the direction $-\mathbf{B}$. We say "classical", although the spin actually makes sense only in quantum mechanics. Classically we can understand the vector $\mathbf{s}$—up to a constant factor—as the magnetic moment of an electron, which explains the form of E_{cl}. The corresponding Hamiltonian operator reads

$$H = \frac{\alpha\hbar}{2}(B_x \sigma_x + B_y \sigma_y + B_z \sigma_z). \qquad (2.149)$$

If **B** depends on time, $H = H(t)$ does too. For the moment, we assume a time-independent **B** and therefore H. The analysis of systems with time-dependent H is much more complicated. We will say more about this at the end of this section.

The eigenvalues E_i of H are the possible energies of the system. The corresponding eigenvalue equation is the so-called

---

Stationary Schrödinger Equation

$$H|v\rangle = E|v\rangle, \tag{2.150}$$

---

also known as time-independent Schrödinger equation. The solutions for E are the eigenvalues E_i; the solutions for $|v\rangle$ are the eigenstates $|E_i\rangle$, which form an orthonormal basis of the Hilbert space (where the same eigenvalue can occur multiple times). If the time-dependent state $|v(t)\rangle$ is an eigenstate of H at the moment $t = 0$, $|v(0)\rangle = |E_i\rangle$, then (2.144) has a simple solution:

$$|v(t)\rangle = e^{-i\frac{E_i}{\hbar}t}|E_i\rangle \tag{2.151}$$

The state thus remains unchanged up to a rotating phase factor. Energy eigenstates are therefore **stationary states**: In all expectation values $\langle v(t)|A|v(t)\rangle$ and probabilities $\langle v(t)|P_\lambda|v(t)\rangle$, the time dependent phase factors in bra and ket vector cancel each other, such that all of these quantities are constant in time. This is a remarkable property. From classical mechanics we are used to the fact that a system with kinetic energy is in motion—that's exactly what the term kinetic energy means. In QM, this is no longer true: A system in a stationary state does not move at all, even if a large part of the energy E_i is kinetic. The solution (2.151) also implies the energy conservation law: $|E_i\rangle$ remains $|E_i\rangle$ for all times. This is, however, only true for a time-independent H. If H depends on time, its eigenvalues do too, and (2.151) is no longer valid. The system then exchanges energy with its environment.

If $|v(0)\rangle$ is not a stationary state, it is still a composition of several stationary states,

$$|v(0)\rangle = \sum_n |E_n\rangle\langle E_n|v(0)\rangle, \tag{2.152}$$

cf. (2.73). The Schrödinger equation is linear, that is, the components of the state vector appear only to first power. Therefore, the sum of several solutions is again a solution. In other words: each term of a linear combination evolves independent of the others. This is the **superposition principle**. Applied to (2.152) it implies:

$$|v(t)\rangle = \sum_n |E_n\rangle\langle E_n|v(0)\rangle e^{-i\frac{E_n}{\hbar}t} \tag{2.153}$$

Here, the component of $|v(0)\rangle$ in the direction of $|E_n\rangle$ oscillates as $e^{-i\frac{E_n}{\hbar}t}$. This is equivalent to saying that

$$|v(t)\rangle = e^{-\frac{i}{\hbar}Ht}|v(0)\rangle. \tag{2.154}$$

We can see that by expressing H in the basis $\{|E_i\rangle\}$, where H is diagonal with the values E_i on the diagonal,

$$H^{(E)} = \mathrm{diag}(E_1, E_2, ...) \quad \Rightarrow \quad \left(e^{-\frac{i}{\hbar}Ht}\right)^{(E)} = \mathrm{diag}(e^{-\frac{i}{\hbar}E_1 t}, e^{-\frac{i}{\hbar}E_2 t}, ...). \tag{2.155}$$

As an example we choose the Hamiltonian operator (2.149) with constant magnetic field in z-direction, $\mathbf{B} = B\mathbf{e}_z$. The eigenvalues are

$$E_1 = \frac{\alpha B\hbar}{2}, \quad E_2 = -\frac{\alpha B\hbar}{2}, \tag{2.156}$$

with eigenstates

$$|E_1\rangle = |z+\rangle, \quad |E_2\rangle = |z-\rangle. \tag{2.157}$$

The lowest energy eigenvalue E_2 is the so-called **zero-point energy**, $|E_2\rangle$ the so-called **ground state** of the system. Let $|v_1(0)\rangle = |E_1\rangle$, $|v_2(0)\rangle = |E_2\rangle$. Then

$$|v_1(t)\rangle = e^{-i\omega t}|E_1\rangle, \quad |v_2(t)\rangle = e^{i\omega t}|E_2\rangle, \quad \omega = \frac{\alpha B}{2}. \tag{2.158}$$

On the other hand, if

$$|v(0)\rangle = |x+\rangle = \frac{1}{\sqrt{2}}(|E_1\rangle + |E_2\rangle), \tag{2.159}$$

then

$$|v(t)\rangle = \frac{1}{\sqrt{2}}\begin{pmatrix} e^{-i\omega t} \\ e^{i\omega t} \end{pmatrix}. \tag{2.160}$$

One therefore has

$$|v(t)\rangle = e^{-i\omega t\sigma_z}|v(0)\rangle. \tag{2.161}$$

We already know from the previous section how the operator $e^{-i\omega t\sigma_z}$ acts on the vector $|x+\rangle$: it rotates $|x+\rangle$ to $|y+\rangle$, then to $|x-\rangle$, to $|y-\rangle$ and finally back to $|x+\rangle$ (in each case up to an irrelevant phase factor). So, the spin oscillates around the axis of the magnetic field, just as one would expect form a magnetic moment in the classical case. The speed of the oscillation is proportional to the magnetic field and to the coupling strength α.

The time evolution of a state between time t_0 and t is, according to (2.154), given by the action of the unitary operator

$$U(t, t_0) = e^{-\frac{i}{\hbar}(t-t_0)H}, \tag{2.162}$$

$$|v(t)\rangle = U(t, t_0)|v(t_0)\rangle. \tag{2.163}$$

Time evolution is thus a continuous unitary transformation of the state. In particular, the norm of the state does not change. A normalized state vector remains normalized. The operator $U(t, t_0)$ is called **time evolution operator** or **propagator**. It has the properties

$$U(t_2, t_0) = U(t_2, t_1)U(t_1, t_0), \qquad U(t_0, t_0) = 1, \tag{2.164}$$

which follow directly from (2.163).

The entire calculation becomes much more complicated if H depends on time, for then there are no time-independent eigenvalues and eigenstates. We can then take (2.163) as a starting point. That is, we define U by (2.163), where (2.162) is no longer valid. Plugging (2.163) into the Schrödinger equation we obtain a differential equation for U,

$$i\hbar \frac{d}{dt}U(t, t_0) = H(t)U(t, t_0), \tag{2.165}$$

with the initial condition $U(t_0, t_0) = 1$.

Here, the derivative of a continuous family $A(t)$ of operators is defined similarly to the derivative of functions:

$$\frac{d}{dt}A(t) = \lim_{\epsilon \to 0} \frac{A(t + \epsilon) - A(t)}{\epsilon} \tag{2.166}$$

Written as a matrix with respect to some basis, each matrix element can be differentiated separately,

$$\left[\frac{d}{dt}A(t)\right]_{ij}^{(e)} = \frac{d}{dt}\left[A(t)_{ij}^{(e)}\right]. \tag{2.167}$$

As the inversion of differentiation, the integral $\int dt\, A(t)$ of an operator is defined correspondingly. Again, in a matrix representation, each matrix element can be integrated separately.

If one is able to solve (2.165), one obtains with $U(t, t_0)$, due to (2.163), the time evolution for any initial state $|v(t_0)\rangle$. Were H and U scalar quantities, the solution of this differential equation would be

$$U(t, t_0) = \exp\left(-\frac{i}{\hbar}\int_{t_0}^{t} dt'\, H(t')\right). \tag{2.168}$$

However, H and U are operators. The problem is that the operators $H(t)$ at different times don't necessarily commute with each other. But the chain rule $\frac{d}{dt}e^{A(t)} = \dot{A}(t)e^{A(t)}$ applies only if A and $\dot{A}$ commute.

Exercise 2.17
Verify this, by differentiating the terms of the exponential power series with the product rule

$$\frac{d}{dt}(A(t)B(t)) = \left(\frac{d}{dt}A(t)\right)B(t) + A(t)\frac{d}{dt}B(t). \tag{2.169}$$

So, why is (2.168) not a solution of (2.165)?

In fact, the correct solution is

$$U(t, t_0) = T\left\{\exp\left(-\frac{i}{\hbar}\int_{t_0}^{t} dt' \, H(t')\right)\right\}, \tag{2.170}$$

where T is the **time ordering operator**, which sorts the operators of a product $A(t_1)A(t_2)\dots A(t_n)$ by decreasing times t_i. For example, the quadratic term of the exponential expansion in (2.170) is

$$-\frac{1}{2\hbar^2}\int_{t_0}^{t} dt_1 \int_{t_0}^{t} dt_2 \, H(t_1)H(t_2). \tag{2.171}$$

The time ordering operator turns this into

$$-\frac{1}{2\hbar^2}\left(\int_{t_0}^{t} dt_1 \int_{t_0}^{t_1} dt_2 \, H(t_1)H(t_2) + \int_{t_0}^{t} dt_1 \int_{t_1}^{t} dt_2 \, H(t_2)H(t_1)\right). \tag{2.172}$$

After thinking about it for a while, one finds that the two terms inside the brackets are identical.

Exercise 2.18
Verify this.

The quadratic term thus becomes

$$-\hbar^{-2}\int_{t_0}^{t} dt_1 \int_{t_0}^{t_1} dt_2 \, H(t_1)H(t_2). \tag{2.173}$$

With similar considerations, one finds the cubic term

$$i\hbar^{-3}\int_{t_0}^{t} dt_1 \int_{t_0}^{t_1} dt_2 \int_{t_0}^{t_2} dt_3 \, H(t_1)H(t_2)H(t_3) \tag{2.174}$$

etc. The term with the k-th power contains, after time ordering, $k!$ summands inside the bracket (cf. 2.172), since there are $k!$ possible orders of k different times. The summands are all identical, thereby canceling the factor of $k!$ in the denominator of the coefficients of the exponential series. Equation (2.170) is therefore equivalent to

$$
\begin{aligned}
U(t, t_0) = {}& 1 - i\hbar^{-1} \int_{t_0}^{t} dt_1\, H(t_1) \\
& - \hbar^{-2} \int_{t_0}^{t} dt_1 \int_{t_0}^{t_1} dt_2\, H(t_1)H(t_2) \\
& + i\hbar^{-3} \int_{t_0}^{t} dt_1 \int_{t_0}^{t_1} dt_2 \int_{t_0}^{t_2} dt_3\, H(t_1)H(t_2)H(t_3) + \cdots
\end{aligned}
\tag{2.175}
$$

We now want to show that (2.175) is indeed a solution of (2.165). For that purpose, we integrate (2.165), transforming it into an integral equation:

$$
i\hbar \int_{t_0}^{t} dt_1\, \frac{d}{dt_1} U(t_1, t_0) = \int_{t_0}^{t} dt_1\, H(t_1)U(t_1, t_0) \tag{2.176}
$$

$$
\Rightarrow i\hbar U(t_1, t_0)|_{t_1=t_0}^{t_1=t} = \int_{t_0}^{t} dt_1\, H(t_1)U(t_1, t_0) \tag{2.177}
$$

$$
\Rightarrow U(t, t_0) = 1 - \frac{i}{\hbar} \int_{t_0}^{t} dt_1\, H(t_1)U(t_1, t_0) \tag{2.178}
$$

This integral equation can be solved iteratively: As a zeroth approximation, we set $U^{(0)}(t, t_0) = 1$ on the right hand side, and obtain as a first approximation on the left hand side

$$
U^{(1)}(t, t_0) = 1 - i\hbar^{-1} \int_{t_0}^{t} dt_1\, H(t_1). \tag{2.179}
$$

Then we plug this first approximation back into the right hand side of (2.178), obtaining a second approximation on the left hand side,

$$
\begin{aligned}
U^{(2)}(t, t_0) = {}& 1 - i\hbar^{-1} \int_{t_0}^{t} dt_1\, H(t_1) \\
& - \hbar^{-2} \int_{t_0}^{t} dt_1 \int_{t_0}^{t_1} dt_2\, H(t_1)H(t_2),
\end{aligned}
\tag{2.180}
$$

etc. In this way, (2.175) is reproduced step by step.

Self-check questions:

1. How do you construct the Hamiltonian operator of a quantum system? What kind of difficulty can occur in this construction?
2. What are stationary states, and what properties do they have?
3. What is a time evolution operator, and what properties does it have?

2.8 Commutator and Uncertainty

In this section, it will be shown that two observables a and b can be measured simultaneously if and only if the corresponding operators A and B commute. On the way, we will introduce the important notion of the commutator and will find what a **complete set of commuting observables** is. Afterwards, we will derive Heisenberg's famous **Uncertainty Relation**. This relation gives a lower bound for the combined "uncertainty" of two observables, if the corresponding operators *don't* commute.

The **commutator** of two operators A and B is defined as

$$[A, B] = AB - BA. \tag{2.181}$$

The commutator of two hermitian operators is antihermitian, and thus not an observable:

$$[A, B]^\dagger = (AB)^\dagger - (BA)^\dagger = B^\dagger A^\dagger - A^\dagger B^\dagger = BA - AB = -[A, B] \tag{2.182}$$

Exercise 2.19
Show that

$$[\sigma_x, \sigma_y] = 2i\sigma_z, \quad [\sigma_y, \sigma_z] = 2i\sigma_x, \quad [\sigma_z, \sigma_x] = 2i\sigma_y \tag{2.183}$$

and hence

$$[S_x, S_y] = i\hbar S_z, \quad [S_y, S_z] = i\hbar S_x, \quad [S_z, S_x] = i\hbar S_y. \tag{2.184}$$

Exercise 2.20
Show that the commutator of two antihermitean operators is again antihermitian.

Exercise 2.21
Show that commutators obey the following rules:

$$[B, A] = -[A, B] \tag{2.185}$$
$$[A, B + C] = [A, B] + [A, C] \tag{2.186}$$
$$[A, BC] = [A, B]C + B[A, C] \tag{2.187}$$
$$[A, [B, C]] + [B, [C, A]] + [C, [A, B]] = 0 \tag{2.188}$$

The last one of these equations is the **Jacobi identity**.

Exercise 2.22

Show that $[A, B]$ can never be a multiple of the unit operator. For this, use the **trace** of the operators, which you may be still familiar with from your Linear Algebra course. The trace is defined as the sum of diagonal entries of a matrix representation

$$\text{tr}(A) = \sum_i A_{ii}. \qquad (2.189)$$

This property of an operator is, as one proves in Linear Algebra, independent of a choice of basis. One has

$$\text{tr}(AB) = \text{tr}(BA) \qquad (2.190)$$

(why?). Interestingly, the statement to be shown is valid only in finite-dimensional Hilbert spaces. In infinite-dimensional Hilbert spaces, the trace is not defined, as the sum (2.189) in general does not converge. Fortunately, for a large part of quantum mechanics is based on the fact that the commutator of position and momentum operators is a multiple of the unit operator, as we will see.

We now want to prove that two observables are simultaneously measurable if and only if the associated operators A and B commute, that is, if $[A, B] = 0$. At first we clarify for ourselves what it means to say that two observables are simultaneously measurable: A measurement is always accompanied by a projection, namely onto an eigenvector corresponding to the measured eigenvalue. If two observables are simultaneously measurable, the state vector after the measurement is an eigenvector of both A and B. This holds for any arbitrary initial state $|v\rangle$ (the simultaneous measurability should not be restricted to specific states, for at the moment of measurement one presumably doesn't know in what state the system is). This in turn implies that there must exist an orthonormal basis consisting only of vectors which are eigenvectors of both A and B. For any vector must be a linear combination of such eigenvectors, such that a common projection is possible. In this orthonormal basis, both A and B are diagonal, with the respective eigenvalues on the diagonal. Therefore, two observables are simultaneously measurable if and only if the associated operators can be diagonalized simultaneously. It thus remains to show:

Two operators can be diagonalized simultaneously if and only if they commute.

Proof ($\Rightarrow$):

Let $|e_i\rangle$ be a basis in which A and B are diagonal. In this basis one has

$$A^{(e)} = \text{diag}(a_1, a_2, \ldots, a_n) \qquad (2.191)$$
$$B^{(e)} = \text{diag}(b_1, b_2, \ldots, b_n), \qquad (2.192)$$

where we denote with diag(...) the entries of a purely diagonal matrix. The a_i and b_i are the eigenvalues of A and B, where the same eigenvalue can occur several times. Multiplying diagonal matrices is rather simple, namely

$$(AB)^{(e)} = \text{diag}(a_1 b_1, a_2 b_2, \ldots, a_n b_n) = (BA)^{(e)}, \qquad (2.193)$$

since we multiplied only eigenvalues, and in the multiplication of numbers, the order does not matter. As a consequence, A and B commute.

Proof ($\Leftarrow$):

We proceed as follows: at first, A is diagonalized (this is always possible for a hermitian operator). Then it is shown that B operates only inside the eigenspaces of A, i.e. that an eigenvector of A is not thrown out of its eigenspace by the operation of B. One can thus consider each eigenspace of A separately. Then B is diagonalized inside each of the eigenspaces of A, and it is shown that this does not break the diagonality of A.

So, let $[A, B] = 0$ and $|e_i\rangle$ an orthonormal basis in which A is diagonal, i.e. the $|e_i\rangle$ are eigenvectors of A with eigenvalues a_i, where again the same eigenvalue can occur several times. From $[A, B] = 0$ follows:

$$0 = \langle e_i|[A, B]|e_j\rangle = \langle e_i|AB|e_j\rangle - \langle e_i|BA|e_j\rangle \qquad (2.194)$$

$$= a_i \langle e_i|B|e_j\rangle - a_j \langle e_i|B|e_j\rangle = (a_i - a_j)\langle e_i|B|e_j\rangle \qquad (2.195)$$

(A acts once to the left, supplying the eigenvalue a_i, and once to the right, supplying the eigenvalue a_j). One can read the result in this way: If $a_i \neq a_j$, then $\langle e_i|B|e_j\rangle = 0$. That is, the vector $B|e_j\rangle$ has no component in any A-eigenspace belonging to an A-eigenvalue different from a_j. So, $B|e_j\rangle$, just as $|e_j\rangle$, is located in the A-eigenspace of the A-eigenvalue a_j. The eigenspace $\mathcal{H}_{a_j}$ is a subspace of the entire Hilbert space $\mathcal{H}$. Restricted to this subspace, A is a multiple of the unit operator,

$$A|_{\mathcal{H}_{a_j}} = a_j \mathbf{1}. \qquad (2.196)$$

Since each A-eigenspace is closed under the action of B,

$$|v\rangle \in \mathcal{H}_{a_j} \quad \Rightarrow \quad B|v\rangle \in \mathcal{H}_{a_j} \qquad (2.197)$$

(for this property is passed from the basis vectors $|e_{j_r}\rangle$ spanning the eigenspace $\mathcal{H}_{a_j}$ to their linear combinations), we can consider each A-eigenspace separately. We can diagonalize B within the subspace $\mathcal{H}_{a_j}$, since $B|_{\mathcal{H}_{a_j}}$ is hermitian (this property is passed from the entire space to the subspace). The unit matrix is not modified by basis transformations, it always remains the unit matrix. Therefore, the diagonality of $A|_{\mathcal{H}_{a_j}}$ is not destroyed by the diagonalization of $B|_{\mathcal{H}_{a_j}}$. We can do that in each A-eigenspace, and in the end we recombine the several subspace bases to an allover basis spanning the entire Hilbert space. In this basis now both A and B are diagonal.

Exercise 2.23
Verify that the matrices

$$A = \begin{pmatrix} 1 & 0 & 0 & 0 \\ 0 & 1 & 0 & 0 \\ 0 & 0 & -1 & 0 \\ 0 & 0 & 0 & -1 \end{pmatrix}, \quad B = \begin{pmatrix} 0 & 1 & 0 & 0 \\ 1 & 0 & 0 & 0 \\ 0 & 0 & 0 & 1 \\ 0 & 0 & 1 & 0 \end{pmatrix} \qquad (2.198)$$

commute, and determine an orthonormal basis of common eigenvectors.

Assume a measurement of a has resulted in the value a_i, but the dimension of the eigenspace $\mathcal{H}_{a_i}$ of this value is larger than 1. Then we still don't know in which state the system is after the measurement. But the measurement can be **refined** by an additional measurement of b (with associated operator B, having $[A, B] = 0$). If the b-measurement results in the value b_j, then we already know that the system after the measurement is in a state belonging to the intersection $\mathcal{H}_{a_i} \cap \mathcal{H}_{b_j}$, where $\mathcal{H}_{b_j}$ is the B-eigenspace of the eigenvalue b_j. It is conceivable that $\mathcal{H}_{a_i} \cap \mathcal{H}_{b_j}$ still has a dimension larger than 1. Then we need to find another operator C which commutes with both A and B, further refining the measurement. We can continue this game until we have a set $\{A, B, C, \ldots\}$ of operators, such that each intersection of eigenspaces of all these operators, $\mathcal{H}_{a_i} \cap \mathcal{H}_{b_j} \cap \cdots$, is only one-dimensional, and the measurement thus cannot be further refined. Then we have found a **complete set of commuting observables**. The result of a simultaneous measurement of all these observables uniquely determines the state after the measurement. The elements of the basis in which all operators $\{A, B, C, \ldots\}$ are diagonal are determined by the eigenvalues. If it is contextually clear which operators are meant, the basis vectors are denoted by the eigenvalues. For example, $|a_i b_j \ldots\rangle$ is the basis vector defined by having A-eigenvalue a_i and B-eigenvalue b_j etc. The values $\{a_i, b_j, \ldots\}$ are also called the **quantum numbers** of the system.

Exercise 2.24
Show that the operators A and B in (2.198) form a complete set of commuting observables.

Exercise 2.25
Show that

$$\mathbf{S}^2 := S_x^2 + S_y^2 + S_z^2 = \frac{3\hbar^2}{4}\mathbf{1}. \qquad (2.199)$$

Conclude that the squared norm and an arbitrary component of the spin are simultaneously measurable. The measurement of $\mathbf{s}^2$ is, however, rather boring:

It has only one possible result, $\frac{3\hbar^2}{4}$. The chosen component of the spin is already by itself a complete set of commuting observables.

Now we want to approach Heisenberg's Uncertainty Relation. For that we need one further object, the **anticommutator**, defined by

$$\{A, B\} = AB + BA, \qquad \cdot \qquad (2.200)$$

just like the commutator, but with a relative plus sign. The anticommutator of two hermitian operators A and B is hermitian:

$$\{A, B\}^\dagger = (AB)^\dagger + (BA)^\dagger = B^\dagger A^\dagger + A^\dagger B^\dagger = BA + AB = \{A, B\} \quad (2.201)$$

Exercise 2.26
Show that

$$\{\sigma_x, \sigma_y\} = \{\sigma_y, \sigma_z\} = \{\sigma_z, \sigma_x\} = 0. \qquad (2.202)$$

As a hermitian operator, $\{A, B\}$ has real eigenvalues. In contrast, $[A, B]$ as an antihermitian operator has purely imaginary eigenvalues. This can be proven in complete analogy with (2.56) and (2.57), only with an additional minus sign: let C be an antihermitian operator. Then

$$\langle v|C|v\rangle = \langle v|C^\dagger|v\rangle^* = -\langle v|C|v\rangle^*. \qquad (2.203)$$

For an eigenvector $|v\rangle$ with eigenvalue λ follows $\lambda = -\lambda^*$, and so λ is imaginary.

Now we have the prerequisites to derive the Uncertainty Relation. We remember (cf. (2.101)) that the uncertainty $(\Delta A)_v$ of an observable with operator A for a given state $|v\rangle$ is defined as the square root of the average quadratic deviation from the expectation value,

$$(\Delta A)_v = \sqrt{\langle v|(A - \langle A\rangle_v)^2|v\rangle}. \qquad (2.204)$$

With the definition

$$\tilde{A} = A - \langle A\rangle_v \qquad (2.205)$$

we can write $(\Delta A)_v$ as the norm of a vector ($\tilde{A}$ is like A hermitian):

$$(\Delta A)_v = \sqrt{\langle v|\tilde{A}^2|v\rangle} = \sqrt{\langle \tilde{A}v|\tilde{A}v\rangle} = ||\tilde{A}v|| \qquad (2.206)$$

The same considerations apply to a second observable B, and we write the product of the uncertainties as

$$(\Delta A)_v(\Delta B)_v = ||\tilde{A}v|| \, ||\tilde{B}v||. \qquad (2.207)$$

Now we apply the Schwarz inequality (2.16) to the right hand side and obtain

$$(\Delta A)_v (\Delta B)_v \geq |\langle \tilde{A}v | \tilde{B}v \rangle| = |\langle v | \tilde{A}\tilde{B} | v \rangle|. \tag{2.208}$$

As a next step, we realize that

$$AB = \frac{1}{2}(\{A, B\} + [A, B]) \tag{2.209}$$

which yields

$$(\Delta A)_v (\Delta B)_v \geq \frac{1}{2} |\langle v | \{\tilde{A}, \tilde{B}\} | v \rangle + \langle v | [\tilde{A}, \tilde{B}] | v \rangle|. \tag{2.210}$$

Since $\{\tilde{A}, \tilde{B}\}$ has only real, $[\tilde{A}, \tilde{B}]$ only imaginary eigenvalues, and since we can compose $|v\rangle$ of eigenvectors of each of these operators, the first expectation value on the right hand side is also real, the second imaginary. (More precisely, the second expression is not an expectation value, since $[\tilde{A}, \tilde{B}]$ as an antihermitian operator does not correspond to any observable. However, we can extend the notion of an expectation value to include any expression of the type $\langle v | C | v \rangle$.) The norm of a sum of a real and an imaginary value is given by the Pythagorean theorem, $|x + iy| = \sqrt{x^2 + y^2}$, thus

$$|\langle v | \{\tilde{A}, \tilde{B}\} | v \rangle + \langle v | [\tilde{A}, \tilde{B}] | v \rangle| = \sqrt{\langle v | \{\tilde{A}, \tilde{B}\} | v \rangle^2 + |\langle v | [\tilde{A}, \tilde{B}] | v \rangle|^2}. \tag{2.211}$$

Finally we find that $[\tilde{A}, \tilde{B}] = [A, B]$ (please verify!), and conclude

$$(\Delta A)_v (\Delta B)_v \geq \frac{1}{2} \sqrt{\langle v | \{\tilde{A}, \tilde{B}\} | v \rangle^2 + |\langle v | [A, B] | v \rangle|^2}. \tag{2.212}$$

And now we make things simple for us: we simply ignore the expression with the anticommutator. After leaving out this expression, the square root is at most as large as before,

$$\sqrt{\langle v | \{\tilde{A}, \tilde{B}\} | v \rangle^2 + |\langle v | [A, B] | v \rangle|^2} \geq |\langle v | [A, B] | v \rangle|, \tag{2.213}$$

and hence we have

Heisenberg's Uncertainty Relation

$$(\Delta A)_v (\Delta B)_v \geq \frac{1}{2} |\langle v | [A, B] | v \rangle|. \tag{2.214}$$

Why does one simply ignore the expression with the anticommutator? For several reasons:

- because it makes the inequality look ugly.
- because it vanishes in many cases, as for example for the Pauli matrices. Another example is the Gaussian wave packet we will get to know later on. While in this case the anticommutator $\{\tilde{X}, \tilde{P}\}$ does not vanish (X is the position operator, P the momentum operator), its expectation value does (cf. Sect. 3.5).
- because the expression with the commutator is easier to handle than the one with the anticommutator. In particular, for the anticommutator the expectation values of A and B need to be determined, but not for the commutator, since we could leave out the tilde there. So, one trades the better estimate (2.212) for a simpler usage.
- The Uncertainty Relation (2.214) in general depends on the state (this is important!), but with an important exception: If $[A, B]$ is a multiple of the unit operator (which is impossible for finite-dimensional Hilbert spaces, as we have seen in Exercise 2.22), $[A, B] = \lambda \mathbf{1}$, then the corresponding expectation value is always λ, independent of the state. The Uncertainty Relation—without the anticommutator term—becomes then independent of state as well. In fact, it was originally conceived for such cases. The most prominent example is given by the position and momentum operators, where $\lambda = i\hbar$, as we will see in Chap. 3.

As an example we use the spin again. We want to estimate $\Delta S_x \Delta S_y$ for two states with the Uncertainty Relation and compare the result with the actual values. One has

$$[S_x, S_y] = \frac{\hbar^2}{4}[\sigma_x, \sigma_y] = \frac{\hbar^2}{2}i\sigma_z = i\hbar S_z. \tag{2.215}$$

At first, let $|v\rangle = |z+\rangle$. Then the expectation value of S_z equals $\hbar/2$ and (2.214) yields, together with (2.215),

$$(\Delta S_x)_v(\Delta S_y)_v \geq \frac{\hbar^2}{4}. \tag{2.216}$$

The expectation values $\langle S_x \rangle_v$ and $\langle S_y \rangle_v$ are both equal to zero, for in the state $|z+\rangle$, the values $+\hbar/2$ and $-\hbar/2$ are equally likely for the spin in x- or y-direction. Hence

$$\langle (S_x - \langle S_x \rangle_v)^2 \rangle_v = \langle S_x^2 \rangle_v = \frac{\hbar^2}{4} \tag{2.217}$$

and thus $(\Delta S_x)_v = \hbar/2$. The same holds for $(\Delta S_y)_v$ and therefore

$$(\Delta S_x)_v(\Delta S_y)_v = \frac{\hbar^2}{4}. \tag{2.218}$$

The estimate of the Uncertainty Relation is thus perfect in this case.

Next we set $|v\rangle = |x+\rangle$. Now the expectation value of S_z vanishes, and the Uncertainty Relation does not yield much:

$$(\Delta S_x)_v(\Delta S_y)_v \geq 0 \qquad\qquad (2.219)$$

However, this is justified, for $(\Delta S_x)_v$ vanishes indeed, since $|v\rangle$ is "sharp" with respect to S_x; the expectation value $\langle S_x\rangle_v$ is $\hbar/2$, and the deviation from it is zero, because $\hbar/2$ is the only possible measurement value for S_x. Thus

$$(\Delta S_x)_v(\Delta S_y)_v = 0, \qquad\qquad (2.220)$$

and again the Uncertainty Relation has provided the exact value.

Some more remarks about the Uncertainty Relation: In the early years of QM, it was considered its quintessence, its most central theorem. Many old textbooks take it as a starting point. As its source, many physicists thought that the measurement of a would cause a "recoil" on the system, thereby "washing out" the value of b, and only thereby making it uncertain. The more precise the measurement of a should be, the stronger the recoil occurring in the measurement, and the more uncertain the value of b. Today, with the results from the violation of Bell's inequalities, we know that this interpretation is only partially correct. In fact, already before the measurement, the state of the system does not allow for a "classical" value for the observables a and b to exist. So, the uncertainty is not to be ascribed to the recoil (although this contributes to a further "uncertification").

Today, the fascination about the Uncertainty Relation has somewhat calmed down. The center of amazement is now the phenomenon of entanglement, on which we got a little preview in the Introduction, and into which we will delve more in Sect. 2.10 on tensor products. In the early years of QM, almost nothing was known about entanglement; in particular, the technical prerequisites for experimenting with it were not yet given. This has changed in the last few decades.

Self-check questions:

1. Under what conditions can two operators be simultaneously diagonalized?
2. What is a complete set of commuting observables?
3. What does Heisenberg's Uncertainty Relation say?

2.9 Schrödinger Picture and Heisenberg Picture

The time dependence of probabilities in QM can be formulated in different ways. So far we have been operating in the so-called **Schrödinger picture**, where the time evolution of a state is determined by the Schrödinger equation, and the operators corresponding to the measured observables are in general time-independent. An exception was the time-dependent Hamiltonian operator $H(t)$ we discussed in

Sect. 2.7. Such time dependencies of operators occur when the observable to be measured is actively modified from outside. For example, we can modify the energy of an electron by switching on or off an external electric field, or a magnetic field which has an influence on the spin. In this case the Hamiltonian operator becomes time-dependent. An operator which is time-dependent in the Schrödinger picture is called **explicitly time-dependent**, in order to distinguish this kind of time dependence from the kind we will meet in the Heisenberg picture.

To get from the Schrödinger picture into the **Heisenberg picture**, we transform the state $|v_S(t)\rangle$ (in this section we supply a state of the Schrödinger picture with the subscript S) at any moment in time in such a way that it is identical to the state at some arbitrary but fixed moment t_0:

$$|v_H\rangle := U^\dagger(t, t_0)|v_S(t)\rangle = U^\dagger(t, t_0)U(t, t_0)|v_S(0)\rangle = |v_S(t_0)\rangle \qquad (2.221)$$

Here, $U(t, t_0)$ is the time evolution operator from the Schrödinger picture. The transformation with $U^\dagger(t, t_0)$ compensates for the time evolution from the Schrödinger picture. The Heisenberg state $|v_H\rangle$ does therefore not depend on time! In order to make the predictions for measurement statistics agree with those from the Schrödinger picture, one has to make the operators time-dependent instead,

$$A_H(t) = U^\dagger(t, t_0)A_S U(t, t_0). \qquad (2.222)$$

Now all scalar products are just as before,

$$\langle u_H|A_H|v_H\rangle = \langle u_S|A_S|v_S\rangle, \qquad (2.223)$$

similar to (2.122) and the reasoning that followed it.

One can understand the switch to the Heisenberg picture in several ways (compare with the discussion of unitary transformations as active or passive transformations in Sect. 2.6):

- as a passive basis transformation. One chooses a time-dependent basis which rotates with the state $|v_S\rangle$ from the Schrödinger picture. The state then still rotates in the Heisenberg picture, but its component representation w.r.t. the new time-dependent basis remains constant. The subscript H is then to be understood similar to the basis superscript (f) after a transformation into a basis $\{|f_i\rangle\}$. Similarly, (2.223) is the transformation of the components of A into the new time-dependent basis.
- as an active transformation of the state. The basis remains the same, but the state $|v_H\rangle$ is kept constant by the transformation. It is thus a state different from $|v_S\rangle$, and therefore the operators must be also different (in particular time-dependent), in order to get the same predictions.
- as a change of vector identification through time. That is, we change the mathematical description in the sense that we no longer speak of a fixed Hilbert space $\mathcal{H}$, but of a family of Hilbert spaces $\mathcal{H}_t$, one for each moment in time t. The Hilbert

spaces at different times are copies of each other. But since the vectors of a Hilbert space are something abstract and apart from scalar products have no inner structure, there is a certain freedom which vector of $\mathcal{H}_t$ to consider a "copy" of a certain vector $|v\rangle$ in $\mathcal{H}_{t_0}$. The choice of how we assign copies in $\mathcal{H}_t$ to a vector $|v\rangle \in \mathcal{H}_{t_0}$ for all times t, is what we call vector identification through time. Schrödinger and Heisenberg picture represent two such choices. We have some $|v(t_0)\rangle \in \mathcal{H}_{t_0}$. The copy of $|v(t_0)\rangle$ in $\mathcal{H}_t$ is denoted $|v_t(t_0)\rangle$. In the Schrödinger picture, $|v_t(t_0)\rangle$ is chosen such that $|v(t)\rangle = U(t, t_0)|v_t(t_0)\rangle$. In the Heisenberg picture, $|v_t(t_0)\rangle$ is chosen such that $|v(t)\rangle = |v_t(t_0)\rangle$.

All three interpretations of the Heisenberg picture are equally justified. In each case, the switch to the Heisenberg picture constitutes an interesting change of perspective. In the Schrödinger picture, the observed system evolves with time, and the surrounding world of the observer, who performs measurements on the system using some operators, remains (relatively) constant. In the Heisenberg picture, the state of the system remains unchanged (in the quantum system time does not exist, so to speak!), and the entire time-dependence lies in the perspective of the observer (like someone sitting in a train, watching the landscape through the window). This perspective changes permanently, since the operators he can use for measurement according to (2.223) permanently "transform through time".

As an example, consider the spin rotating in a magnetic field from Sect. 2.7. In the Schrödinger picture, the state of the spin rotates in the (xy)-plane, i.e. between the states $|x+\rangle$, $|y+\rangle$, $|x-\rangle$ and $|y-\rangle$. In the Heisenberg picture, the state remains still, say, at $|x+\rangle$, which is the Schrödinger picture state at $t = 0$. Instead, the operators which are used to measure the spin in x- or y-direction are changing. $(S_x)_H$ is not just given by $\frac{\hbar}{2}\sigma_x$. Instead of σ_x we have a time-dependent matrix, passing through σ_x, $-\sigma_y$, $-\sigma_x$ and σ_y successively. A similar statement holds for $(S_y)_H$.

Exercise 2.27

Compute $(S_x)_H$ and $(S_y)_H$ as functions of time. Use (2.222) and the results from Sect. 2.7. Compute the expectation value $\langle s_x(t)\rangle_v$ of the spin in x-direction at time t in both Schrödinger and Heisenberg picture. The two results must be identical.

In Newtonian mechanics there are independent criteria to determine whether the observer or the observed system rotates. Does the sun orbit around the earth (and the observer sitting on it) once a day? Or is it the earth which rotates and thereby changes the perspective of the observer, while the sun stands still? When I have a small rotating gyroscope in front of me, is it then really the gyroscope that rotates? Or is it me, the observer, who rotates with the entire world around the gyroscope? In Newtonian mechanics, this can be clearly decided by means of centrifugal and Coriolis forces. In the first case, it is the earth with its resident observers, which rotates, in the second case it is the gyroscope, while the observer is at rest. In QM, there are no such centrifugal forces to distinguish between Schrödinger and Heisenberg picture. On the other hand, in QM the observed systems are in most cases very small, with rotations fast as with a gyroscope (even much faster). Therefore the Schrödinger-Bild appears more correct intuitively. Accordingly, it is the one which is used more often.

In Sect. 11.2 about time-dependent perturbation theory we will meet another picture which is located half-way between Schrödinger- and Heisenberg picture: the Dirac picture.

Some nice properties of QM are easier to be proved in the Heisenberg picture. Let's compute the time derivative of an operator A_H in the Heisenberg picture, using (2.165) and the hermitian conjugate of this equation:

$$i\hbar \frac{d}{dt} A_H(t) = i\hbar \frac{dU^\dagger}{dt} A_S U + i\hbar U^\dagger \frac{dA_S}{dt} U + i\hbar U^\dagger A_S \frac{dU}{dt} \qquad (2.224)$$

$$= -U^\dagger H_S A_S U + i\hbar U^\dagger \frac{dA_S}{dt} U + U^\dagger A_S H_S U \qquad (2.225)$$

$$= U^\dagger [A_S, H_S] U + i\hbar U^\dagger \frac{dA_S}{dt} U \qquad (2.226)$$

Here, dA_S/dt is a possible explicit time dependence of the operator in the Schrödinger picture. The first term can be written as:

$$U^\dagger [A_S, H_S] U = U^\dagger A_S H_S U - U^\dagger H_S A_S U \qquad (2.227)$$

$$= U^\dagger A_S U U^\dagger H_S U - U^\dagger H_S U U^\dagger A_S U \qquad (2.228)$$

$$= A_H H_H - H_H A_H = [A_H, H_H] \qquad (2.229)$$

Additionally, one uses the notation

$$\frac{\partial A_H}{\partial t} := U^\dagger \frac{dA_S}{dt} U. \qquad (2.230)$$

This results in the

Heisenberg Equation

$$i\hbar \frac{d}{dt} A_H(t) = [A_H, H_H] + i\hbar \frac{\partial A_H}{\partial t}. \qquad (2.231)$$

It is for the Heisenberg picture what the Schrödinger equation is for the Schrödinger picture.

Scalar products and expectation values are independent of in which picture they were determined,

$$\langle A \rangle_v = \langle v_S | A_S | v_S \rangle = \langle v_H | A_H | v_H \rangle. \qquad (2.232)$$

The time dependence of an expectation value can be calculated more easily in the Heisenberg picture, since the state is time-independent there. Let's assume that A

does not explicitly depend on time. Then the Heisenberg equation yields

$$\frac{d\langle A\rangle_v}{dt} = \frac{d}{dt}\langle v_H | A_H | v_H\rangle = \langle v_H | \frac{dA_H}{dt} | v_H\rangle \qquad (2.233)$$

$$= \frac{1}{i\hbar}\langle v_H | [A_H, H_H] | v_H\rangle = \frac{1}{i\hbar}\langle [A, H]\rangle_v. \qquad (2.234)$$

In the last step we could drop the subscript H, since the expectation value is picture independent. In summary: the expectation value of a not explicitly time-dependent operator A obeys the

---

Ehrenfest Theorem

$$\frac{d\langle A\rangle_v}{dt} = \frac{1}{i\hbar}\langle [A, H]\rangle_v. \qquad (2.235)$$

As a consequence, the expectation value does not change if A commutes with the Hamiltonian operator. An observable commuting with H is therefore a **conserved quantity**.

If we consider again the Hamiltonian operator responsible for the spin rotation,

$$H = \frac{\alpha\hbar B}{2}\sigma_z, \qquad (2.236)$$

we see that it commutes with S_z. The spin in z-direction is thus a conserved quantity in this system. Only the spin components in the (xy)-plane rotate.

A consequence of the Ehrenfest theorem is the so-called **Energy-Time Uncertainty Relation**. While this relation looks like an example of Heisenberg's Uncertainty Relation (2.212), it is formally something different, and also regarding its interpretation. We start with Heisenberg's Uncertainty Relation, applied to A and H (both again not explicitly time-dependent):

$$(\Delta A)_v(\Delta E)_v \geq \frac{1}{2}|\langle [A, H]\rangle_v| \qquad (2.237)$$

(the uncertainty of H is always denoted ΔE instead of ΔH), and with (2.235) follows:

$$(\Delta A)_v(\Delta E)_v \geq \frac{\hbar}{2}\frac{d\langle A\rangle_v}{dt} \qquad (2.238)$$

Now we define

$$(\Delta\tau)_v = \frac{(\Delta A)_v}{\frac{d\langle A\rangle_v}{dt}}. \qquad (2.239)$$

This is, to first approximation, the time needed by the expectation value of A to change by the amount $(\Delta A)_v$. For a given moment t, the measured value of A with a relatively large probability lies in the interval $[\langle A \rangle_v - (\Delta A)_v, \langle A \rangle_v + (\Delta A)_v]$ (all quantities evaluated at t). At the time $t + (\Delta \tau)_v$, the interval has moved forward by half its width; so, around this time there is a significant change of the expected measurement values. With the definition (2.239), (2.238) yields

$$(\Delta \tau)_v (\Delta E)_v \geq \frac{\hbar}{2}, \tag{2.240}$$

which is the **Energy-Time Uncertainty Relation**. It can be interpreted in the following way: the more sharply the energy of a system is determined, the slower is the change of the expectation values of its observables. In an energy eigenstate, $(\Delta E)_v = 0$, the system is stationary (as we've seen in Sect. 2.7), and nothing ever changes in it; $(\Delta \tau)_v$ is not well-defined and (2.240) is invalid. On the other hand, the faster a system changes, the broader its state must be regarding its contained energy components.

A rather adventurous, but often expressed interpretation is that the system is allowed to violate the classical energy conservation for a time $(\Delta \tau)_v$ by a maximal amount of $(\Delta E)_v = \frac{\hbar}{2}(\Delta \tau)_v$. That is, one imagines that the system "borrows" the amount $(\Delta E)_v$ and has to return it after the loan period $(\Delta \tau)_v$. This interpretation should be better not taken too literally.

Self-check questions:

1. What is the difference between the Schrödinger and the Heisenberg picture? What ensures that all predictions are the same in both pictures?
2. What does the Ehrenfest theorem say? When is an observable a conserved quantity?
3. What does the Energy-Time Uncertainty Relation say?

2.10 Tensor Products

In classical mechanics, the motion of a particle can be described via a trajectory in six-dimensional phase space (three dimensions for position, three for momentum). If one adds another particle, the dimension of the phase space is increased by 6, since for an additional particle three further position and momentum coordinates are added. For n particles, phase space has $6n$ dimensions. In other words: If two systems S_1 and S_2 are combined into one system S, where S_1 is described by a d_1-dimensional phase space P_1, S_2 by a d_2-dimensional phase space P_2, then the phase space P of S has $d = d_1 + d_2$ dimensions. For P is the **direct sum** of P_1 and P_2,

$$P = P_1 \oplus P_2. \tag{2.241}$$

This means: if P_1 has a basis $B_1 = \{e_1^{(1)}, \ldots, e_r^{(1)}\}$ and P_2 a basis $B_2 = \{e_1^{(2)}, \ldots, e_s^{(2)}\}$, then

$$B = B_1 \cup B_2 = \{e_1^{(1)}, \ldots e_r^{(1)}, e_1^{(2)}, \ldots, e_s^{(2)}\} \tag{2.242}$$

is a basis of P.

The expression "direct sum" is used for the spaces as well as for the corresponding vectors and operators. The direct sum $u \oplus v \in P$ of $u \in P_1$ and $v \in P_2$ is in components defined as

$$u \oplus v = (u, v) = (u_1, \ldots, u_r, v_1, \ldots, v_s). \tag{2.243}$$

The direct sum $A \oplus B$ of a linear operator A on P_1 and a linear operator B on P_2 is (in matrix form, with respect to the above mentioned basis)

$$A \oplus B = \begin{pmatrix} A & 0 \\ 0 & B \end{pmatrix}, \tag{2.244}$$

such that

$$(A \oplus B)(u \oplus v) = Au \oplus Bv. \tag{2.245}$$

In quantum mechanics, things are different. If a quantum object is described by a d-dimensional Hilbert space, then the Hilbert space for n such objects is d^n-dimensional. If two quantum systems S_1 and S_2 are combined into one system S, where S_1 is described by a d_1-dimensional Hilbert space $\mathcal{H}_1$, S_2 by a d_2-dimensional Hilbert space $\mathcal{H}_2$, then the Hilbert space $\mathcal{H}$ of S has $d = d_1 d_2$ dimensions. For $\mathcal{H}$ is the **tensor product** of $\mathcal{H}_1$ and $\mathcal{H}_2$,

$$\mathcal{H} = \mathcal{H}_1 \otimes \mathcal{H}_2. \tag{2.246}$$

This means: if $\mathcal{H}_1$ has a basis $B_1 = \{e_1^{(1)}, \ldots, e_r^{(1)}\}$ and $\mathcal{H}_2$ a basis $B_2 = \{e_1^{(2)}, \ldots, e_s^{(2)}\}$ (for the sake of clarity we deviate from the bra/ket notation for a moment), then

$$B = \{e_i^{(1)} \otimes e_j^{(2)} \mid i = 1, \ldots, r; \; j = 1, \ldots, s\} \tag{2.247}$$

is a basis of $\mathcal{H}$. Each basis vector of $\mathcal{H}$ is thus a combination of a basis vector $e_i^{(1)}$ of $\mathcal{H}_1$ and a basis vector $e_j^{(2)}$ of $\mathcal{H}_2$, where the combination is expressed via the tensor symbol $\otimes$.

The tensor product $u \otimes v$ of two vectors

$$u = \sum_{i=1}^{r} u_i e_i^{(1)} \quad \in \mathcal{H}_1 \tag{2.248}$$

$$\mathbf{v} = \sum_{j=1}^{s} v_j \mathbf{e}_j^{(2)} \quad \in \mathcal{H}_2 \tag{2.249}$$

is defined as

$$\mathbf{u} \otimes \mathbf{v} = \sum_{i=1}^{r} \sum_{j=1}^{s} u_i v_j \, \mathbf{e}_i^{(1)} \otimes \mathbf{e}_j^{(2)} \tag{2.250}$$

and the distributive property holds,

$$\mathbf{u} \otimes (\alpha \mathbf{v} + \beta \mathbf{w}) = \alpha \mathbf{u} \otimes \mathbf{v} + \beta \mathbf{u} \otimes \mathbf{w}. \tag{2.251}$$

Here, an essential difference between QM and classical mechanics becomes evident. In a direct sum of vector spaces, $V = V_1 \oplus V_2$, *each* vector $\mathbf{w} \in V$ can be written as a direct sum of vectors $\mathbf{u} \in V_1$ and $\mathbf{v} \in V_2$: With respect to the basis (2.242), the first r components of $\mathbf{w}$ form the vector $\mathbf{u}$, the remaining s components the vector $\mathbf{v}$. In a tensor product of vector spaces, however, only *some* vectors $\mathbf{w} \in V$ can be written as a tensor product of vectors $\mathbf{u} \in V_1$ and $\mathbf{v} \in V_2$. For example,

$$\mathbf{w} = \mathbf{e}_1^{(1)} \otimes \mathbf{e}_1^{(2)} + \mathbf{e}_1^{(1)} \otimes \mathbf{e}_2^{(2)} - \mathbf{e}_2^{(1)} \otimes \mathbf{e}_1^{(2)} - \mathbf{e}_2^{(1)} \otimes \mathbf{e}_2^{(2)} \tag{2.252}$$

can be written as $\mathbf{u} \otimes \mathbf{v}$ with

$$\mathbf{u} = \mathbf{e}_1^{(1)} - \mathbf{e}_2^{(1)}, \quad \mathbf{v} = \mathbf{e}_1^{(2)} + \mathbf{e}_2^{(2)}. \tag{2.253}$$

For

$$\mathbf{w} = \mathbf{e}_1^{(1)} \otimes \mathbf{e}_1^{(2)} - \mathbf{e}_2^{(1)} \otimes \mathbf{e}_2^{(2)} \tag{2.254}$$

however, there is no such representation.

Exercise 2.28
Verify this.

This implies that the phase space trajectories of the individual particles in classical mechanics can always be described independently. At any given moment, the phase space P can be decomposed into P_1 and P_2, and one can say that S_1 is now in the state $\mathbf{u} \in P_1$, S_2 in the state $\mathbf{v} \in P_2$. In the equations determining the trajectories, the systems may be coupled to each other. But in the description of the momentary state, they can always be considered separately.

In QM, this works only if $\mathbf{w} = \mathbf{u} \otimes \mathbf{v}$. In all other cases, as for example in (2.254), we have an **entangled state**: The state of system S_1 cannot be described independently of the state of S_2. This leads to the correlated probabilities and "spooky action at a distance" we've talked about in the introductory chapter on Bell's inequality.

In a moment, we will demonstrate this again, after we've discussed operators and measurements on a tensor product $\mathcal{H} = \mathcal{H}_1 \otimes \mathcal{H}_2$.

The tensor product $A \otimes B$ of a linear operator A on $\mathcal{H}_1$ and a linear operator B on $\mathcal{H}_2$ is defined as

$$(A \otimes B)(\mathbf{u} \otimes \mathbf{v}) = A\mathbf{u} \otimes B\mathbf{v} \tag{2.255}$$

for all $\mathbf{u} \in \mathcal{H}_1$ and $\mathbf{v} \in \mathcal{H}_2$. In a matrix representation w.r.t a basis (2.247) one then has

$$(A \otimes B)_{(ij)(kl)} = A_{ik}B_{jl}. \tag{2.256}$$

Here we have used double indices: (ij) identifies a row corresponding to the basis vector $\mathbf{e}_i^{(1)} \otimes \mathbf{e}_j^{(2)}$, and similarly for the column (kl). The components of $\mathbf{w} = \mathbf{u} \otimes \mathbf{v}$ are

$$w_{ij} = u_i v_j \tag{2.257}$$

and it follows

$$[(A \otimes B)\mathbf{w}]_{(ij)} = \sum_{k=1}^{r} \sum_{l=1}^{s} (A \otimes B)_{(ij)(kl)} w_{kl} \tag{2.258}$$

$$= \sum_{k=1}^{r} \sum_{l=1}^{s} A_{ik}B_{jl}u_k v_l \tag{2.259}$$

$$= (A\mathbf{u})_i (B\mathbf{v})_j \tag{2.260}$$

$$= [(A\mathbf{u}) \otimes (B\mathbf{v})]_{(ij)}, \tag{2.261}$$

as required by (2.255).

As an example we consider a system of two spins/qubits. The corresponding Hilbert space is four-dimensional. We evaluate the tensor product of two vectors

$$|x+\rangle \otimes |x-\rangle = \frac{1}{2}\begin{pmatrix} 1 \\ 1 \end{pmatrix} \otimes \begin{pmatrix} 1 \\ -1 \end{pmatrix} = \frac{1}{2}\begin{pmatrix} 1 \\ -1 \\ 1 \\ -1 \end{pmatrix} \tag{2.262}$$

(where we write the components in the order $(z+, z+)$, $(z+, z-)$, $(z-, z+)$, $(z-, z-)$), and of two Pauli matrices:

$$\sigma_z \otimes \sigma_y = \begin{pmatrix} 1 & 0 \\ 0 & -1 \end{pmatrix} \otimes \begin{pmatrix} 0 & -i \\ i & 0 \end{pmatrix} = \begin{pmatrix} 0 & -i & 0 & 0 \\ i & 0 & 0 & 0 \\ 0 & 0 & 0 & i \\ 0 & 0 & -i & 0 \end{pmatrix} \tag{2.263}$$

Exercise 2.29
Evaluate $\sigma_x \otimes \sigma_x$, $\sigma_x \otimes \sigma_y$, $\sigma_y \otimes \sigma_x$ and $\sigma_y \otimes \sigma_y$.

If a measurement is performed on only one of the systems S_1 or S_2, the associated hermitian operator $A^{(1)}$ or $B^{(2)}$ acts only on the corresponding Hilbert space $\mathcal{H}_1$ or $\mathcal{H}_2$. The other system is not affected by the operator. $A^{(1)}$ and $B^{(2)}$ thus have the form

$$A^{(1)} = A \otimes 1 \quad \text{and} \quad B^{(2)} = 1 \otimes B. \tag{2.264}$$

We are using the same symbol (in this case A and B) for the operator in the combined system as for the operator in the subsystem, only adding a superscript for the operator in the combined system which denotes to which subsystem the operator refers. For example, if we want to measure the z-component $s_z^{(1)}$ of the first spin in the two-spin system, the associated operator is

$$S_z^{(1)} = S_z \otimes 1 = \frac{\hbar}{2} \sigma_z \otimes 1. \tag{2.265}$$

Exercise 2.30
Show, using (2.255):

$$[A^{(1)}, B^{(2)}] = 0, \tag{2.266}$$

i.e. observables referring to different subsystems are simultaneously measurable.

Another combination which occurs quite often is

$$C = A^{(1)} + B^{(2)} = A \otimes 1 + 1 \otimes B, \tag{2.267}$$

that is, we have the sum of an operator operating only on $\mathcal{H}_1$ and one operating only on $\mathcal{H}_2$. For example, the total spin in z-direction, s_z, is given by the sum of the individual spins:

$$S_z^{(tot)} = S_z^{(1)} + S_z^{(2)} = S_z \otimes 1 + 1 \otimes S_z \tag{2.268}$$

Another example is the Hamiltonian operator H of the total system. If the systems don't interact with each other, then H is the sum of the Hamiltonian operators H_1 and H_2 of the subsystems,

$$H = H_1^{(1)} + H_2^{(2)} = H_1 \otimes 1 + 1 \otimes H_2. \tag{2.269}$$

From now on we use the bra/ket notation again. We write $|u^{(1)}v^{(2)}\rangle$ or simply $|uv\rangle$ for $|u\rangle \otimes |v\rangle = \mathbf{u} \otimes \mathbf{v}$, and $\langle u^{(1)}v^{(2)}|$ or simply $\langle uv|$ for the associated vector in the dual space. For the basis vectors $\mathbf{e}_i^{(1)} \otimes \mathbf{e}_j^{(2)}$ we write $|ij\rangle$, and $\langle ij|$ for the associated bra vector. The scalar product in the tensor product space is defined by

$$\langle u^{(1)} v^{(2)} | w^{(1)} x^{(2)} \rangle = \langle u|w \rangle \langle v|x \rangle. \tag{2.270}$$

In particular, basis vectors obey (assuming the bases of $\mathcal{H}_1$ and $\mathcal{H}_2$ to be orthonormal)

$$\langle ij|kl \rangle = \delta_{ik} \delta_{jl}. \tag{2.271}$$

The eigenvalues of a tensor product $C = A \otimes B$ are, due to (2.255), simply the *products* of the eigenvalues of A and B. Let $|u\rangle \in \mathcal{H}_{\lambda_1}^{(1)}$ and $|v\rangle \in \mathcal{H}_{\lambda_2}^{(2)}$, hence

$$A|u\rangle = \lambda_1|u\rangle, \quad B|v\rangle = \lambda_2|v\rangle. \tag{2.272}$$

Then

$$C|uv\rangle = (A|u\rangle) \otimes (B|v\rangle) = \lambda_1 \lambda_2 |uv\rangle. \tag{2.273}$$

The eigenspace of eigenvalue $\lambda = \lambda_1 \lambda_2$ is the tensor product of the eigenspaces of the eigenvalues $\lambda_{1/2}$,

$$\mathcal{H}_\lambda = \mathcal{H}_{\lambda_1}^{(1)} \otimes \mathcal{H}_{\lambda_2}^{(2)}. \tag{2.274}$$

This holds if there is only one possibility to get λ as a product of two eigenvalues λ_1 and λ_2 of A and B, respectively. If there are several possibilities, then each vector of the form

$$|w\rangle = \sum_{\lambda_1, \lambda_2 | \lambda_1 \lambda_2 = \lambda} \alpha_{\lambda_1 \lambda_2} |\lambda_1\rangle \otimes |\lambda_2\rangle, \tag{2.275}$$

with complex coefficients $\alpha_{\lambda_1 \lambda_2}$, is an eigenvector of eigenvalue λ (where $|\lambda_1\rangle \in \mathcal{H}_{\lambda_1}^{(1)}$ and $|\lambda_2\rangle \in \mathcal{H}_{\lambda_2}^{(2)}$). The eigenspace of eigenvalue λ is then the direct sum of several tensor products:

$$\mathcal{H}_\lambda = \bigoplus_{\lambda_1, \lambda_2 | \lambda_1 \lambda_2 = \lambda} \mathcal{H}_{\lambda_1}^{(1)} \otimes \mathcal{H}_{\lambda_2}^{(2)} \tag{2.276}$$

An example: the operator

$$C = \sigma_z \otimes \sigma_z \tag{2.277}$$

in our two-spin system has the eigenvalues ± 1. Each vector of the form

$$|w_+\rangle = \alpha|z+, z+\rangle + \beta|z-, z-\rangle \tag{2.278}$$

is an eigenvector of eigenvalue $+1$, each vector of the form

$$|w_-\rangle = \alpha|z+, z-\rangle + \beta|z-, z+\rangle \tag{2.279}$$

an eigenvector of eigenvalue -1. The eigenspace $\mathcal{H}_{+1}$ of eigenvalue $+1$ is therefore

$$\mathcal{H}_{+1} = \left(\mathcal{H}_{+1}^{(1)} \otimes \mathcal{H}_{+1}^{(2)}\right) \oplus \left(\mathcal{H}_{-1}^{(1)} \otimes \mathcal{H}_{-1}^{(2)}\right). \tag{2.280}$$

Here, $\mathcal{H}_{+1}^{(1)} \otimes \mathcal{H}_{+1}^{(2)}$ is the one-dimensional vector space spanned by $|z+, z+\rangle = |z+\rangle \otimes |z+\rangle$, and similarly for $\mathcal{H}_{-1}^{(1)} \otimes \mathcal{H}_{-1}^{(2)}$. Equation (2.278) says that each element of $\mathcal{H}_{+1}$ lies in the two-dimensional direct sum of these two spaces.

The eigenvalues of an operator C of the form (2.267) are the *sums* of the eigenvalues of A and B. Let again $|u\rangle$ and $|v\rangle$ be as in (2.272). Then

$$C|uv\rangle = (A|u\rangle) \otimes |v\rangle + |u\rangle \otimes (B|v\rangle) = (\lambda_1 + \lambda_2)|uv\rangle. \tag{2.281}$$

In a similar way as in the previous case we find

$$\mathcal{H}_\lambda = \bigoplus_{\lambda_1,\lambda_2|\lambda_1+\lambda_2=\lambda} \mathcal{H}_{\lambda_1}^{(1)} \otimes \mathcal{H}_{\lambda_2}^{(2)}. \tag{2.282}$$

An example: the operator

$$C = \sigma_z \otimes \mathbf{1} + \mathbf{1} \otimes \sigma_z \tag{2.283}$$

in our two-spin system has the eigenvalues $\{-2, 0, 2\}$. These are the possible sums of the eigenvalues of the individual spins. The corresponding eigenspaces are

$$\mathcal{H}_2 = \mathcal{H}_{+1}^{(1)} \otimes \mathcal{H}_{+1}^{(2)} \tag{2.284}$$

$$\mathcal{H}_0 = \left(\mathcal{H}_{+1}^{(1)} \otimes \mathcal{H}_{-1}^{(2)}\right) \oplus \left(\mathcal{H}_{-1}^{(1)} \otimes \mathcal{H}_{+1}^{(2)}\right) \tag{2.285}$$

$$\mathcal{H}_{-2} = \mathcal{H}_{-1}^{(1)} \otimes \mathcal{H}_{-1}^{(2)}. \tag{2.286}$$

We now want to demonstrate the **correlation** in an entangled state by means of an example. Given a two-spin system in the state

$$|w\rangle = \frac{1}{\sqrt{2}} (|z+, z+\rangle + |z-, z-\rangle) = \frac{1}{\sqrt{2}} \begin{pmatrix} 1 \\ 0 \\ 0 \\ 1 \end{pmatrix}. \tag{2.287}$$

Exercise 2.31
Show that this vector can be also written as

$$|w\rangle = \frac{1}{\sqrt{2}} (|x+, x+\rangle + |x-, x-\rangle) = \frac{1}{\sqrt{2}} (|y+, y-\rangle + |y-, y+\rangle). \tag{2.288}$$

For a simultaneous measurement of the spin on z-direction on each of the two systems, the probability for spin-up or spin-down is $\frac{1}{2}$ in each case. Mathematically this follows from the projection operators

$$
\begin{aligned}
P_{z+}^{(1)} &= (|z+\rangle\langle z+|) \otimes \mathbf{1} = \operatorname{diag}(1, 1, 0, 0), \\
P_{z+}^{(2)} &= \mathbf{1} \otimes (|z+\rangle\langle z+|) = \operatorname{diag}(1, 0, 1, 0), \\
P_{z-}^{(1)} &= (|z-\rangle\langle z-|) \otimes \mathbf{1} = \operatorname{diag}(0, 0, 1, 1), \\
P_{z-}^{(2)} &= \mathbf{1} \otimes (|z-\rangle\langle z-|) = \operatorname{diag}(0, 1, 0, 1),
\end{aligned}
$$

and hence

$$
\langle w| P_{z+}^{(1)} |w\rangle = \langle w| P_{z+}^{(2)} |w\rangle = \langle w| P_{z-}^{(1)} |w\rangle = \langle w| P_{z-}^{(2)} |w\rangle = \frac{1}{2}. \tag{2.289}
$$

However, the probabilities in system 1 are not independent of the measurement result in system 2. The following statement (A1) holds: if spin-up is measured in system 2, then spin-up will be also measured in system 1, with hundred percent probability. For the probability to have spin-up in system 2 and spin-down in system 1 is

$$
\langle w| P_{z-}^{(1)} P_{z+}^{(2)} |w\rangle = 0. \tag{2.290}
$$

Note that the order of the projection operators is irrelevant here: they commute, because they act on different systems (one measurement takes place on system 1, the other one on system 2).

The following also holds (A2): if spin-down is measured in system 2, then spin-down will be also measured in system 1, with hundred percent probability. According to Exercise 2.31, similar statements hold for measurements of the spin in x-direction. The situation is similar to the entangled photons in the introduction. For an arbitrary direction $\mathbf{r}$ in the (xz)-plane it is true that the spin in $\mathbf{r}$-direction is measured to be the same for both systems, either positive or negative. For the spin in y-direction, according to Exercise 2.31, the opposite is true: Here the measurement yields opposite results for the two systems.

Statement (A1) can also be derived with conditional probabilities Let $p_{z+^{(2)}}(z+^{(1)})$ be the probability for measuring spin-up in system 1, under the condition that spin-up was measured in system 2. In general, conditional probabilities obey $p_X(Y) = p(X, Y)/p(X)$. It follows

$$
p_{z+^{(2)}}(z+^{(1)}) = \frac{p(z+, z+)}{p(z+^{(2)})} = \frac{\langle w| P_{z+}^{(1)} P_{z+}^{(2)} |w\rangle}{\langle w| P_{z+}^{(2)} |w\rangle} = \frac{1/2}{1/2} = 1. \tag{2.291}
$$

The statements (A1) and (A2) together imply that the spins in z-direction are maximally correlated.

One speaks of a **correlation** of two simultaneously measurable observables A and B if their measurement results are not independent. A measure for this is the expectation value of their product: If A and B are uncorrelated, then

$$\langle AB \rangle = \langle A \rangle \langle B \rangle. \tag{2.292}$$

If $\langle AB \rangle$ is larger or smaller than this value, this is called a **positive** or **negative correlation**, respectively, and in the second case also an **anticorrelation**.

In our example, the relevant quantity is

$$\langle S_z^{(1)} S_z^{(2)} \rangle_w = \langle w | S_z^{(1)} S_z^{(2)} | w \rangle = \frac{\hbar^2}{4} \langle w | \sigma_z \otimes \sigma_z | w \rangle = \frac{\hbar^2}{4}. \tag{2.293}$$

On the other hand, one has

$$\langle S_z^{(1)} \rangle_w = \langle S_z^{(2)} \rangle_w = 0, \tag{2.294}$$

and so we have a positive correlation.

In order to define the correlation with a value between -1 und 1, we can take the expectation value of the tensor product $\sigma_z \otimes \sigma_z$ as a measure. The eigenvalues of $\sigma_z \otimes \sigma_z$ are the possible products of the eigenvalues of σ_z in each of the systems, i.e. ± 1. If

$$\langle \sigma_z \otimes \sigma_z \rangle_v = 0 \tag{2.295}$$

then the two z-spins are **uncorrelated**. If

$$0 < \langle \sigma_z \otimes \sigma_z \rangle_v \leq 1, \tag{2.296}$$

then the two z-spins are **correlated**. If

$$-1 \leq \langle \sigma_z \otimes \sigma_z \rangle_v < 0, \tag{2.297}$$

then the two z-spins are **anticorrelated**. In the case of our example,

$$\langle \sigma_z \otimes \sigma_z \rangle_w = \langle w | \sigma_z \otimes \sigma_z | w \rangle = 1, \tag{2.298}$$

that is, we have **maximal correlation**.

Exercise 2.32
Compute the conditional probabilities and the correlation for the case that in system 1 the spin in z-direction is measured, but in system 2 the spin in x-direction. Result: the measurements are uncorrelated.

Exercise 2.33
Compute the correlation for the case that in system 1 the spin in z-direction is measured, but in system 2 the spin in direction $\frac{1}{\sqrt{2}}(\mathbf{e}_x + \mathbf{e}_z)$. Result: the measurements are correlated, but not maximally correlated. Hint: Use the operator (2.141).

Given at time $t = 0$ a non-entangled state $|w\rangle = |u\rangle \otimes |v\rangle$ of the total system S and a time-independent Hamiltonian operator H. Under what conditions does the state remain non-entangled? There are two cases worth mentioning:

1. $|w\rangle$ is an eigenstate of H. In this case

$$|w(t)\rangle = e^{-i\frac{E}{\hbar}t}|u\rangle \otimes |v\rangle. \tag{2.299}$$

2. H is of the form (2.269). Then, since $H_1^{(1)}$ and $H_2^{(2)}$ commute,

$$|w(t)\rangle = e^{-i\frac{H}{\hbar}t}|w(0)\rangle = e^{-i\frac{H_1^{(1)}+H_2^{(2)}}{\hbar}t}|w(0)\rangle \tag{2.300}$$

$$= e^{-i\frac{H^{(1)}}{\hbar}t}e^{-i\frac{H^{(2)}}{\hbar}t}|w(0)\rangle \tag{2.301}$$

$$= \left(e^{-i\frac{H_1}{\hbar}t} \otimes \mathbf{1}\right)\left(\mathbf{1} \otimes e^{-i\frac{H_2}{\hbar}t}\right)(|u\rangle \otimes |v\rangle) \tag{2.302}$$

$$= \left(e^{-i\frac{H_1}{\hbar}t}|u\rangle\right) \otimes \left(e^{-i\frac{H_2}{\hbar}t}|v\rangle\right). \tag{2.303}$$

Entanglement occurs when systems interact with each other. In the case of the two spins, the classical energy of a typical interaction is given by

$$E = \alpha \mathbf{s}^{(1)} \cdot \mathbf{s}^{(2)} = \alpha(s_x^{(1)}s_x^{(2)} + s_y^{(1)}s_y^{(2)} + s_z^{(1)}s_z^{(2)}), \tag{2.304}$$

i.e. there is a force between the two systems which tries to align the spins antiparallel to each other (assuming $\alpha > 0$). The associated Hamiltonian operator reads

$$H = \frac{\alpha\hbar^2}{4}\left(\sigma_x \otimes \sigma_x + \sigma_y \otimes \sigma_y + \sigma_z \otimes \sigma_z\right) \tag{2.305}$$

$$= \frac{\alpha\hbar^2}{4}\begin{pmatrix} 1 & 0 & 0 & 0 \\ 0 & -1 & 2 & 0 \\ 0 & 2 & -1 & 0 \\ 0 & 0 & 0 & 1 \end{pmatrix}. \tag{2.306}$$

H has the eigenvalues

$$E_1 = \frac{\alpha\hbar^2}{4}, \quad E_2 = -3\frac{\alpha\hbar^2}{4}. \tag{2.307}$$

The three-dimensional eigenspace of eigenvalue E_1 is spanned by the three symmetric states (invariant under exchange of the two spins)

$$|s_+\rangle = |z+, z+\rangle, \quad |s_0\rangle = \frac{1}{\sqrt{2}}(|z+, z-\rangle + |z-, z+\rangle), \quad |s_-\rangle = |z-, z-\rangle$$

$$\tag{2.308}$$

The one-dimensional eigenspace of eigenvalue E_2 is spanned by the three antisymmetric states (acquiring a minus sign under exchange of the two spins)

$$|a_0\rangle = \frac{1}{\sqrt{2}}(|z+, z-\rangle - |z-, z+\rangle). \tag{2.309}$$

Exercise 2.34
Verify all that: the matrix for H, the eigenvalues and eigenvectors.

Let's take a moment to think about these eigenvalues and eigenvectors. In an uncorrelated, homogeneously distributed statistical mixture of two-spin states, the average for the measured values of the scalar product $\mathbf{s}^{(1)} \cdot \mathbf{s}^{(2)}$ is, as expected, $\hbar^2(1+1+1+(-3))/4 = 0$. But what may be surprising is the eigenvectors $|s_0\rangle$ and $|a_0\rangle$. Intuitively, one would expect that for parallel spins $\mathbf{s}^{(1)} \cdot \mathbf{s}^{(2)} = \hbar^2/4$, and for antiparallel spins $\mathbf{s}^{(1)} \cdot \mathbf{s}^{(2)} = -\hbar^2/4$. Instead, the subspace of antiparallel spin states is split into a symmetric state with $\mathbf{s}^{(1)} \cdot \mathbf{s}^{(2)} = +\hbar^2/4$, and an antisymmetric state with $\mathbf{s}^{(1)} \cdot \mathbf{s}^{(2)} = -3\hbar^2/4$. In particular, the eigenvalue $-3\hbar^2/4$ surprises, which seems to exceed the product of $\mathbf{s}^{(1)}$ and $\mathbf{s}^{(2)}$. A similarly remarkable behavior is already found for a single spin. For the state $|z+\rangle$ one has

$$\langle S_x \rangle = \langle S_y \rangle = 0, \quad \langle S_z \rangle = \frac{\hbar}{2}, \tag{2.310}$$

but

$$\langle \mathbf{S}^2 \rangle = \frac{\hbar^2}{4} \langle \sigma_x^2 + \sigma_y^2 + \sigma_z^2 \rangle \tag{2.311}$$

$$= \frac{3\hbar^2}{4} \langle \mathbf{1} \rangle = \frac{3\hbar^2}{4} = 3 \left(\langle S_x \rangle, \langle S_y \rangle, \langle S_z \rangle \right)^2. \tag{2.312}$$

We will come back to this topic in the Chaps. 7 and 9.

Returning to our discussion of entanglement via interaction: let the system described by the Hamiltonian operator (2.305) be in the non-entangled initial state (at $t = 0$)

$$|w(0)\rangle = |z+, z-\rangle = \frac{1}{\sqrt{2}}(|s_0\rangle + |a_0\rangle). \tag{2.313}$$

With (2.307), one immediately gets

$$|w(t)\rangle = \frac{1}{\sqrt{2}} \left(e^{i\omega_1 t}|s_0\rangle + e^{i\omega_2 t}|a_0\rangle \right), \quad \omega_1 = \frac{\alpha\hbar}{4}, \quad \omega_2 = -\frac{3\alpha\hbar}{4}. \quad (2.314)$$

At any time, $|w(t)\rangle$ is of the form

$$|w(t)\rangle = \alpha(t)|z+, z-\rangle + \beta(t)|z-, z+\rangle. \quad (2.315)$$

If $e^{i\omega_1 t} = e^{i\omega_2 t}$, i.e. $t = 2\pi n/(\omega_1 - \omega_2)$, $n \in \mathbb{Z}$, then $\beta = 0$. If $e^{i\omega_1 t} = -e^{i\omega_2 t}$, i.e. $t = (2n + 1)\pi/(\omega_1 - \omega_2)$, then $\alpha = 0$. At all other times, the system is entangled.

Self-check questions:

1. What are the differences between a tensor product and a direct sum?
2. What are entangled states?
3. How do you calculate conditional probabilities and correlations?

Chapter 3
Formalism II: Infinite-Dimensional Hilbert Spaces

Abstract The weird formalism of QM is extended to function spaces. Wave functions and the Schrödinger equation in position space are introduced. On the way, we explain why basis vectors don't need to be elements of the space they are a basis of.

In QM, state vectors mostly appear in the form of **wave functions** $\psi(\mathbf{r})$. We therefore have to figure out first in what sense sets of functions are vector spaces. The corresponding field in mathematics is **functional analysis**. We will see that operations such as differentiation can be understood as linear operators that can be represented by matrices with infinitely many entries. On our way we will meet lots of new features that don't occur in the finite-dimensional case.

The transition from classical mechanics to QM is undertaken by means of the Hamilton formalism, where positions and momenta of particles are the fundamental variables.

The **position observable** describes the position (x, y, z) of a particle according to a previously chosen cartesian coordinate system. Associated to them are the **position operators** (X, Y, Z) in QM, which act on the particle's wave function. We will find that these operators act via multiplication with the coordinate, e.g. $(X\psi)(\mathbf{r}) = x\psi(\mathbf{r})$.

However, there is a snag to it: Although X is hermitian, it has no eigenvalues and eigenvectors in the considered Hilbert space. Only with the use of a trick we will be able to construct a so-called **pseudo-basis** of **pseudo-eigenvectors**, which can be practically used for calculations, and which gives us a connection between the three-dimensional position space and the infinite-dimensional Hilbert space.

Similar statements hold for the **momentum operators** (P_x, P_y, P_z), which are associated to the three components of momentum. They act via differentiation, e.g. $(P_x\psi)(\mathbf{r}) = -i\hbar\frac{\partial}{\partial x}\psi(\mathbf{r})$. These operators also don't have eigenvalues and eigenvectors in the Hilbert space, and again we have to resort to pseudo-vetors.

Just as we recapitulated some Linear Algebra in the beginning of Chap. 2, we will now begin by acquiring the mathematical equipment needed to understand the QM of positions and momenta. This is done in the Sects. 3.1–3.3.

A particular focus will be on the operator X which multiplies functions of one variable with that variable, and on the operator D which takes the derivative of a function. For these two operators are fundamental for position and momentum in QM. Only then, in Sect. 3.4, we will return to physics. Then we will meet the

© Springer International Publishing Switzerland 2016

J.-M. Schwindt, *Conceptual Basis of Quantum Mechanics*,
Undergraduate Lecture Notes in Physics, DOI 10.1007/978-3-319-24526-3_3

Schrödinger equation in position space, the **Position-Momentum Uncertainty**, the **wave-particle duality**, and some beautiful connections between QM and classical mechanics. This will complete our work on the formalism of QM, and we will then be ready to solve some concrete problems in the remaining parts of this book.

But now it is time to roll up our sleeves and understand some math!

3.1 Sets of Functions as Vector Spaces

Let M be an arbitrary set and V a vector space over the field K, where $K = \mathbb{R}$ or $\mathbb{C}$. Then the set of functions $F(M, V) = \{f : M \to V\}$ also forms a vector space over K. The addition of two functions, $f + g$, and the multiplication of a function with a number $\alpha \in K$ are defined in the following way:

$$(f + g)(x) = f(x) + g(x), \qquad (\alpha f)(x) = \alpha \, f(x) \tag{3.1}$$

That is to say, these operations are reduced to the corresponding operations in the target space V. They are well-defined, because the corresponding operations on V are well-defined.

It is thus crucial that V is a vector space. If V were just an interval $[a, b]$, for instance, then one could always find two functions whose image is in $[a, b]$, but the image of their sum is not.

The zero function $0_{M,V}$ is the function that maps all elements of M on the zero vector 0 in V:

$$0_{M,V}(x) = 0 \tag{3.2}$$

For each function $f \in F(M, V)$ there is a function $-f \in F(M, V)$ such that $f + (-f) = 0_{M,V}$. For that, set $(-f)(x) = -f(x)$. So, the zero and the inverse with respect to addition in $F(M, V)$ are also reduced to the corresponding properties in V. Again it is crucial that V is a vector space.

In QM one considers wave functions which are elements of $F(\mathbb{R}^n, \mathbb{C})$. Here, the elements of $\mathbb{R}^n$ represent n real spatial coordinates of one or several "particles" (the quotation marks are supposed to indicate that the notion of particles in QM means something different than in classical mechanics). The target space $\mathbb{C}$ can be understood as a one-dimensional vector space over itself. Therefore, $F(\mathbb{R}^n, \mathbb{C})$ is also a vector space over $\mathbb{C}$.

Instead of the entire space $F(M, V)$, one can also consider subspaces of functions with specific "nice" properties, e.g.

- the set $C^0(M, V)$ of continuous functions
- the set $C^1(M, V)$ of differentiable functions
- the set $L^1(M, V)$ of integrable functions, where $\int_M f$ is well-defined and finite
- the set of functions $f : M \to V$ having a zero at a specific point $a \in M$, $f(a) = 0$.

Of course, these subsets can only be constructed if M allows for such a definition. If M, for instance, is the set of animals in the zoo of Heidelberg, it is hard to say what a "continuous" function $M \to V$ would be.

It is easy to convince yourself that the respective property is not lost if one constructs a **finite** linear combination of functions having this property. For example: If f and g are continuous, then $\alpha f + \beta g$ is continuous too. All the mentioned subspaces are indeed vector spaces.

The notion of a **linear combination** is used for function spaces just as for any other vector spaces, namely for a sum $\sum_{i=1}^{n} \alpha_i f_i$, with $\alpha_i \in K$ and all f_i in the considered function space. The vector space properties ensure that linear combinations don't leave the vector space **as long as they are finite**, i.e. have only finitely many terms. A sum of infinitely many functions does not need to converge, and even if it converges, it is not ensured that the properties of the summands (e.g. continuity) is conserved.

As an example for a function space, we choose the set of real polynomials,

$$\text{Pol}(\mathbb{R}, \mathbb{R}) = \{f : \mathbb{R} \to \mathbb{R} \mid f(x) = \sum_{i=0}^{n} \alpha_i x^i, \alpha_i \in \mathbb{R}, n \in \mathbb{N}\}. \tag{3.3}$$

Even infinite-dimensional vector spaces have a basis, i.e. an (infinite) set of basis vectors $\{e_i\}$, such that all elements of the vector space can be written as a **finite** linear combination of basis vectors, in a unique way.

For the polynomials, there is an obvious basis: the monomials, $e_i(x) = x^i$. Each polynomial can then be expressed in components w.r.t. this basis:

$$f = (\alpha_0, \alpha_1, \alpha_2, \dots) \tag{3.4}$$

means

$$f = \alpha_0 e_0 + \alpha_1 e_1 + \alpha_2 e_2 + \cdots, \tag{3.5}$$

which is

$$f(x) = (\alpha_0 e_0 + \alpha_1 e_1 + \alpha_2 e_2 + \cdots)(x) \tag{3.6}$$
$$= \alpha_0 e_0(x) + \alpha_1 e_1(x) + \alpha_2 e_2(x) + \cdots \tag{3.7}$$
$$= \alpha_0 + \alpha_1 x + \alpha_2 x^2 + \cdots \tag{3.8}$$

Since each polynomial terminates at a finite power of x, it is ensured that in the component representation (3.4) only finitely many values are different from 0.

But what if instead of polynomials we choose the space of all functions that can be expressed as a power series in all of $\mathbb{R}$,

$$\text{PR}(\mathbb{R}, \mathbb{R}) = \{f : \mathbb{R} \to \mathbb{R} \mid f(x) = \sum_{i=0}^{\infty} \alpha_i x^i \text{ converges in all of } \mathbb{R}\}. \tag{3.9}$$

In contrast to polynomials, power series in general don't terminate at finite n, i.e., they are *infinite* linear combinations of the monomials. Fortunately, mathematicians have an understanding here and still allow for the monomials as a basis. It is, however, not a basis according to the original definition, where the linear combinations of basis vectors have to be *finite*.

A basis that requires infinite linear combinations is called **Schauder basis**. The monomials thus form a Schauder basis of the space $PR(\mathbb{R}, \mathbb{R})$. There *is* also a basis which requires only finite linear combinations. However, one can show that such a basis consists not only of infinitely many but even *uncountably infinitely many* basis vectors. Such a basis is called **Hamel basis** and is completely useless for our purposes.

The problem with a Schauder basis is that basis vectors can no longer be combined in arbitrary ways. As mentioned above, an infinite linear combination of functions in a function space is not necessarily again an element of that space. In our case: Not every power series (infinite linear combination of monomials) converges, i.e. is in $PR(\mathbb{R}, \mathbb{R})$. On one hand, every function in $PR(\mathbb{R}, \mathbb{R})$ can be written as a finite or infinite linear combination of the basis vectors, but on the other hand, not every infinite linear combination of basis vectors leads to an element of $PR(\mathbb{R}, \mathbb{R})$!

Operators on a function space $F(M, V)$ are functions $T : F(M, V) \to F(M, V)$ mapping functions to functions. An operator T is **linear** if

$$T(\alpha f + \beta g) = \alpha T(f) + \beta T(g) \tag{3.10}$$

for any $\alpha, \beta \in K$, $f, g \in F(M, V)$. One usually writes just Tf instead of $T(f)$.

Linear functions (**automorphisms**) on a vector space can be represented as matrices with respect to a given basis. This also holds for operators on function spaces. Since function spaces are infinite-dimensional, the matrices are infinitely large.

We want to consider two examples for operators on $Pol(\mathbb{R}, \mathbb{R})$ and $PR(\mathbb{R}, \mathbb{R})$, respectively, which play an important role in QM, namely the operator X which multiplies each function with x, and the derivative operator D which differentiates each function:

$$(Xf)(x) = xf(x), \qquad (Df)(x) = f'(x) \tag{3.11}$$

The operators act on the monomials $\mathbf{e}_i$ as follows:

$$X\mathbf{e}_n = \mathbf{e}_{n+1}, \qquad D\mathbf{e}_0 = 0_{\mathbb{R}, \mathbb{R}}, \qquad D\mathbf{e}_n = n\mathbf{e}_{n-1} \text{ for } n > 0 \tag{3.12}$$

Thus, X and D have the following matrix representation in the monomial basis:

$$X^{(\mathbf{e})} = \begin{pmatrix} 0 & 0 & 0 & 0 & \cdots \\ 1 & 0 & 0 & 0 & \cdots \\ 0 & 1 & 0 & 0 & \cdots \\ 0 & 0 & 1 & 0 & \cdots \\ \vdots & \vdots & \vdots & \vdots & \ddots \end{pmatrix}, \qquad D^{(\mathbf{e})} = \begin{pmatrix} 0 & 1 & 0 & 0 & \cdots \\ 0 & 0 & 2 & 0 & \cdots \\ 0 & 0 & 0 & 3 & \cdots \\ 0 & 0 & 0 & 0 & \cdots \\ \vdots & \vdots & \vdots & \vdots & \ddots \end{pmatrix} \tag{3.13}$$

What about eigenvalues and eigenvectors of X and D? Since the matrices are infinitely large, we cannot determine a characteristic polynomial and have to get by otherwise. In the case of X we search for a power series f with the property $Xf = \lambda f$, i.e.

$$xf(x) = \lambda f(x) \tag{3.14}$$

for all $x \in \mathbb{R}$ and fixed $\lambda \in \mathbb{R}$. One immediately sees that this cannot work: x varies over the entire real line, while λ remains fixed. The only solution is $f = 0_{\mathbb{R},\mathbb{R}}$. Hence, there are no eigenvalues and no eigenvectors. (As a resort, we will later find the delta distribution. The function space has to be extended for that.)

In the case of D we look for a polynomial or power series f with the property $Df = \lambda f$, i.e.

$$f'(x) = \lambda f(x) \tag{3.15}$$

for all $x \in \mathbb{R}$ and fixed $\lambda \in \mathbb{R}$. In $\mathrm{Pol}(\mathbb{R}, \mathbb{R})$ there is obviously no solution, since the derivative decreases the degree of each polynomial by 1. In $\mathrm{PR}(\mathbb{R}, \mathbb{R})$ however there are indeed solutions: The functions $f(x) = e^{\lambda x}$ can be expressed as power series in all of $\mathbb{R}$ and obey (3.15). In $\mathrm{PR}(\mathbb{R}, \mathbb{R})$, D thus has a **continuous spectrum** of eigenvalues: every real number λ is an eigenvalue of D. The corresponding eigenspace PR_λ is the one-dimensional space spanned by the function $f(x) = e^{\lambda x}$, i.e.

$$\mathrm{PR}_\lambda = \{f \in \mathrm{PR}(\mathbb{R}, \mathbb{R}) \mid f(x) = \alpha e^{\lambda x}, \alpha \in \mathbb{R}\}. \tag{3.16}$$

Self-check questions:

1. Do the functions $f : \mathbb{R} \to \mathbb{R}$ with $f(0) \geq 0$ form a vector space?
2. What is a Schauder basis?

3.2 Scalar Product and Orthonormal Basis

QM takes place in Hilbert spaces. In order to turn function spaces into Hilbert spaces, we need to define a scalar product. For this we consider function spaces $F(\mathbb{R}^n, K)$, where $K = \mathbb{R}$ or $\mathbb{C}$. On these spaces (more precisely, on subspaces where the following integrals are well-defined) we can define a scalar product via

$$\langle f|g \rangle = \int_{\mathbb{R}^n} d^n x \; f^*(\mathbf{r}) g(\mathbf{r}), \tag{3.17}$$

with $\mathbf{r} = (x_1, \ldots, x_n)$, and a norm via

$$\|f\| = \sqrt{\int_{\mathbb{R}^n} d^n x \; f^*(\mathbf{r}) f(\mathbf{r})}. \tag{3.18}$$

Of course, the star for complex conjugation is only required if $K = \mathbb{C}$. With this definition, the scalar product has all the required properties mentioned in Sect. 2.2. It is actually similar to the scalar product on finite-dimensional Hilbert spaces: Only the sum $\sum_i u_i^* v_i$ was replaced by the integral $\int f^*(\mathbf{r}) g(\mathbf{r})$. The spatial vector $\mathbf{r}$ has, in a sense, taken the place of the index i.

We want to allow only finite values for the scalar product. This is fulfilled (as we shall see in a moment) in the **space of square-integrable functions** $L^2(\mathbb{R}^n, K)$,

$$L^2(\mathbb{R}^n, K) = \{f \in F(\mathbb{R}^n, K) \mid f \text{ measurable, } \int_{\mathbb{R}^n} d^n x \ f^*(\mathbf{r}) f(\mathbf{r}) < \infty\}.$$

$$(3.19)$$

The notion of measurability shall not bother us here any further. It ensures that the integrals are well-defined. It has nothing to do with measurement in the physical sense! The second condition says that $\|f\|$ is finite. For elements of the Hilbert space $\mathcal{H} = L^2(\mathbb{R}^n, K)$ we will again use the bra/ket notation.

The Schwarz inequality

$$|\langle f|g \rangle| \leq \|f\| \, \|g\|$$

$$(3.20)$$

and the triangle inequality

$$\|f + g\| \leq \|f\| + \|g\|$$

$$(3.21)$$

hold also in function spaces (their proof didn't make any use of the vector space being finite-dimensional). From (3.20) follows that $\langle f|g \rangle$ with f, g in $L^2(\mathbb{R}^n, K)$ is finite: the left hand side of (3.20) is finite, because the right hand side is. It also follows from (3.21) that $L^2(\mathbb{R}^n, K)$ is a vector space:

$$\|\alpha f + \beta g\| \leq \|\alpha f\| + \|\beta g\| = |\alpha| \, \|f\| + |\beta| \, \|g\|$$

$$(3.22)$$

The left hand side is finite, because the right hand side is, and thus $\alpha f + \beta g$ is in $L^2(\mathbb{R}^n, K)$.

In the following we will construct some orthonormal bases for several L^2-spaces. We learn how to calculate with these infinite-dimensional vector spaces, and find some useful things on the way: the Legendre, Hermite, and Laguerre polynomials, all of which play a role in some later parts of this book, and the Fourier series (real and complex). In this way we will also meet our first pseudo-basis.

In $L^2(\mathbb{R}, \mathbb{R})$, the derivative operator D again has no eigenvectors, since the functions $f(x) = e^{\alpha x}$ are obviously not square-integrable. In $L^2(\mathbb{R}, \mathbb{C})$, where α can be complex, it doesn't get better:

$$\langle f|f \rangle = \int_{-\infty}^{\infty} dx \ f^*(x) f(x) = \int_{-\infty}^{\infty} dx \ e^{\alpha^* x} e^{\alpha x} = \int_{-\infty}^{\infty} dx \ e^{2\text{Re}(\alpha)x} = \infty$$

$$(3.23)$$

The polynomials in $\text{Pol}(\mathbb{R}, \mathbb{R})$ are also not square-integrable. We therefore want to restrict them to the interval $[-1, 1]$, that is, we consider $\text{Pol}([-1, 1], \mathbb{R})$. In the interval $[-1, 1]$, all polynomials are square-integrable, i.e. $\text{Pol}([-1, 1], \mathbb{R})$ is a subspace of $L^2([-1, 1], \mathbb{R})$. The monomials form a basis of $\text{Pol}([-1, 1], \mathbb{R})$, but not an orthonormal basis; for instance,

$$\langle e_0 | e_2 \rangle = \int_{-1}^{1} dx \, 1 \cdot x^2 = \frac{2}{3}. \tag{3.24}$$

As an exercise, we want to take the basis $\{|e_i\rangle\}$ of monomials and orthogonalize and normalize it to an orthonormal basis $\{|P_i\rangle\}$, at least for $i = 0$ to 3. For this, we use the **Gram-Schmidt process**, which should be known from Linear Algebra, but let's recall it in an exercise:

Exercise 3.1
Show that the following process constructs an orthonormal basis $|f_i\rangle$ out of an arbitrary basis $|e_i\rangle$:
(i) Set

$$|f_1\rangle = \frac{|e_1\rangle}{\||e_1\||}. \tag{3.25}$$

(ii) Assume the first k basis vectors $|f_1\rangle, \ldots, |f_k\rangle$ are already known. Set

$$|\tilde{f}_{k+1}\rangle = |e_{k+1}\rangle - \sum_{i=1}^{k} |f_i\rangle \langle f_i | e_{k+1}\rangle \tag{3.26}$$

and then

$$|f_{k+1}\rangle = \frac{|\tilde{f}_{k+1}\rangle}{\||\tilde{f}_{k+1}\||}. \tag{3.27}$$

Now apply this to the basis of monomials (please follow all the steps of the calculation): It is $\langle e_0 | e_0 \rangle = 2$, hence we normalize $|e_0\rangle$ to $|P_0\rangle$ with

$$P_0(x) = \frac{1}{\sqrt{2}}. \tag{3.28}$$

Due to $\langle P_0 | e_1 \rangle = 0$, $|e_1\rangle$ is already orthogonal to $|P_0\rangle$ and only has to be normalized, $\langle e_1 | e_1 \rangle = \frac{2}{3}$, to $|P_1\rangle$ with

$$P_1(x) = \sqrt{\frac{3}{2}} x. \tag{3.29}$$

The next monomial, $|e_2\rangle$, is already orthogonal to $|P_1\rangle$, but not to $|P_0\rangle$: $\langle P_0 | e_2 \rangle = \frac{\sqrt{2}}{3}$. We thus have to subtract $|P_0\rangle \langle P_0 | e_2 \rangle$ from $|e_2\rangle$ an obtain the vector $|\tilde{P}_2\rangle$ orthogonal to $|P_0\rangle$ with

$$\tilde{P}_2(x) = x^2 - \frac{1}{\sqrt{2}} \frac{\sqrt{2}}{3} = x^2 - \frac{1}{3}.$$ (3.30)

This needs to be normalized:

$$\langle \tilde{P}_2 | \tilde{P}_2 \rangle = \int_{-1}^{1} dx \left(x^2 - \frac{1}{3} \right)^2 = \frac{8}{45},$$ (3.31)

from which we get $|P_2\rangle$:

$$P_2(x) = \frac{\sqrt{5}}{2\sqrt{2}} (3x^2 - 1).$$ (3.32)

Finally, $|e_3\rangle$ is already orthogonal to $|P_0\rangle$ and $|P_2\rangle$, but not to $|P_1\rangle$:

$$\langle P_1 | e_3 \rangle = \int_{-1}^{1} dx \sqrt{\frac{3}{2}} x \cdot x^3 = \frac{\sqrt{6}}{5}.$$ (3.33)

We subtract $|P_1\rangle\langle P_1 | e_3 \rangle$ from $|e_3\rangle$ and obtain the vector $|\tilde{P}_3\rangle$ orthogonal to $|P_1\rangle$ with

$$\tilde{P}_3(x) = x^3 - \sqrt{\frac{3}{2}} x \cdot \frac{\sqrt{6}}{5} = x^3 - \frac{3}{5}x.$$ (3.34)

This needs to be normalized:

$$\langle \tilde{P}_3 | \tilde{P}_3 \rangle = \int_{-1}^{1} dx \left(x^3 - \frac{3}{5}x \right)^2 = \frac{8}{175},$$ (3.35)

from which we get $|P_3\rangle$:

$$P_3(x) = \frac{\sqrt{7}}{2\sqrt{2}} (5x^3 - 3x).$$ (3.36)

This procedure can be continued forever. The result are the polynomials $P_i(x)$ (where P_i is a polynomial of degree i) which are orthonormal to each other on the interval $[-1, 1]$:

$$\int_{-1}^{1} dx\, P_i(x) P_j(x) = \delta_{ij}$$ (3.37)

The P_i are (up to a constant factor) the **Legendre polynomials**, and we will meet them again when we study spherically symmetric problems in three dimensions, where they will show up in the form $P_i(\cos\theta)$ ($\cos\theta$ runs from -1 to $+1$!).

If we want to find an orthonormal system in $L^2(\mathbb{R}, \mathbb{R})$ instead of $L^2([-1, 1], \mathbb{R})$, we should not begin with the monomials, for these are in $L^2([-1, 1], \mathbb{R})$, but not in $L^2(\mathbb{R}, \mathbb{R})$. However, we can slightly change our starting point if we multiply the monomials by a factor (e.g. $e^{-x^2/2}$) which damps them sufficiently in the limit $x \to \pm\infty$ such that they become square integrable. We thus start with the basis functions

$$\mathbf{e}_n(x) = e^{-x^2/2} x^n, \qquad (3.38)$$

which span a subspace of $L^2(\mathbb{R}, \mathbb{R})$, and then orthogonalize and normalize them with the same procedure as above. The result are the functions

$$\tilde{H}_n(x) = e^{-x^2/2} H_n(x), \qquad (3.39)$$

where H_n are polynomials of degree n and, due to the orthonormalization, obey

$$\int_{-\infty}^{\infty} e^{-x^2} H_i(x) H_j(x) = \delta_{ij}. \qquad (3.40)$$

They are (up to a constant factor) the **Hermite polynomials**, and we will meet them again when we study the harmonic oscillator.

If we look for an orthonormal system in $L^2([0, \infty), \mathbb{R})$, we need again a damping factor for the monomials to begin with, but this time it needs to damp only in the direction of $+\infty$, for instance $e^{-x/2}$. We thus start with

$$\mathbf{e}_n(x) = e^{-x/2} x^n, \qquad (3.41)$$

which span a subspace of $L^2([0, \infty), \mathbb{R})$, and then orthonormalize them. The result are the functions

$$\tilde{L}_n(x) = e^{-x/2} L_n(x), \qquad (3.42)$$

where L_n are polynomials of degree n and, due to the orthonormalization, obey

$$\int_0^{\infty} e^{-x} L_i(x) L_j(x) = \delta_{ij}. \qquad (3.43)$$

They are the **Laguerre polynomials**, and we will meet them again when we study the hydrogen atom, as functions of the radial coordinate (which runs from 0 to ∞!).

Let's come back to the two bases of $\mathrm{Pol}([-1, 1], \mathbb{R})$: the monomial basis $|e_i\rangle$ and the orthonormal basis of the Legendre polynomials $|P_i\rangle$. We want to do some "finger exercises" with them.

Associated with a transformation between two bases is a transformation matrix. For infinite-dimensional vectors spaces, this is an $\infty \times \infty$-matrix. In our case the task is somewhat simplified by the fact that polynomials of degree n contain only the basis vectors $|e_0\rangle, \ldots, |e_n\rangle$, and with the other basis they also contain only the basis vectors $|P_0\rangle, \ldots, |P_n\rangle$. Hence we can restrict ourselves to polynomials up to 3rd degree to obtain a 4×4-matrix. The matrix A which leads us from the components of a vector in the P-basis to the components of this vector in the $\mathbf{e}$-basis, can be directly read of from the polynomials $P_0(x)$ to $P_3(x)$:

$$A = \begin{pmatrix} \frac{1}{\sqrt{2}} & 0 & -\frac{\sqrt{5}}{2\sqrt{2}} & 0 \\ 0 & \sqrt{\frac{3}{2}} & 0 & -\frac{3\sqrt{7}}{2\sqrt{2}} \\ 0 & 0 & \frac{3\sqrt{5}}{2\sqrt{2}} & 0 \\ 0 & 0 & 0 & \frac{5\sqrt{7}}{2\sqrt{2}} \end{pmatrix} \tag{3.44}$$

Is this clear to you? If not, convince yourself by means of examples, for instance

$$\begin{pmatrix} -\frac{\sqrt{5}}{2\sqrt{2}} \\ 0 \\ \frac{3\sqrt{5}}{2\sqrt{2}} \\ 0 \end{pmatrix}^{(\mathbf{e})} = \begin{pmatrix} \frac{1}{\sqrt{2}} & 0 & -\frac{\sqrt{5}}{2\sqrt{2}} & 0 \\ 0 & \sqrt{\frac{3}{2}} & 0 & -\frac{3\sqrt{7}}{2\sqrt{2}} \\ 0 & 0 & \frac{3\sqrt{5}}{2\sqrt{2}} & 0 \\ 0 & 0 & 0 & \frac{5\sqrt{7}}{2\sqrt{2}} \end{pmatrix} \begin{pmatrix} 0 \\ 0 \\ 1 \\ 0 \end{pmatrix}^{(P)} . \tag{3.45}$$

The column vector to the left corresponds to the polynomial $-\sqrt{5}/(2\sqrt{2}) + 3\sqrt{5}/(2\sqrt{2})x^2$ in the $\mathbf{e}$ basis, the column vector to the right corresponds to the the polynomial $P_2(x)$ in the P-basis, which is exactly the same.

For the other direction one has:

$$1 = \sqrt{2}P_0(x) \tag{3.46}$$

$$x = \sqrt{\frac{2}{3}}P_1(x) \tag{3.47}$$

$$x^2 = \frac{\sqrt{2}}{3}P_0(x) + \frac{2\sqrt{2}}{3\sqrt{5}}P_2(x) \tag{3.48}$$

$$x^3 = \frac{\sqrt{6}}{5}P_1(x) + \frac{2\sqrt{2}}{5\sqrt{7}}P_3(x), \tag{3.49}$$

from where we can read off the inverse of A, A^{-1}, that is the transformation matrix which takes us from the components of a vector in the $\mathbf{e}$-basis to the components of the same vector in the P-basis:

$$A^{-1} = \begin{pmatrix} \sqrt{2} & 0 & \frac{\sqrt{2}}{3} & 0 \\ 0 & \sqrt{\frac{2}{3}} & 0 & \frac{\sqrt{6}}{5} \\ 0 & 0 & \frac{2\sqrt{2}}{3\sqrt{5}} & 0 \\ 0 & 0 & 0 & \frac{2\sqrt{2}}{5\sqrt{7}} \end{pmatrix} \tag{3.50}$$

Next, we consider the operators X and D in the P basis. Let's begin with D. One has

$$P_0'(x) = 0 \tag{3.51}$$

$$P_1'(x) = \sqrt{\frac{3}{2}} = \sqrt{3}P_0(x) \tag{3.52}$$

$$P_2'(x) = \frac{3\sqrt{5}}{\sqrt{2}}x = \sqrt{15}P_1(x) \tag{3.53}$$

$$P_3'(x) = \frac{15\sqrt{7}}{2\sqrt{2}}x^2 - \frac{3\sqrt{7}}{2\sqrt{2}} = \sqrt{7}P_0(x) + \sqrt{35}P_2(x). \tag{3.54}$$

This gives the components of D:

$$D^{(P)} = \begin{pmatrix} 0 & \sqrt{3} & 0 & \sqrt{7} & \cdots \\ 0 & 0 & \sqrt{15} & 0 & \cdots \\ 0 & 0 & 0 & \sqrt{35} & \cdots \\ 0 & 0 & 0 & 0 & \cdots \\ \vdots & \vdots & \vdots & \vdots & \ddots \end{pmatrix} \tag{3.55}$$

For now we also note that D is neither hermitian nor anti-hermitian (since the P-basis is orthonormal, as we can directly infer from the matrix).

Exercise 3.2
Verify that $D^{(P)} = A^{-1}D^{(e)}A$, as it has to be.

For the X-operator we find

$$x P_0(x) = \frac{1}{\sqrt{2}}x = \frac{1}{\sqrt{3}}P_1(x) \tag{3.56}$$

$$x P_1(x) = \sqrt{\frac{3}{2}}x^2 = \frac{2}{\sqrt{15}}P_2(x) + \frac{1}{\sqrt{3}}P_0(x) \tag{3.57}$$

$$x P_2(x) = \frac{\sqrt{5}}{2\sqrt{2}}(3x^3 - x) = \frac{3}{\sqrt{35}}P_3(x) + \frac{2}{\sqrt{15}}P_1(x) \tag{3.58}$$

$$x P_3(x) = \frac{\sqrt{7}}{2\sqrt{2}}(5x^4 - 3x^2) = ? \tag{3.59}$$

We cannot provide a linear combination for the last vector, $X|P_3\rangle$, since we haven't determined $|P_4\rangle$. In contrast to the derivative operator, the X-operator leads to polynomials of higher degree. If we denote by $\mathrm{Pol}_3([-1, 1], \mathbb{R})$ the subspace of polynomials of degree ≤ 3, then $D\,\mathrm{Pol}_3([-1, 1], \mathbb{R}) \subseteq \mathrm{Pol}_3([-1, 1], \mathbb{R})$, but $X\,\mathrm{Pol}_3([-1, 1], \mathbb{R}) \not\subseteq \mathrm{Pol}_3([-1, 1], \mathbb{R})$.

So far the matrix of the X-operator looks like this:

$$X^{(P)} = \begin{pmatrix} 0 & \frac{1}{\sqrt{3}} & 0 & ? & \cdots \\ \frac{1}{\sqrt{3}} & 0 & \frac{2}{\sqrt{15}} & ? & \cdots \\ 0 & \frac{2}{\sqrt{15}} & 0 & ? & \cdots \\ 0 & 0 & \frac{3}{\sqrt{35}} & ? & \cdots \\ \vdots & \vdots & \vdots & \vdots & \ddots \end{pmatrix} \tag{3.60}$$

The question marks stand for the unknown coefficients of $|P_0\rangle$ to $|P_3\rangle$ in $X|P_3\rangle$. Can we determine these coefficients without knowing $|P_4\rangle$? The answer is yes.

You have certainly noticed that P_n for even n contains only even powers of x, for odd n only odd powers. This can be proven by induction from the way in which we construct the P_n: we begin with the monomial x^n and orthogonalize this to the already determined P_k with $k < n$. For even n, x^n is already orthogonal to all odd powers of x, for odd n orthogonal to all even powers of x, since $\int_{-1}^{1} dx\, x^n x^p = 0$ for $n + p$ odd. By the induction hypothesis, P_k for $k < n$ already consists only of even/odd powers of x, if k is even/odd, respectively. Hence P_n needs to be orthogonalized only to those P_k which are like n even/odd. Hereby only such powers of x are added/subtracted to x^n which are like n even/odd. It follows the induction step that also P_n consists only of even/odd powers of x.

The polynomial $X|P_3\rangle$ consists only of even powers of x and therefore gets no contribution from $|P_1\rangle$ and $|P_3\rangle$. We can thus replace the second and fourth question mark with zero. But what about the contributions of $|P_0\rangle$ and $|P_2\rangle$ (first and third question mark)? Here it comes handy that we are operating with an orthonormal basis, for this implies the relation $|v\rangle = \sum_n |P_n\rangle\langle P_n|v\rangle$. In particular, one has for the (ij) entry in the matrix $X^{(P)}$:

$$X_{ij}^{(P)} = \langle P_i|X|P_j\rangle = \int_{-1}^{1} dx\, P_i(x)x P_j(x) \tag{3.61}$$

$$= \int_{-1}^{1} dx\, P_j(x)x P_i(x) = \langle P_j|X|P_i\rangle = X_{ji}^{(P)} \tag{3.62}$$

Hence, the X-operator is symmetric (as a matrix in an orthonormal basis) and thus also hermitian (a basis-independent property). Now we can also replace the remaining question marks with values:

$$X^{(P)} = \begin{pmatrix} 0 & \frac{1}{\sqrt{3}} & 0 & 0 & \cdots \\ \frac{1}{\sqrt{3}} & 0 & \frac{2}{\sqrt{15}} & 0 & \cdots \\ 0 & \frac{2}{\sqrt{15}} & 0 & \frac{3}{\sqrt{35}} & \cdots \\ 0 & 0 & \frac{3}{\sqrt{35}} & 0 & \cdots \\ \vdots & \vdots & \vdots & \vdots & \ddots \end{pmatrix}. \tag{3.63}$$

Note that we were able to determine the composition of $X|P_3\rangle$ even without knowing its leading ingredient $P_4(x)$. "Normally" it would have been our job to determine $P_4(x)$ first, then subtract the $|P_4\rangle$-part from $X|P_3\rangle$, see which coefficients of x^0 and x^2 remain, and from that find the contributions of $|P_0\rangle$ and $|P_2\rangle$. By making use of our knowledge of Linear Algebra, we could avoid this calculation. The strength of this algebraic perspective will show again and again, most impressively perhaps for the harmonic oscillator.

Exercise 3.3
Calculate $P_4(x)$ by orthogonalizing x^4 with respect to P_0 and P_2, and determine the composition of $X|P_3\rangle$ from that.

Exercise 3.4
Calculate $A^{-1}X^{(e)}A$ (where all matrices are restricted to the first four rows and columns, analogous to the corresponding exercise with the D-operator.) Why does the result not match with $X^{(P)}$ this time?

A different basis for a space of functions $[-1, 1] \rightarrow \mathbb{R}$ is given by the **Fourier expansion**. For that, we consider the space Four$([-1, 1], \mathbb{R})$ of functions $f : [-1, 1] \rightarrow \mathbb{R}$ with the following properties (the so-called **Dirichlet conditions**):

- $f(-1) = f(1)$
- f has only finitely many maxima and minima.
  (A counterexample would be $f(x) = \sin(1/x)$ for $x \neq 0$, $f(0) = 0$.)
- f has only finitely many discontinuities.
- At each discontinuity x_0 one has

$$f(x_0) = \frac{1}{2}\left(\lim_{x \nearrow x_0} f(x) + \lim_{x \searrow x_0} f(x)\right), \tag{3.64}$$

i.e. $f(x_0)$ is the average of the limits from left and right. (If $x_0 = \pm 1$, the right hand side must be replaced with $\lim_{x \nearrow 1} f(x)$ and $\lim_{x \searrow -1} f(x)$.) An example is the **Heaviside step function**, $\theta(x) = 0$ for $x < 0$, $\theta(x) = \frac{1}{2}$ for $x = 0$, $\theta(x) = 1$ for $x > 0$.

Exercise 3.5
Convince yourself that the functions with these properties form a vector space.

A theorem by Dirichlet says that any of these functions can be expanded in a **Fourier series**,

$$f(x) = a_0 \frac{1}{\sqrt{2}} + \sum_{n=1}^{\infty} a_n \cos n\pi x + b_n \sin n\pi x. \tag{3.65}$$

The functions $|v_i\rangle$ with

$$v_0(x) = \frac{1}{\sqrt{2}}, \quad v_{2n}(x) = \cos n\pi x, \quad v_{2n-1}(x) = \sin n\pi x, \quad n \in \mathbb{N}, \tag{3.66}$$

already form an orthonormal basis (an orthonormal Schauder basis, to be precise!) of the space Four$([-1, 1], \mathbb{R})$, since

$$\langle v_0 | v_0 \rangle = \int_{-1}^{1} dx \, \frac{1}{2} = 1 \tag{3.67}$$

$$\langle v_0 | v_{2n} \rangle = \int_{-1}^{1} dx \, \frac{1}{\sqrt{2}} \cos n\pi x = 0 \tag{3.68}$$

$$\langle v_0 | v_{2n-1} \rangle = \int_{-1}^{1} dx \, \frac{1}{\sqrt{2}} \sin n\pi x = 0 \tag{3.69}$$

$$\langle v_{2m} | v_{2n} \rangle = \int_{-1}^{1} dx \, \cos m\pi x \cos n\pi x = \delta_{mn} \tag{3.70}$$

$$\langle v_{2m-1} | v_{2n-1} \rangle = \int_{-1}^{1} dx \, \sin m\pi x \sin n\pi x = \delta_{mn} \tag{3.71}$$

$$\langle v_{2m} | v_{2n-1} \rangle = \int_{-1}^{1} dx \, \cos m\pi x \sin n\pi x = 0. \tag{3.72}$$

Therefore, coefficients can be determined via scalar products again, $|f\rangle = \sum_{i=0}^{\infty} |v_i\rangle \langle v_i | f \rangle$, and so

$$a_0 = \langle v_0 | f \rangle = \int_{-1}^{1} dx \, \frac{1}{2} f(x) \tag{3.73}$$

$$a_n = \langle v_{2n} | f \rangle = \int_{-1}^{1} dx \, \cos n\pi x f(x) \tag{3.74}$$

$$b_n = \langle v_{2n-1} | f \rangle = \int_{-1}^{1} dx \, \sin n\pi x f(x). \tag{3.75}$$

The derivative operator D in this basis results from the relations

$$v'_{2n}(x) = -n\pi \sin n\pi x, \quad v'_{2n-1}(x) = n\pi \cos n\pi x, \tag{3.76}$$

hence

$$D^{(v)} = \begin{pmatrix} 0 & 0 & 0 & 0 & 0 & \cdots \\ 0 & 0 & -\pi & 0 & 0 & \cdots \\ 0 & \pi & 0 & 0 & 0 & \cdots \\ 0 & 0 & 0 & 0 & -2\pi & \cdots \\ 0 & 0 & 0 & 2\pi & 0 & \cdots \\ \vdots & \vdots & \vdots & \vdots & \vdots & \ddots \end{pmatrix}.$$

(3.77)

Note that D is antisymmetric in this basis. Therefore, D is antihermitian on the function space Four$([-1, 1], \mathbb{R})$ (a basis-independent property). What distinguishes Four$([-1, 1], \mathbb{R})$ from PR$([-1, 1], \mathbb{R})$, that D has this property in one space but not in the other? In the Fourier case,

$$D_{ij}^{(v)} = \langle v_i | D | v_j \rangle = \int_{-1}^{1} dx\, v_i(x) v_j'(x)$$

(3.78)

$$= -\int_{-1}^{1} dx\, v_i'(x) v_j(x) + v_i(x) v_j(x)|_{-1}^{1}$$

(3.79)

$$= -\langle v_j | D | v_i \rangle + 0 = -D_{ji}^{(v)}.$$

(3.80)

The boundary term in the second row vanishes, because the functions are periodic: One always has $v_i(1) = v_i(-1)$. This is exactly the relevant difference to the space of power series PR$([-1, 1], \mathbb{R})$.

The result can be generalized: **On a function space $F([a, b], K)$ the derivative operator D is antihermitian if and only if $f(a) = f(b)$ for all $f \in F([a, b], K)$.**

This also holds in the limit $a \to -\infty$, $b \to \infty$, i.e. for function spaces $F(\mathbb{R}, K)$. Here the condition for D to be antihermitian is $\lim_{x \to \infty} f(x) = \lim_{x \to -\infty} f(x)$. The condition holds for the space of square-integrable functions, $L^2(\mathbb{R}, K)$, since these functions converge to 0 for $x \to \pm\infty$.

The X-operator looks quite complicated in the Fourier basis. Its components $X_{ij}^{(v)} = \langle v_i | X | v_j \rangle$ are given by the corresponding integrals, which we are not going to compute here though. As in any orthonormal basis, $X^{(v)}$ is symmetric. You can determine the first row/column, exemplarily:

Exercise 3.6
Show via partial integration that

$$X | v_0 \rangle = \sum_{n=1}^{\infty} \frac{(-1)^n \sqrt{2}}{n\pi} | v_{2n-1} \rangle.$$

(3.81)

To what extent can the Schauder bases $|P_i\rangle$ of $PR([-1, 1], \mathbb{R})$ and $|v_i\rangle$ of Four$([-1, 1], \mathbb{R})$ be transformed into each other? It is clear that this basis transformation can work only on the intersection of the two function spaces. There are elements of $PR([-1, 1], \mathbb{R})$ which cannot be written as Fourier series, since they don't obey all of the mentioned requirements from Dirichlet's theorem. In particular, this includes all functions f with $f(1) \neq f(-1)$. Vice versa, there are Fourier series which cannot be written as power series. In particular, this includes the non-continuous functions, as for example the above-mentioned Heaviside step function.

The basis vectors $|v_i\rangle$ can be written as power series,

$$\sin n\pi x = \sum_{k=0}^{\infty} (-1)^k \frac{(n\pi x)^{2k+1}}{(2k+1)!}, \quad \cos n\pi x = \sum_{k=0}^{\infty} (-1)^k \frac{(n\pi x)^{2k}}{(2k)!}, \qquad (3.82)$$

i.e. as an infinite linear combination of the monomials. For each monomial, one can calculate how to represent it as a combination of Legendre polynomials. From that, one can in principle determine the infinite transformation matrix that takes us from the v-basis into the P-basis. But can it actually be applied?

The problem originates from the impossibility to reorder a series in a desired way. The series $1 - \frac{1}{2} + \frac{1}{3} - \frac{1}{4} \pm \cdots$ is known to converge to $\ln 2$. However, if one tries to reorder the terms, the result will not be the same, or there may not even be a result at all. If one, for instance, pulls all positive terms to the front, $(1 + \frac{1}{3} + \cdots) - (\frac{1}{2} + \frac{1}{4} + \cdots)$, one gets two divergent series which are completely useless.

In case of the Heaviside function, we already know that something has to go wrong, for a corresponding power series cannot exist. One easily verifies that $\theta(x)$ is in the interval $[-1, 1]$ given by the Fourier series

$$\theta(x) = \frac{1}{2} + \frac{2}{\pi} \left(\frac{\sin \pi x}{1} + \frac{\sin 3\pi x}{3} + \frac{\sin 5\pi x}{5} + \cdots \right). \qquad (3.83)$$

If we now try to turn this into a power series, by writing each single sine function as a power series, adding them up, and then sorting them by powers of x, we realize that it won't work. Already the coefficient of x^1, $2(1 + 1 + 1 + \cdots)$, diverges, and the coefficients of the higher monomials diverge too.

The process doesn't even work on the intersections of $PR([-1, 1], \mathbb{R})$ and Four$([-1, 1], \mathbb{R})$. From (3.81) we know that in $[-1, 1]$

$$x = \frac{2}{\pi} \left(\frac{\sin \pi x}{1} - \frac{\sin 2\pi x}{2} + \frac{\sin 3\pi x}{3} - \frac{\sin 4\pi x}{4} \pm \cdots \right). \qquad (3.84)$$

If one now tries to write the right hand side as a sum of power series and to sort them by powers of x, one might expect to retrieve the left hand side. But that is not the case. The coefficient for x^1 is $2(1 - 1 + 1 - 1 \pm \cdots)$, and for all higher odd powers of x the coefficients diverge too.

This is the problem with Schauder bases. One always has to deal with infinite series that cannot be reordered, and with basis vectors that cannot be arbitrarily combined. In this way, some of the rules know from the Linear Algebra of finite-dimensional vector spaces are invalidated. In the conversion between the e- and the P-basis of $PR([-1, 1], \mathbb{R})$, the problem did not occur, since here each element of one basis had only finitely many contributions from elements of the other.

Considering complex Fourier series, Four$([-1, 1], \mathbb{C})$, there is a basis which is a bit simpler to use than the v-basis: the new basis $|w_i\rangle$ is given by

$$w_0(x) = \frac{1}{\sqrt{2}}, \quad w_{2n}(x) = \frac{1}{\sqrt{2}} e^{in\pi x}, \quad w_{2n-1}(x) = \frac{1}{\sqrt{2}} e^{-in\pi x}. \quad (3.85)$$

All complex-valued functions on the interval $[-1, 1]$ obeying the Dirichlet conditions can be expressed as a series in this basis. Again, it is an orthonormal basis. Please note that you now have to complex conjugate the first factor in scalar products, e.g.

$$\langle w_{2n}|w_{2n}\rangle = \int_{-1}^{1} dx \, w_{2n}^*(x) w_{2n}(x) = \int_{-1}^{1} dx \, \frac{1}{2} e^{-in\pi x} e^{in\pi x} = 1. \quad (3.86)$$

Exercise 3.7
Verify that the w-basis is orthonormal, $\langle w_i|w_j\rangle = \delta_{ij}$.

The v-Basis is still a basis of Four$([-1, 1], \mathbb{C})$, now with complex-valued coefficients. The two bases can be easily converted into each other. Since

$$e^{ix} = \cos x + i \sin x \Rightarrow |w_{2n}\rangle = \frac{1}{\sqrt{2}} (|v_{2n}\rangle + i|v_{2n-1}\rangle), \quad (3.87)$$

$$e^{-ix} = \cos x - i \sin x \Rightarrow |w_{2n}\rangle = \frac{1}{\sqrt{2}} (|v_{2n}\rangle - i|v_{2n-1}\rangle), \quad (3.88)$$

the transformation matrix from the w-basis to the v-basis reads

$$T = \begin{pmatrix} 1 & 0 & 0 & 0 & 0 & \cdots \\ 0 & -\frac{i}{\sqrt{2}} & \frac{i}{\sqrt{2}} & 0 & 0 & \cdots \\ 0 & \frac{1}{\sqrt{2}} & \frac{1}{\sqrt{2}} & 0 & 0 & \cdots \\ 0 & 0 & 0 & -\frac{i}{\sqrt{2}} & \frac{i}{\sqrt{2}} & \cdots \\ 0 & 0 & 0 & \frac{1}{\sqrt{2}} & \frac{1}{\sqrt{2}} & \cdots \\ \vdots & \vdots & \vdots & \vdots & \vdots & \ddots \end{pmatrix}. \quad (3.89)$$

Due to its block diagonal form, this infinite matrix can be easily inverted. The transformation matrix from the v-basis to the w-basis therefore reads

$$
T^{-1} = \begin{pmatrix}
1 & 0 & 0 & 0 & 0 & \cdots \\
0 & \frac{i}{\sqrt{2}} & \frac{1}{\sqrt{2}} & 0 & 0 & \cdots \\
0 & -\frac{i}{\sqrt{2}} & \frac{1}{\sqrt{2}} & 0 & 0 & \cdots \\
0 & 0 & 0 & \frac{i}{\sqrt{2}} & \frac{1}{\sqrt{2}} & \cdots \\
0 & 0 & 0 & -\frac{i}{\sqrt{2}} & \frac{1}{\sqrt{2}} & \cdots \\
\vdots & \vdots & \vdots & \vdots & \vdots & \ddots
\end{pmatrix}.
\tag{3.90}
$$

This matches the known relations

$$
\sin x = \frac{i}{2}\left(e^{-ix} - e^{ix}\right) \Rightarrow |v_{2n-1}\rangle = \frac{1}{\sqrt{2}}\left(i|w_{2n-1}\rangle - i|w_{2n}\rangle\right), \tag{3.91}
$$

$$
\cos x = \frac{1}{2}\left(e^{-ix} + e^{ix}\right) \Rightarrow |v_{2n}\rangle = \frac{1}{\sqrt{2}}\left(|w_{2n-1}\rangle + |w_{2n}\rangle\right). \tag{3.92}
$$

If you don't believe the integrals of (3.70)–(3.72), you can derive the orthonormality of the v-basis from the orthonormality of the w-basis:

Exercise 3.8
Show, using the orthonormality of the w-basis and the relations (3.91) and (3.92), that $\langle v_i | v_j \rangle = \delta_{ij}$ for $i, j > 0$. Don't compute any integrals, only use already known scalar products.

The derivative operator D is, thanks to the simple rule $\frac{d}{dx}e^{\alpha x} = \alpha e^{\alpha x}$, already diagonal in the w-basis,

$$
D^{(w)} = \begin{pmatrix}
0 & 0 & 0 & 0 & 0 & \cdots \\
0 & -i\pi & 0 & 0 & 0 & \cdots \\
0 & 0 & i\pi & 0 & 0 & \cdots \\
0 & 0 & 0 & -2i\pi & 0 & \cdots \\
0 & 0 & 0 & 0 & 2i\pi & \cdots \\
\vdots & \vdots & \vdots & \vdots & \vdots & \ddots
\end{pmatrix}.
\tag{3.93}
$$

The w-basis is thus an eigenbasis of D. As it is right and proper for an antihermitian operator, D has only imaginary eigenvalues.

Exercise 3.9
Show that $D^{(w)} = T^{-1} D^{(v)} T$.

The w-basis is "simpler" than the v-basis, because exponential functions are more easy to calculate with than sine and cosine. The real functions in Four$([-1, 1], \mathbb{R})$ can also be written as linear combinations of functions from the w-basis: Setting

$$f(x) = \alpha_0 + \sum_{n=1}^{\infty} \alpha_n e^{in\pi x} + \beta_n e^{-in\pi x} \tag{3.94}$$

(where the factors $1/\sqrt{2}$ have been absorbed into the coefficients α_n and β_n), we see that f is real if and only if α_0 is real and $\beta_n = \alpha_n^*$ for all n. **Although the w-basis consists of functions which do not lie in** Four$([-1, 1], \mathbb{R})$ **(because they are complex), we can use them as a Schauder basis for** Four$([-1, 1], \mathbb{R})$. Only the coefficients have to fulfill certain relations for a linear combination of basis vectors to be in Four$([-1, 1], \mathbb{R})$. But this holds for Schauder bases anyway: Not every infinite linear combination of basis vectors gives an element of the function space. This restriction gets only somewhat tightened when we restrict ourselves to a space which does not even include the basis vectors themselves.

Since the basis vectors are not elements of the space they serve as a basis for, they are better called a **pseudo-basis**, and the basis vectors themselves denoted as **pseudo-vectors** or **improper vectors**. In the next section we will construct two important pseudo-bases for the space $L^2(\mathbb{R}, \mathbb{C})$ of square-integrable functions.

Self-check questions:

1. How are scalar product and norm defined in $L^2(\mathbb{R}, \mathbb{C})$?
2. On which function spaces is the derivative operator D antihermitian?
3. What is the meaning of the following statement: "The functions $e^{in\pi x}$, $n \in \mathbb{Z}$, form a pseudo-basis of Four$([-1, 1], \mathbb{R})$"?

3.3 Pseudo-Vectors and Fourier Transformation

In this section we are dealing with the space $L^2(\mathbb{R}, \mathbb{C})$ of square-integrable functions. We want to construct pseudo-bases in which the X- or the D-operator is diagonal, respectively, i.e. in which the pseudo-basis vectors are pseudo-eigenvectors of the corresponding operator. These two bases are of fundamental importance for the QM of wave functions.

We remind ourselves of the eigenvalue (3.14) for the X-operator. The problem was that the factor x on the left hand side varies over the entire real line, while the factor of λ on the right hand side is fixed. In PR$(\mathbb{R}, \mathbb{R})$ there was no solution. A solution would have to be a function which contributes only at a single value of x. In $L^2(\mathbb{R}, \mathbb{C})$, such functions exist for any real value of λ: Set $f_\lambda(x) = 0$ for $x \neq \lambda$, and $f_\lambda(\lambda) = a$ with an arbitrary constant a. The functions f_λ are eigenvectors of eigenvalue λ, (3.14) is fulfilled. But there is a problem: The scalar product of f_λ with each function g in $L^2(\mathbb{R}, \mathbb{C})$ vanishes: $\int_{-\infty}^{\infty} dx \, f_\lambda^*(x) g(x) = 0$. Therefore f_λ is ineligible as a basis vector.

The way out is to replace the functions f_λ with the **Dirac delta distributions** $\delta_\lambda(x) := \delta(x - \lambda)$. These are defined as

$$\delta_\lambda(x) = 0 \text{ for } x \neq \lambda, \quad \int_{-\infty}^{\infty} dx\, \delta_\lambda(x) g(x) = g(\lambda). \quad (3.95)$$

With $g(x) = 1$ one has in particular $\int_{-\infty}^{\infty} dx\, \delta_\lambda(x) = 1$. Loosely speaking, δ_λ is a version of f_λ where the value a was "multiplied up to infinity" in such a way that the scalar product does not vanish. δ_λ is not a function in the strict sense, since the function value $\delta_\lambda(\lambda)$ is undefined. It is insufficient to say "$\delta_\lambda(\lambda) = \infty$", since neither is ∞ in $\mathbb{C}$, nor is the integral determined this way. At least we can say that δ_λ is real in the following formal sense: If one complex conjugates the integral in (3.95),

$$\int_{-\infty}^{\infty} dx\, \delta_\lambda^*(x) g^*(x) = g^*(\lambda), \quad (3.96)$$

one gets the same as if one lets δ_λ act on g^* directly,

$$\int_{-\infty}^{\infty} dx\, \delta_\lambda(x) g^*(x) = g^*(\lambda). \quad (3.97)$$

The effects of δ_λ^* and δ_λ under the integral are thus the same, and in this sense we can say $\delta_\lambda^* = \delta_\lambda$ and therefore

$$\langle \delta_\lambda | f \rangle = \int_{-\infty}^{\infty} dx\, \delta_\lambda^*(x) f(x) = \int_{-\infty}^{\infty} dx\, \delta_\lambda(x) f(x) = f(\lambda). \quad (3.98)$$

As a pseudo-vector, δ_λ is not square-integrable:

$$\langle \delta_\lambda | \delta_\lambda \rangle = \int_{-\infty}^{\infty} dx\, \delta_\lambda(x) \delta_\lambda(x) = \delta_\lambda(\lambda) = \text{undefined} \quad (3.99)$$

Although the δ_λ are not contained in $L^2(\mathbb{R}, \mathbb{C})$, they can be used as a pseudo-basis of $L^2(\mathbb{R}, \mathbb{C})$. In the following, we write x_0 (or x_1 etc. or simply x) instead of λ, since it represents a certain value of x where δ_{x_0} gives a contribution, and $|x_0\rangle$ for the vector δ_{x_0}. In this way, we follow the convention of the ket notation to use the eigenvalue itself as the identifier of the vector.

So, the set $\{|x_0\rangle,\ x_0 \in \mathbb{R}\}$ is a pseudo-basis of $L^2(\mathbb{R}, \mathbb{C})$. It is orthogonal,

$$\langle x_0 | x_1 \rangle = \int_{-\infty}^{\infty} dx\, \delta_{x_0}(x) \delta_{x_1}(x) = \delta_{x_1}(x_0) = 0 \text{ for } x_1 \neq x_0. \quad (3.100)$$

It is complete, since f is completely determined by the values $f(x) = \langle x | f \rangle$ for all $x \in \mathbb{R}$. It is not a Schauder basis (not even a Schauder pseudo-basis), for it is not countable, but continuous and therefore uncountable. Hence, one cannot write f as a sum (not even an infinite sum) of the basis vectors, but only as an integral

$$|f\rangle = \int_{-\infty}^{\infty} dx \, |x\rangle\langle x|f\rangle = \int_{-\infty}^{\infty} dx \, f(x)|x\rangle \tag{3.101}$$

We will drop the prefix "pseudo" in the following from time to time. It should be clear by now that these basis vectors are not contained in $L^2(\mathbb{R}, \mathbb{C})$. And in fact the $|x\rangle$ *are* vectors, only in a different vector space, e.g. in the vector space of distributions. So, the vectors $|x_0\rangle$ are eigenvectors of the X-operator, obeying $X|x_0\rangle = x_0|x_0\rangle$. One says the X-operator is "diagonal" in the x-basis, although with an uncountable basis one can no longer represent an operator by a matrix consisting of rows and columns. But formally one can still write:

$$X_{x_1 x_2}^{(x)} = \langle x_1|X|x_2\rangle = x_2 \delta(x_1 - x_2) \tag{3.102}$$

One can understand the Dirac delta distribution in different ways:

- as the density distribution of a point particle. The density vanishes everywhere except for a single point where it is infinite. The integral over the density yields a certain value (for example the mass of the particle if we consider the mass density).
- as a continuous version of basis vector components. Just as in the countable case the i-th basis vector $\mathbf{e}^{(i)}$ has the components $e_j^{(i)} = \delta_{ij}$, in the continuous case the basis vectors $\mathbf{e}^{(x_0)} := |x_0\rangle$ have the "components"

$$\mathbf{e}^{(x_0)}(x) = \delta_{x_0}(x) = \delta(x - x_0). \tag{3.103}$$

- as matrix entries of the unit matrix. Just as in the countable case one has $\mathbf{1}_{ij} = \langle i|\mathbf{1}|j\rangle = \delta_{ij}$, in the continuous case this becomes

$$\mathbf{1}_{xx'} = \langle x|\mathbf{1}|x'\rangle = \delta(x - x') \tag{3.104}$$

and thus

$$\mathbf{1} = \int_{-\infty}^{\infty} dx \int_{-\infty}^{\infty} dx' \, |x\rangle \mathbf{1}_{xx'} \langle x'| = \int_{-\infty}^{\infty} dx \, |x\rangle\langle x|. \tag{3.105}$$

In other words, the unit operator now is the integral over all projection operators $|x\rangle\langle x|$. The definition of the scalar product on $L^2(\mathbb{R}, \mathbb{C})$ can be obtained from the abstract scalar product by inserting a one:

$$\langle f|g\rangle = \langle f|\mathbf{1}|g\rangle = \int_{-\infty}^{\infty} dx \, \langle f|x\rangle\langle x|g\rangle = \int_{-\infty}^{\infty} dx \, f^*(x)g(x) \tag{3.106}$$

- as a linear **functional** (a linear form on a function space is called functional) which maps each function f to a number: $\hat{\delta}_{x_0}(f) := \langle x_0|f\rangle = f(x_0)$.

Nerd's Corner 3.1

Although we've already discussed the mathematical background in much more detail than most other QM books, many aspects could be only briefly touched. For example, one might ask about the dual space of $L^2(\mathbb{R}, \mathbb{C})$. For an infinite-dimensional vector space V, the dual space V' no longer needs to be isomorphic to V. What are the consequences for our bra/ket notation, where each bra vector is an element of the dual space of $L^2(\mathbb{R}, \mathbb{C})$? Fortunately one can prove that the dual space of $L^2(\mathbb{R}, \mathbb{C})$ is again $L^2(\mathbb{R}, \mathbb{C})$, hence there are no additional complications.

A thoughtful reader may be surprised. Didn't we just say that the Dirac delta distribution is a linear functional? But now it should nevertheless not be contained in the dual space of $L^2(\mathbb{R}, \mathbb{C})$? There are two reasons for this:

1. As a functional the delta distribution is not continuous. Were $\hat{\delta}_\lambda$ continuous, this would imply: For each $\epsilon > 0$ there is a $\delta > 0$ such that $\hat{\delta}_\lambda(f) - \hat{\delta}_\lambda(0_{\mathbb{R},\mathbb{C}}) < \epsilon$ if $||f - 0_{\mathbb{R},\mathbb{C}}|| < \delta$. But if we set $f = f_\lambda$ with f_λ as above: $f_\lambda(x) = 0$ for $x \neq \lambda$, $f_\lambda(\lambda) = a$ and $a > \epsilon$, then

$$||f - 0_{\mathbb{R},\mathbb{C}}|| = \sqrt{\langle f_\lambda | f_\lambda \rangle} = 0 < \delta, \qquad (3.107)$$

but

$$\hat{\delta}_\lambda(f) - \hat{\delta}_\lambda(0_{\mathbb{R},\mathbb{C}}) = a > \epsilon. \qquad (3.108)$$

In finitely many dimensions, each linear function and each linear functional is continuous, but not in infinitely many dimensions.

In order to make certain desirable theorems hold, the dual space is only defined as the set of *continuous* linear functionals. Distributions however need not be continuous.

2. That the f_λ-functions have norm zero poses a problem in mathematics. For by definition, only the null vector—in our case $0_{\mathbb{R},\mathbb{C}}$—is supposed to have zero norm. One therefore defines functions in $L^2(\mathbb{R}, \mathbb{C})$ as *equivalent* if they differ only on a null set (a set of points with zero extension). By this definition, f_λ is "the same element" of $L^2(\mathbb{R}, \mathbb{C})$ as $0_{\mathbb{R},\mathbb{C}}$. This makes also sense from a physical perspective: we will interpret the functions of $L^2(\mathbb{R}, \mathbb{C})$ (more precisely, the norm squared of their values) as probability densities. In order to compute probabilities from that, one always has to perform an integral. Functions differing only on a null set yield the same result under integration.

But with this definition, the delta distribution is not even a functional on $L^2(\mathbb{R}, \mathbb{C})$ any more. For if f_λ is "the same element" of $L^2(\mathbb{R}, \mathbb{C})$ as $0_{\mathbb{R},\mathbb{C}}$, but $\hat{\delta}_\lambda$ yields two different results (namely once a and once 0), then $\hat{\delta}_\lambda$ is no longer well-defined.

In fact, in QM one mostly assumes that functions are not only square-integrable but also continuous. This means we are working with $L^2(\mathbb{R}, \mathbb{C}) \cap C^0(\mathbb{R}, \mathbb{C})$, the intersection of square-integrable and continuous functions. Here, $\hat{\delta}_\lambda$ is well-defined again, since continuous functions never differ only on a null set. In exchange one gets other problems, for the dual space of $L^2(\mathbb{R}, \mathbb{C}) \cap C^0(\mathbb{R}, \mathbb{C})$ is no longer isomorphic to $L^2(\mathbb{R}, \mathbb{C}) \cap C^0(\mathbb{R}, \mathbb{C})$. And whether $\hat{\delta}_\lambda$ is continuous—and therefore belongs to the dual space—depends on which norm one uses. With the L^2-norm, defined via the integral, $\hat{\delta}_\lambda$ is not continuous. With the C^0-Norm, defined as the supremum of the absolute value of the function,

$$\|f\|_{C^0} := \sup(|f|), \tag{3.109}$$

$\hat{\delta}_\lambda$ *is* continuous.

This mixture of several function spaces has led to the notion of the **Gelfand triple** $S(\mathbb{R}, \mathbb{C}) \subset L^2(\mathbb{R}, \mathbb{C}) \subset S^*(\mathbb{R}, \mathbb{C})$. Here S is a subspace of L^2, namely the space of "physically meaningful" functions, for which additional properties hold (continuity, differentiability, maybe more). The dual space S^* of S is *larger* than L^2, in particular it contains distributions like the delta distribution. In bra/ket expressions $\langle f|g \rangle$, $\langle f|$ can be an element of S^* if $|g\rangle$ is an element of S. Via $\langle g|f \rangle = \langle f|g \rangle^*$ the opposite constellation is also defined.

As you see, we easily get into hell's kitchen if we try to take into account each mathematical subtlety. Further problems arise with the question whether or not we want to allow for functions with poles (like for example $f(x) = |x|^{-1/2}e^{-x^2}$). Such functions are not elements of $L^2(\mathbb{R}, \mathbb{C})$, for there *any* real value must be mapped to a complex value (and not ∞). Now if we include functions with poles, $\hat{\delta}_\lambda$ is again not defined for functions having a pole in λ etc.

As physicists, we rarely spend much time thinking about such subtle issues. And this is for good reason. Not so much because it plays no role for calculations in practice. (We want to understand what we are doing, not just apply cooking recipes.) Rather because in physics, any theory has to be considered as an approximation in the first place, valid within a certain range of scales, but losing its validity at some point. For example, we don't know if space on a scale of 10^{-33} cm still has the same three-dimensional continuous quality as on the scales we are familiar with. We cannot even say whether space contains infinitely many points or if continuity is only an approximation used by us. On the side of large scales, we don't know if space is infinite. Therefore, many of the distinctions which have to be made in an exact manner in mathematics are not so relevant for physics. For instance, we calculate with the delta distribution as a density distribution when we consider objects smaller than the resolution of our best measurement devices. Or smaller than all other objects occurring

in our calculation. Whether the object is indeed infinitely small or just tremendously small is something we often cannot judge. Similarly, we don't know
whether the Hilbert space of our universe is really infinite-dimensional or just
of tremendously large but finite dimension. For these reasons, we are well
advised not to get lost in mathematical subtleties.

Nevertheless I find it important for a physicist to understand the mathematical foundations too, and not just to sweep over everything vaguely. Accordingly, this part of the book takes quite some space.

What does the derivative operator D look like in the x-basis? For that, we make
use of the derivative of the delta distribution, which is formally defined via its effect
on a differentiable function under the integral. One defines via partial integration

$$\int_{-\infty}^{\infty} dx\, \delta'_{x_0}(x) f(x) := - \int_{-\infty}^{\infty} dx\, \delta_{x_0}(x) f'(x) = -f'(x_0). \tag{3.110}$$

It can then be conclude that

$$D_{xx'}^{(x)} = \langle x|D|x'\rangle = \delta'(x - x') \tag{3.111}$$

(we write $\delta'(x-x')$ synonymously to $\delta'_{x'}(x)$ and $\frac{d}{dx}\delta(x-x')$), because this is needed
for

$$\langle x|D|f\rangle = \int_{-\infty}^{\infty} dx'\, \langle x|D|x'\rangle\langle x'|f\rangle \tag{3.112}$$

$$= \int_{-\infty}^{\infty} dx'\, \frac{d}{dx}\delta(x - x') f(x') \tag{3.113}$$

$$= - \int_{-\infty}^{\infty} dx'\, \frac{d}{dx'}\delta(x - x') f(x') \tag{3.114}$$

$$= - \int_{-\infty}^{\infty} dx'\, \frac{d}{dx'}\delta(x' - x) f(x') \tag{3.115}$$

$$= \frac{d}{dx'} f(x')|_{x'=x} = f'(x). \tag{3.116}$$

This is how it has to be, since $\langle x|D|f\rangle$ means the derivative of f, evaluated at the
point x. In the above calculation we used the fact that

$$\frac{d}{dx}\delta(x - x') = -\frac{d}{dx'}\delta(x - x'), \qquad \delta(x - x') = \delta(x' - x). \tag{3.117}$$

The first is a standard rule for derivatives, which formally needs to hold also for
distributions. The second follows from $\delta(x - x')$ being real, in the sense mentioned
above, and thus $\langle x|x'\rangle = \langle x'|x\rangle$.

We have shown that the derivative operator D is antihermitian on $L^2(\mathbb{R}, \mathbb{C})$. It therefore has only imaginary eigenvalues. The eigenfunctions are obviously

$$g_k(x) = \frac{1}{\sqrt{2\pi}} e^{ikx} \qquad (3.118)$$

with arbitrary k (the prefactor will become clear in a moment). Compare the discussion of the w-Basis in the previous section: there, only discrete values of k were allowed, namely multiples of π, for we were operating on the finite interval $[-1, 1]$. Now we don't have such a restriction any more.

Unfortunately, the g_k are not contained in $L^2(\mathbb{R}, \mathbb{C})$ though, since

$$\int_{-\infty}^{\infty} dx \, g_k^*(x) g_k(x) = \int_{-\infty}^{\infty} dx \, \frac{1}{2\pi} = \infty. \qquad (3.119)$$

They are again pseudo-vectors. As with the eigenvectors of X, one can however construct square-integrable functions as linear combinations of the g_k. The (continuous) coefficients $\tilde{f}(k)$ of a function $f(x)$ are defined for any L^2-function and are given by the **Fourier transformation** of f:

$$\tilde{f}(k) = \langle k|f \rangle = \frac{1}{\sqrt{2\pi}} \int_{\infty}^{\infty} dx \, e^{-ikx} f(x) \qquad (3.120)$$

Here we denoted the vector g_k as $|k\rangle$, similarly to $|x\rangle$ in the x-basis.

Do the $\{|k\rangle, \; k \in \mathbb{R}\}$ form a pseudo-basis? Does $\{|k\rangle, \; k \in \mathbb{R}\}$ span the entire space $L^2(\mathbb{R}, \mathbb{C})$? This would imply that each L^2-function f can be written as

$$|f\rangle = \int_{-\infty}^{\infty} dk \, |k\rangle\langle k|f\rangle, \qquad (3.121)$$

in particular

$$f(x) = \langle x|f \rangle = \int_{-\infty}^{\infty} dk \, \langle x|k\rangle\langle k|f\rangle \qquad (3.122)$$

$$= \frac{1}{\sqrt{2\pi}} \int_{-\infty}^{\infty} dk \, e^{ikx} \tilde{f}(k). \qquad (3.123)$$

This is just the inverse Fourier transformation. The Fourier integral theorem says that this condition is fulfilled if $f(x)$ obeys the Dirichlet conditions on each finite interval. Functions where this is not the case are again such a mathematical subtlety, but don't play any role in QM, and so we don't consider them any further. With a more or less clear conscience, we thus treat the $\{|k\rangle, \; k \in \mathbb{R}\}$ as a pseudo-basis of $L^2(\mathbb{R}, \mathbb{C})$.

Plugging (3.120) into (3.123), we obtain

$$f(x) = \frac{1}{2\pi} \int_{-\infty}^{\infty} dk \, e^{ikx} \int_{-\infty}^{\infty} dx' \, e^{-ikx'} f(x') \tag{3.124}$$

$$= \int_{-\infty}^{\infty} dx' \left(\frac{1}{2\pi} \int_{-\infty}^{\infty} dk \, e^{ik(x-x')} \right) f(x'). \tag{3.125}$$

Apparently, the expression in brackets has the same effect on a function f as the delta distribution; hence we formally set

$$\frac{1}{2\pi} \int_{-\infty}^{\infty} dk \, e^{ik(x-x')} = \delta(x - x'). \tag{3.126}$$

This is true only in a very formal sense, for $e^{ik(x-x')}$ is actually not integrable in the limits $\pm\infty$, since the function does not converge there.

With this relation we can show that $\{|k\rangle, \ k \in \mathbb{R}\}$ is an orthonormal basis:

$$\langle k|k'\rangle = \frac{1}{2\pi} \int_{-\infty}^{\infty} dx \, e^{ix(k'-k)} = \delta(k' - k) \tag{3.127}$$

We say orthonormal, not just orthogonal, for we *define* orthonormality in a continuous (pseudo-)basis just by (3.127), as a generalization of the discrete $\langle i|j\rangle = \delta_{ij}$. We have thus met two continuous orthonormal bases of $L^2(\mathbb{R}, \mathbb{C})$: $\{|x\rangle, \ x \in \mathbb{R}\}$ and $\{|k\rangle, \ k \in \mathbb{R}\}$, which are related via Fourier transformation. The elements of the first basis are the pseudo-eigenvectors of the X-operator, those of the second basis are the pseudo-eigenvectors of the D-operator.

Similar to the x-basis, we can represent the unit operator in the k-Basis,

$$\mathbf{1}_{kk'} = \langle k|\mathbf{1}|k'\rangle = \delta(k - k'), \tag{3.128}$$

$$\mathbf{1} = \int_{-\infty}^{\infty} dk \int_{-\infty}^{\infty} dk' \, |k\rangle \mathbf{1}_{kk'} \langle k'| = \int_{-\infty}^{\infty} dk \, |k\rangle\langle k|, \tag{3.129}$$

and hence derive the scalar product in the k-basis by inserting a one:

$$\langle f|g\rangle = \langle f|\mathbf{1}|g\rangle = \int_{-\infty}^{\infty} dk \, \langle f|k\rangle\langle k|g\rangle = \int_{-\infty}^{\infty} dk \, \tilde{f}^*(k)\tilde{g}(k) \tag{3.130}$$

The scalar product $\langle f|g\rangle$ is, of course, basis-independent, implying

$$\int_{-\infty}^{\infty} dk \, \tilde{f}^*(k)\tilde{g}(k) = \int_{-\infty}^{\infty} dx \, f^*(x)g(x). \tag{3.131}$$

In particular, $\tilde{f}$ is square-integrable as a function of k if f is as a function of x, and vice versa.

The D-operator is diagonal in the k-basis,

$$D|k\rangle = ik|k\rangle, \tag{3.132}$$

$$D_{kk'}^{(k)} = \langle k|D|k'\rangle = ik'\delta(k - k'). \tag{3.133}$$

On a function $\tilde{f}(k)$, D acts by multiplication with ik:

$$(D\tilde{f})(k) = \langle k|D|f\rangle = \int_{-\infty}^{\infty} dk' \, \langle k|D|k'\rangle\langle k'|f\rangle \tag{3.134}$$

$$= \int_{-\infty}^{\infty} dk' \, ik'\delta(k - k')\tilde{f}(k') = ik\,\tilde{f}(k) \tag{3.135}$$

What does the X-operator look like in this basis?

$$X_{kk'}^{(k)} = \langle k|X|k'\rangle = \int_{-\infty}^{\infty} dx \int_{-\infty}^{\infty} dx' \, \langle k|x\rangle\langle x|X|x'\rangle\langle x'|k'\rangle \tag{3.136}$$

$$= \int_{-\infty}^{\infty} dx \int_{-\infty}^{\infty} dx' \, e^{-ikx} x' \delta(x - x') e^{ik'x'} \tag{3.137}$$

$$= \frac{1}{2\pi} \int_{-\infty}^{\infty} dx \, e^{-ikx} \, x \, e^{ik'x} \tag{3.138}$$

$$= i\frac{d}{dk}\left(\frac{1}{2\pi}\int_{\infty}^{\infty} dx \, e^{i(k'-k)x}\right) = i\frac{d}{dk}\delta(k - k') \tag{3.139}$$

The action on $\tilde{f}(k)$ is thus, analogous to the D-operator in the x-basis (cf. 3.112–3.116):

$$X\tilde{f}(k) = \langle k|X|f\rangle = \int_{-\infty}^{\infty} dk' \, \langle k|X|k'\rangle\langle k'|f\rangle = i\tilde{f}'(k) \tag{3.140}$$

Between the X- and the D-operator and between the x- and the k-basis therefore holds an almost miraculous relation: The X-operator acts via multiplication in the x-basis and via differentiation in the k-basis. The D-operator acts via differentiation in the x-basis and via multiplication in the k-basis. This is all very nice!

By its effect on a function f, we determine the commutator of X and D, which we will need later for the Position-Momentum Uncertainty:

$$[X, D]f(x) = x\frac{d}{dx}f(x) - \frac{d}{dx}(x\,f(x)) \tag{3.141}$$

$$= x\frac{d}{dx}f(x) - \left(\frac{d}{dx}x\right)f(x) - x\frac{d}{dx}f(x) = -f(x) \tag{3.142}$$

Hence

$$[X, D] = -\mathbf{1}. \tag{3.143}$$

We will now show that

$$[V(X), D] = -V'(X) \tag{3.144}$$

for a power series

$$V(X) = \sum \alpha_n X^n. \tag{3.145}$$

This commutator will become relevant when we discuss the momentum operator of a system in a position-dependent potential $V(x)$. By induction we first show

$$[X^n, D] = -nX^{n-1}. \tag{3.146}$$

For $n = 1$, this statement is identical to (3.143). Assume the statement is valid for $n - 1$. Then

$$\begin{aligned}
[X^n, D] &= X^n D - D X^n \\
&= X^{n-1} X D - X^{n-1} D X + X^{n-1} D X - D X^{n-1} X \\
&= X^{n-1}[X, D] + [X^{n-1}, D]X \\
&= -X^{n-1} - (n - 1)X^{n-2} X \\
&= -nX^{n-1}.
\end{aligned}$$

With (3.146) follows

$$[V(X), D] = \sum \alpha_n [X^n, D] = -\sum \alpha_n (n - 1) X^{n-1} = -V'(X). \tag{3.147}$$

We summarize the most important properties of X, D, $|x\rangle$ and $|k\rangle$:

Properties of X, D, $|x\rangle$ and $|k\rangle$

- Eigenvalues:

$$X|x\rangle = x|x\rangle, \quad D|k\rangle = ik|k\rangle \tag{3.148}$$

- Orthonormality:

$$\langle x|x'\rangle = \delta(x - x'), \quad \langle k|k'\rangle = \delta(k - k') \tag{3.149}$$

- Completeness:

$$1 = \int_{-\infty}^{\infty} dx \, |x\rangle\langle x| = \int_{-\infty}^{\infty} dx \, |k\rangle\langle k| \qquad (3.150)$$

- Fourier transformation:

$$\tilde{f}(k) = \langle k|f \rangle = \frac{1}{\sqrt{2\pi}} \int_{-\infty}^{\infty} dx \, e^{-ikx} f(x) \qquad (3.151)$$

$$f(x) = \langle x|f \rangle = \frac{1}{\sqrt{2\pi}} \int_{-\infty}^{\infty} dk \, e^{ikx} \tilde{f}(k) \qquad (3.152)$$

- Scalar product:

$$\langle f|g \rangle = \int_{-\infty}^{\infty} dx \, f^*(x)g(x) = \int_{-\infty}^{\infty} dk \, \tilde{f}^*(k)\tilde{g}(k) \qquad (3.153)$$

- Action on functions:

$$(Xf)(x) = \langle x|X|f \rangle = xf(x) \qquad (3.154)$$
$$(Df)(x) = \langle x|D|f \rangle = f'(x) \qquad (3.155)$$
$$(X\tilde{f})(k) = \langle k|X|f \rangle = i\,\tilde{f}'(k) \qquad (3.156)$$
$$(D\tilde{f})(k) = \langle k|D|f \rangle = ik\,\tilde{f}(k) \qquad (3.157)$$

- Commutator:

$$[X, D] = -\mathbf{1}, \quad [V(X), D] = -V'(X) \qquad (3.158)$$

Self-check questions:

1. What are the pseudo-eigenvectors of X and D?
2. In what sense do they form an orthonormal pseudo-basis of the space of square-integrable functions?
3. How is the Fourier transformed function $\tilde{f}(k)$ of $f(x)$ defined? How do X and D act on $\tilde{f}$?

3.4 Position and Momentum Operator, Correspondence Principle

After such a lengthy preparation of tools, we finally return to quantum mechanics. We begin with the Hamilton formalism of classical mechanics for a point particle moving in one dimension in a potential $V(x)$. The associated **Hamiltonian function** h is a function of the variables x and p, where x is the position and p the momentum of the particle (we only use the Hamiltonian function in this most simple form),

$$h(x, p) = \frac{p^2}{2m} + V(x).$$ (3.159)

This function obeys **Hamilton's equations**,

$$\frac{dx}{dt} = \frac{\partial h}{\partial p}, \qquad \frac{dp}{dt} = -\frac{\partial h}{\partial x}.$$ (3.160)

The first equation implies that the velocity dx/dt of the particle equals momentum divided by mass. The second equation implies that the force dp/dt accelerating the particle equals the negative derivative of the potential.

In QM the observables x and p have to be replaced by operators, and the state of a particle is described by a vector $|\psi\rangle$ in the Hilbert space $L^2(\mathbb{R}, \mathbb{C})$. The function $\psi(x)$ is called **wave function**, for reasons that will become clear in a moment.

Associated with the observable x is obviously the X-operator we have extensively discussed in the previous sections: The set of eigenstates are the elements $\{|x\rangle, \ x \in \mathbb{R}\}$ of the x-basis, where the particle is "sharply" localized on one point in space. The corresponding eigenvalue is x, the position of the particle in this state, the result of a position measurement. Since $|x\rangle$ is not square-integrable, it is only a pseudo-state. The particle can never be in such a state, because it is not contained in the Hilbert space $L^2(\mathbb{R}, \mathbb{C})$. But by the second postulate of QM it should be, since a measurement always projects on an eigenstate. We conclude that an exact measurement of position is simply impossible. In practice, a position measurement always has a finite resolution. After such a measurement the particle has a new state $|\psi_n\rangle$, a square-integrable wave function, for example a Gaussian function whose width corresponds to the resolution of the measurement.

In the following, we always use normalized wave functions, i.e.

$$\langle\psi|\psi\rangle = \int_{-\infty}^{\infty} dx \, \psi^*(x)\psi(x) = 1.$$ (3.161)

The probability to find the particle in the interval $[x_0, x_1]$ is

$$\int_{x_0}^{x_1} dx \, \langle\psi|x\rangle\langle x|\psi\rangle = \int_{x_0}^{x_1} dx \, \psi^*(x)\psi(x).$$ (3.162)

We thus interpret the norm squared $\psi^*(x)\psi(x)$, of ψ as a **probability density** of the particle. Here $\int_{x_0}^{x_1} dx\, |x\rangle\langle x|$ is the projection operator which restricts the wave function to the interval $[x_0, x_1]$.

The expectation value for the position of the particle is

$$\langle X \rangle_\psi = \langle \psi | X | \psi \rangle = \int_{-\infty}^{\infty} dx\, x\, |\psi(x)|^2. \tag{3.163}$$

Now what about the momentum operator P associated with the observable p? Since classical mechanics is quite a good approximation on macroscopic scales, we demand that Hamilton's equations are at least retained as equations for the expectation values. This requirement leads to the

---

Ehrenfest Equations

$$\frac{d\langle X \rangle}{dt} = \langle \frac{\partial H}{\partial P} \rangle, \quad \frac{d\langle P \rangle}{dt} = -\langle \frac{\partial H}{\partial X} \rangle. \tag{3.164}$$

Here $\partial H/\partial P$ means the operator associated with the quantity $\partial h/\partial p$, and similarly for $\partial H/\partial X$. For the value of the Hamiltonian function h is the total energy of the particle (kinetic plus potential energy), and this observable is in QM associated with the Hamiltonian operator H. This operator is made of combinations of the operators X and P, in the same way as h is made of combinations of x and p. This last statement is also known as the **Correspondence Principle**.

In the case $h = p^2/(2m) + V(x)$, one has

$$H = \frac{P^2}{2m} + V(X) \tag{3.165}$$

For $V(X)$ to be well-defined, we have to assume that the potential V is a polynomial or power series in x, so that we can define $V(X)$ as the same polynomial or power series in X (since, for instance, the square root of an operator is in general undefined). In practice, we will have to extend the definition beyond power series though: on the level of wave functions we simply define $V(X)$ as the operator which multiplies $\psi(x)$ with $V(x)$. For polynomials and power series, this definition is obviously equivalent to the previous one. In fact, only a single one of the potentials we will discuss can be represented as a polynomial or power series: the potential of the harmonic oscillator, $V(x) = \alpha x^2$.

The Ehrenfest equations now read

$$\frac{d\langle X \rangle}{dt} = \frac{\langle P \rangle}{m}, \quad \frac{d\langle P \rangle}{dt} = -\langle V'(X) \rangle. \tag{3.166}$$

Here $V'(X)$ is the operator associated with the quantity $V'(x)$, where similar considerations hold as for $V(X)$. With $d\langle X \rangle / dt$ we identify the velocity of the particle, with $d\langle P \rangle / dt$ the force accelerating it.

Comparing the Ehrenfest equations (3.166) with the Ehrenfest theorem (2.235), we see that for an arbitrary potential and for each state $|\psi\rangle$ the following equations have to hold:

$$\langle [X, H] \rangle = \frac{i\hbar}{m} \langle P \rangle, \quad \langle [P, H] \rangle = -i\hbar \langle V'(X) \rangle \tag{3.167}$$

Due to $[X, V(X)] = 0$, the left hand side of the first equation becomes

$$\langle [X, H] \rangle = \frac{1}{2m} \langle [X, P^2] \rangle = \frac{1}{2m} \langle [X, P]P + P[X, P] \rangle. \tag{3.168}$$

Hence, the first equation in (3.167) is obviously fulfilled if

$$[X, P] = i\hbar \mathbf{1}, \tag{3.169}$$

a constant multiple of the unit operator. This commutator relation holds, due to (3.143), if we set

$$P = -i\hbar D. \tag{3.170}$$

We choose this as our ansatz for P and check whether this leads to a proper result in the second equation of (3.167). The left hand side is then (since $[P, P^2] = 0$) equal to $-i\hbar \langle [D, V(X)] \rangle$, and from (3.144) follows that this equals $-i\hbar \langle V'(X) \rangle$. So, we found our momentum operator! Since D is hermitian on $L^2(\mathbb{R}, \mathbb{C})$, P is hermitian, as required.

The logic of our derivation was as follows: We know that in macroscopic physics, where one can almost always replace quantum mechanical quantities with their average values, the Hamiltonian equations of classical mechanics hold. The quantum mechanical quantities therefore have to obey these equations in terms of their expectation values, that is, the Ehrenfest equations. In combination with the Ehrenfest theorem, which follows directly from the QM postulates, one can deduce the momentum operator from that, $P = -i\hbar D$.

Most other textbooks proceed differently to obtain P. They typically use one of these two ways:

1. They *postulate* the momentum operator in the above-mentioned form, usually as an additional fundamental postulate of QM, and then show that the Ehrenfest equations hold.
2. They start from the observed matter waves (we will discuss them in a moment) and derive the P-operator from that. The Ehrenfest equations again follow as a consequence.

In the end, all arguments lead to the same result, and it is a matter of taste which one to follow.

Self-check questions:

1. What do the Ehrenfest equations say and why should they hold?
2. How does the momentum operator act on a wave function?
3. What is a probability density?

3.5 Matter Waves and Wave-Particle Duality

The pseudo-eigenstates of the momentum operator are those of the derivative operator D, but with eigenvalues multiplied by the factor $-i\hbar$. The pseudo-eigenstate $|p\rangle$ for the momentum eigenvalue p is, written as a wave function, a wave,

$$\psi_p(x) = \langle x|p\rangle = \frac{1}{\sqrt{2\pi\hbar}} e^{i\frac{p}{\hbar}x}, \tag{3.171}$$

and for the **wavenumber** k one obtains the **de Broglie relation**

$$p = \hbar k. \tag{3.172}$$

The normalization factor $1/\sqrt{2\pi\hbar}$ was chosen such that the p-basis is orthonormal,

$$\langle p|p'\rangle = \frac{1}{2\pi\hbar} \int_{-\infty}^{\infty} dx \, e^{i\frac{p'-p}{\hbar}x} = \delta(p' - p). \tag{3.173}$$

To prove this, substitute $y = x/\hbar$ and compare with (3.127). The wave function $\tilde{\psi}$ in the momentum basis can be written as a function of p or k, incorporating the factor $\hbar$ in the correct way:

$$\psi(x) = \langle x|\psi\rangle = \int_{-\infty}^{\infty} dp \, \langle x|p\rangle\langle p|\psi\rangle = \frac{1}{\sqrt{2\pi\hbar}} \int_{-\infty}^{\infty} dp \, e^{i\frac{p}{\hbar}x} \tilde{\psi}(p) \tag{3.174}$$

$$= \sqrt{\frac{\hbar}{2\pi}} \int_{-\infty}^{\infty} dk \, e^{ikx} \tilde{\psi}(k) \tag{3.175}$$

In the second line $dp = \hbar\, dk$ was used. Similarly one has

$$\tilde{\psi}(p) = \langle p|\psi\rangle = \int_{-\infty}^{\infty} dx \, \langle p|x\rangle\langle x|\psi\rangle = \frac{1}{\sqrt{2\pi\hbar}} \int_{-\infty}^{\infty} dx \, e^{-i\frac{p}{\hbar}x} \psi(x), \tag{3.176}$$

or, expressed as a function of k,

$$\tilde{\psi}(k) = \frac{1}{\sqrt{2\pi\hbar}} \int_{-\infty}^{\infty} dx \, e^{-ikx} \psi(x). \tag{3.177}$$

The last equation says, in accordance with (3.175), that $\tilde{\psi}(k)$ is the Fourier transform of $\psi(x)$, multiplied by $1/\sqrt{\hbar}$.

In the momentum basis, the momentum operator acts via multiplication with p. From

$$P|p\rangle = p|p\rangle, \quad \langle p|P = p\langle p| \tag{3.178}$$

follows

$$(P\tilde{\psi})(p) = \langle p|P|\psi\rangle = p\langle p|\psi\rangle = p\,\tilde{\psi}(p). \tag{3.179}$$

The position operator in the momentum basis is derived along the lines of equations (3.136)–(3.140), one only has to take care of factors $\hbar$:

$$\langle p|X|p'\rangle = \frac{1}{2\pi\hbar} \int_{-\infty}^{\infty} dx\, e^{-i\frac{p}{\hbar}x} x\, e^{i\frac{p'}{\hbar}x} \tag{3.180}$$

$$= \frac{\hbar}{2\pi} \int_{-\infty}^{\infty} dy\, e^{-ipy} y\, e^{ipy} \tag{3.181}$$

$$= i\hbar\frac{d}{dp}\left(\frac{1}{2\pi} \int_{-\infty}^{\infty} dy\, e^{i(p'-p)y}\right) = i\hbar\frac{d}{dp}\delta(p - p'), \tag{3.182}$$

where we substituted $y = x/\hbar$ in the second step, and thus

$$(X\tilde{\psi})(p) = \langle p|X|\psi\rangle = \int_{-\infty}^{\infty} dp'\,\langle p|X|p'\rangle\langle p'|f\rangle = i\hbar\frac{d}{dp}\tilde{\psi}(p). \tag{3.183}$$

Altogether we have:

Action of X and P in Position and Momentum Space

$$X\psi(x) = x\,\psi(x), \quad P\psi(x) = -i\hbar\frac{d}{dx}\psi(x) \tag{3.184}$$

$$X\tilde{\psi}(p) = i\hbar\frac{d}{dp}\tilde{\psi}(p), \quad P\tilde{\psi}(p) = p\,\tilde{\psi}(p) \tag{3.185}$$

Just as with position eigenstates $|x\rangle$, the pure momentum eigenstates $|p\rangle$ cannot occur in nature, since they are not square-integrable and therefore not contained in the Hilbert space. Momentum measurements always have, like position measurements, some uncertainty, and the result of an actual measurement is a square-integrable momentum wave function with a non-vanishing width around the measurement result p_0.

Due to the Heisenberg Uncertainty Relation (2.214) and the known value $[X, P] = i\hbar$ we have the

---

Position-Momentum Uncertainty

$$(\Delta X)_\psi (\Delta P)_\psi \geq \frac{\hbar}{2}. \qquad (3.186)$$

It says: the smaller the width of a wave function $\psi(x)$ in position space, the larger is the width of $\tilde{\psi}(p)$ in momentum space, and vice versa. Applying the Position-Momentum Uncertainty to the state *after* a position or momentum measurement, one gets the following statement: the more precise the position of a particle is measured, the fuzzier its momentum *is made by this*, and vice versa.

We want to comprehend the Uncertainty Relation for a **Gaussian wave packet** i.e. for a momentum wave function of the form

$$\tilde{\psi}(p) = \frac{\sqrt{\sigma}}{\sqrt{\hbar}\pi^{1/4}} e^{-\frac{\sigma^2 (p-p_0)^2}{2\hbar^2}}. \qquad (3.187)$$

This function describes a Gaussian distribution peaked at p_0, with a prefactor that normalizes it to 1 (Fig. 3.1),

$$\langle \psi | \psi \rangle = \int_{-\infty}^{\infty} dp \, |\tilde{\psi}(p)|^2 = 1, \qquad (3.188)$$

since

$$\int_{-\infty}^{\infty} dy \, e^{-\frac{y^2}{a}} = \sqrt{\pi a}. \qquad (3.189)$$

Fig. 3.1 Gaussian wave packet in momentum space, peaked at p_0

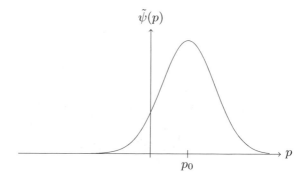

We first compute $\langle P \rangle_\psi$ and $\langle P^2 \rangle_\psi$:

$$\langle P \rangle_\psi = \langle \psi | P | \psi \rangle = \int_{-\infty}^{\infty} dp \, p \, |\tilde{\psi}(p)|^2 \tag{3.190}$$

$$= \frac{\sigma}{\hbar\sqrt{\pi}} \int_{-\infty}^{\infty} dp \, p \, e^{-\frac{\sigma^2 (p-p_0)^2}{\hbar^2}} \tag{3.191}$$

$$= \frac{\sigma}{\hbar\sqrt{\pi}} \int_{-\infty}^{\infty} dq \, (q + p_0) \, e^{-\frac{\sigma^2 q^2}{\hbar^2}} \tag{3.192}$$

with $q = p - p_0$. The first term amounts to 0, as an integral over an odd function, and the second term leads to

$$\langle P \rangle_\psi = p_0. \tag{3.193}$$

Similarly for $\langle P^2 \rangle_\psi$:

$$\langle P^2 \rangle_\psi = \frac{\sigma}{\hbar\sqrt{\pi}} \int_{-\infty}^{\infty} dp \, p^2 \, e^{-\frac{\sigma^2 (p-p_0)^2}{\hbar^2}} \tag{3.194}$$

$$= \frac{\sigma}{\hbar\sqrt{\pi}} \int_{-\infty}^{\infty} dq \, (q^2 + 2qp_0 + p_0^2) e^{-\frac{\sigma^2 q^2}{\hbar^2}} \tag{3.195}$$

The second term vanishes (integral over an odd function), the third one gives p_0^2. For the first term we use

$$\int_{-\infty}^{\infty} dy \, y^2 \, e^{-\frac{y^2}{a}} = \sqrt{\pi a} \, \frac{a}{2} \tag{3.196}$$

and finally obtain

$$\langle P^2 \rangle_\psi = p_0^2 + \frac{\hbar^2}{2\sigma^2}. \tag{3.197}$$

This results in the momentum uncertainty

$$(\Delta P)_\psi = \sqrt{\langle P^2 \rangle_\psi - \langle P \rangle_\psi^2} = \frac{\hbar}{\sigma\sqrt{2}}. \tag{3.198}$$

Next, we calculate the wave function in position space

$$\psi(x) = \frac{1}{2\pi\hbar} \int_{-\infty}^{\infty} dp \, e^{i\frac{p}{\hbar}x} \tilde{\psi}(p) \tag{3.199}$$

$$= \frac{\sqrt{\sigma}}{\hbar\pi^{3/4}} \int_{-\infty}^{\infty} dp \, e^{-\frac{\sigma^2 (p-p_0)^2}{2\hbar^2} + i\frac{p}{\hbar}x} \tag{3.200}$$

$$= \frac{\sqrt{\sigma}}{\hbar \pi^{3/4}} e^{i \frac{p_0}{\hbar} x} \int_{-\infty}^{\infty} dq \, e^{-\frac{\sigma^2 q^2}{2\hbar^2} + i \frac{q}{\hbar} x} \tag{3.201}$$

$$= \frac{\sqrt{\sigma}}{\hbar \pi^{3/4}} e^{i \frac{p_0}{\hbar} x} e^{-\frac{x^2}{2\sigma^2}} \int_{-\infty}^{\infty} dq \, e^{-\frac{\sigma^2}{2\hbar^2} \left(q - \frac{i\hbar x}{\sigma^2} \right)^2} \tag{3.202}$$

$$= \frac{1}{\sqrt{\sigma} \pi^{1/4}} e^{i \frac{p_0}{\hbar} x} e^{-\frac{x^2}{2\sigma^2}}. \tag{3.203}$$

In the last step we have used that the Gaussian integral (3.189) still amounts to $\sqrt{\pi a}$ if the line of integration is shifted in the imaginary direction. The position wave function is thus again a Gauss function, but with an additional oscillation $\exp(ixp_0/\hbar)$. This oscillation is just the one of the momentum pseudo-eigenstate $|p_0\rangle$.

Since in the expectation values $\langle X \rangle_\psi$ and $\langle X^2 \rangle_\psi$ only the square of the absolute value of ψ appears, e.g.

$$\langle X \rangle_\psi = \langle \psi | X | \psi \rangle = \int_{-\infty}^{\infty} dx \, x \, |\psi(x)|^2, \tag{3.204}$$

the oscillation factor is irrelevant for them. One gets, similarly as for P,

$$\langle X \rangle_\psi = 0, \tag{3.205}$$

$$\langle X^2 \rangle_\psi = \frac{\sigma^2}{2}, \tag{3.206}$$

$$(\Delta X)_\psi = \frac{\sigma}{\sqrt{2}} \tag{3.207}$$

and hence

$$(\Delta X)_\psi (\Delta P)_\psi = \frac{\hbar}{2}. \tag{3.208}$$

So, for the Gaussian wave packet the minimal uncertainty allowed is obtained. In this case the Uncertainty Relation goes back to the fact that the width of a Gauss distribution and the width of its Fourier transform are inverse to each other.

Exercise 3.10
Compute $\langle X \rangle_\psi$ and $\langle X^2 \rangle_\psi$ using the momentum wave function $\tilde{\psi}(p)$.

Exercise 3.11
Compute $\langle P \rangle_\psi$ and $\langle P^2 \rangle_\psi$ using the position wave function $\psi(x)$.

Exercise 3.12
Repeat the calculation of expectation values and uncertainties for the case that $\tilde{\psi}(p)$ contains an additional factor $\exp(-ipx_0/\hbar)$.

It is the parameter σ which determines whether the quantum object (an electron or a photon, for instance) represented by $|\psi\rangle$ behaves more like a **wave** or a **particle** (or something in between). If σ is small (where "small" is to be understood in relation to the experimental conditions, for example small compared to the resolution of the measurement device), the position wave function $\psi(x)$ therefore narrow, then the object resembles a property of a classical particle: It is approximately localized at a certain position, i.e. the probability to find it somewhere else almost vanishes.

Some other property of a classical particle is missing though: It doesn't have a certain momentum. The momentum wave function $\tilde{\psi}(p)$ is broad, i.e. when the momentum is measured, a broad range of results is possible. A state with a sharp position is a state with fuzzy momentum.

If σ is large, the object behaves more like a wave: It is spread over a large amount of space and oscillates (with wavenumber $p/\hbar$), so that interference patterns can occur. When this behavior is observed for objects which we are classically used to consider as matter particles (electrons, for instance), one speaks of **matter waves**, in contrast to, say, electromagnetic waves, which have revealed their wave properties much earlier in the history of physics. Instead, the surprise with electromagnetic waves was that they also consist of quanta, the photons, and therefore have particle properties too.

The distinction from classical physics between force fields and matter particles is abolished in QM. Everything consists of quanta (however, the exact meaning of this "consists of" is quite complicated and some aspects of it quite controversial), be it matter or light or force fields of any kind. All of these quanta can appear in a spatially localized, particle-like way, or in a wave-like way, or in different ways where they are neither sharply localized in position nor momentum space, as for example the photons of an ordinary magnetic field (the magnetic field is neither restricted to one position in space, nor does it have the properties of a wave).

Only for gravitation, the quantum nature is not yet verified. Here we have the additional complication that gravity is, according to general relativity, not really a force field in the usual sense, but a side effect of spacetime geometry. Whether it is possible to construct a quantum theory of geometry in the same way as for ordinary force fields, is still an open question.

But what is it that oscillates in a matter wave? It is the probability amplitude $\psi(x)$, whose squared absolute value represents the probability density of the object. In an electromagnetic wave, on the other hand, electric and magnetic fields oscillate. The probability amplitude of the photons constituting the wave oscillates in accordance with these fields. For the **double slit experiment**, the bright stripes on the screen have therefore two meanings:

1. The field amplitudes interfere constructively, i.e. the wave is particularly strong there, the light particularly bright.
2. The probability amplitudes of the photons interfere constructively. The probability to find photons in this place is thus larger than elsewhere, i.e. particularly many photons hit the screen there. Each bright point on the screen represents a position measurement of a photon. The screen is the measurement device of the position operator, where the position uncertainty σ after the measurement is given by the width of the "pixels" on the screen. Through this measurement, the wave-like state of the photon (wavelength and therefore momentum is sharp) is turned into a particle-like state (position is sharp).

The experiment becomes especially impressive, when at every moment only one photon is one the way. Then it is only the wave function of this single photon which interferes with itself. One part of the **probability wave** "streams" through the one slit, another one through the other slit, and both parts interfere with each other. In the position measurement at the end (photon hits screen) only a single point gets bright. This cannot be explained with the first of the two mentioned interpretations (interference of field amplitudes) Only after a large amount of photons have hit the screen, the interference pattern emerges, step by step.

Unfortunately there is a little semantic confusion: Often the quantum object itself is denoted as a "particle". One says the electron or the photon is a "particle". This language convention leads to the confusing statement that a "particle" has both wave and particle properties. This blunder is due to different understandings what exactly is to be considered as a particle property. In our discussion above we meant particle in the sense of an object that is localized within a small spatial volume, i.e. the spatial localization was crucial for us to speak of a particle. For the pioneers of QM, however, already the fact that light comes in quanta, i.e. in certain portions (the photons), meant to them the "particle aspect" of light, which lead to denoting a photon as a "particle".

Self-check questions:

1. What are matter waves? What is the de Broglie Relation? What does it mean?
2. What does the Position-Momentum Uncertainty Relation say?
3. How do you determine expectation values in position and momentum space?

3.6 Schrödinger Equation in One-Dimensional Position Space

For the moment, we are concerned with a single quantum object in one spatial dimension with a time-independent potential $V(x)$. In the x-basis, the effect of the Hamiltonian operator on a state $|\psi\rangle$ is given by

$$\langle x|H|\psi\rangle = (H\psi)(x) = \left[\left(\frac{P^2}{2m} + V(X)\right)\psi\right](x)$$

$$= -\frac{\hbar^2}{2m}\frac{d^2}{dx^2}\psi(x) + V(x)\psi(x). \qquad (3.209)$$

If we now want to determine the time dependency of the state, i.e. consider the family $|\psi(t)\rangle$ instead of the fixed $|\psi\rangle$, we use the **Schrödinger equation in one-dimensional position space**, which follows directly from (3.209) and (2.144):

$$i\hbar\frac{\partial}{\partial t}\psi(x, t) = -\frac{\hbar^2}{2m}\frac{\partial^2}{\partial x^2}\psi(x, t) + V(x)\psi(x, t) \qquad (3.210)$$

The associated eigenvalue equation (**stationary Schrödinger equation in one-dimensional position space**) reads

$$-\frac{\hbar^2}{2m}\frac{d^2}{dx^2}\psi(x) + V(x)\psi(x) = E\psi(x). \qquad (3.211)$$

In general, we will find two kinds of solutions of this equation:

1. **Discrete spectrum**: The eigenvalues E are discrete. The corresponding eigenfunctions ψ are elements of the Hilbert space, and thus square-integrable.
2. **Continuous spectrum**: The eigenvalues E are continuous. The corresponding eigenfunctions ψ are not elements of the Hilbert space, they are not square-integrable. But by integration, square-integrable wave functions can be constructed out of them.

Often we will find that there are two limit energies E_{min} and E_f, such that:

- there are no eigenvalues with $E < E_{min}$.
- the spectrum for $E_{min} \leq E < E_f$ is discrete (**bound states**), with E_{min} being the energy of the **ground state**.
- the spectrum for $E_f < E$ is continuous (**free states** or **scattered states**).

This behavior is well-known from atomic physics. Free electrons which are not bound to an atom have energies $E > E_f = 0$. They can escape "to infinity" or they come from infinity and are scattered by the atom. Then there is a spectrum of negative binding energies corresponding to the several shells of an atom. The innermost shell represents the minimal energy E_{min} an electron can have in this atom. For the hydrogen atom, we will solve the Schrödinger equation exactly. For all higher atoms, numerical approximations are required.

The stationary Schrödinger equation (3.211) is real in the following sense: If the entire equation is complex conjugated, the only effect is that ψ becomes ψ^*. It follows that the complex conjugate $\psi^*(x)$ of any solution $\psi(x)$ of the stationary Schrödinger equation is also a solution. This has the pleasant consequence that we can restrict our search for solutions to real functions. This is because searching for ψ and ψ^* is equivalent to searching for the real functions $\text{Re}(\psi) = \frac{1}{2}(\psi^* + \psi)$ and

$\text{Im}(\psi) = \frac{1}{2i}(\psi^* - \psi)$. The Schrödinger equation is linear, i.e. linear combinations of solutions are again solutions. This applies in particular to $\text{Re}(\psi)$ and $\text{Im}(\psi)$.

The corresponding time-dependent wave function $\psi(x, t)$ is then a product of a phase factor depending only on t and a real function depending only on x:

$$\psi(x, t) = e^{-i\frac{E}{\hbar}t} \psi(x), \tag{3.212}$$

where we have taken the freedom to denote both time-dependent and time-independent wave function with the same letter ψ.

An example: For the **free particle**, $V(x) = 0$ in all of $\mathbb{R}$, the energy eigenfunctions with eigenvalue E are obviously identical to the momentum eigenfunctions with eigenvalue p, where $E = p^2/(2m)$, or $p = \pm\sqrt{2mE}$. Hence, for the eigenvalue E there are two solutions, one for positive and one for negative momentum, related by complex conjugation:

$$\psi(x) = \frac{1}{\sqrt{2\pi\hbar}} e^{ikx}, \quad \psi^*(x) = \frac{1}{\sqrt{2\pi\hbar}} e^{-ikx}, \quad k = \frac{\sqrt{2mE}}{\hbar} \tag{3.213}$$

The real functions

$$\psi_1(x) = \frac{1}{\sqrt{2}} \text{Re}(\psi(x)) = \frac{1}{2\sqrt{\pi\hbar}} \cos kx, \tag{3.214}$$

$$\psi_2(x) = \frac{1}{\sqrt{2}} \text{Im}(\psi(x)) = \frac{1}{2\sqrt{\pi\hbar}} \sin kx \tag{3.215}$$

(the factor $1/\sqrt{2}$ is required for normalization) are also solutions for the same energy eigenvalue E. They are, however, not eigenstates of momentum, for each of the two functions have equal contributions from both $p = +\sqrt{2mE}$ and $p = -\sqrt{2mE}$. The corresponding time-dependent wave functions describe standing waves (peaks and troughs don't move):

$$\psi_1(x, t) = \frac{1}{2\sqrt{\pi\hbar}} e^{-i\frac{E}{\hbar}t} \cos kx, \quad \psi_2(x, t) = \frac{1}{2\sqrt{\pi\hbar}} e^{-i\frac{E}{\hbar}t} \sin kx \tag{3.216}$$

Each solution of eigenvalue E is a linear combination of these two solutions,

$$\psi(x, t) = e^{-i\frac{E}{\hbar}t} \left(\alpha \cos \frac{\sqrt{2mE}}{\hbar} x + \beta \sin \frac{\sqrt{2mE}}{\hbar} x \right) \tag{3.217}$$

with $\alpha, \beta \in \mathbb{C}$.

The norm squared of the wave function,

$$\rho(x, t) = |\psi(x, t)|^2, \tag{3.218}$$

is the spatial probability density of a quantum object. A system is in motion if ρ changes with time. For energy eigenstates, the only time dependency is in the phase factor $\exp(-iEt/\hbar)$, which is irrelevant for ρ. Hence, there is no real motion. At any place, the probability density is unchanged with time. Even if the expectation value of the kinetic energy is positive, there is no motion at all in the system! Due to the Ehrenfest equations (3.164), $\langle P \rangle_\psi$ therefore has to vanish for energy eigenstates, otherwise $\langle X \rangle_\psi$ would change with time. This statement holds for the discrete eigenstaes, for only these are elements of the Hilbert space. The continuous pseudo-eigenstates, like $\psi \sim \exp(ikx)$ for the free particle, do not have well-defined expectation values. Even for a momentum eigenstate $|p\rangle$, it would be misleading to say that its momentum expectation value $\langle P \rangle$ is p: the Ehrenfest equations would then suggest that there is a motion, $\langle X \rangle = p/m$. However, this is not the case: eigenstates of momentum are motionless.

Now let's consider generic states. What can we say about the time evolution of ρ? With help of the Schrödinger equation and its complex conjugate we find:

$$\frac{\partial}{\partial t}(\psi^*\psi) = \left(\frac{\partial}{\partial t}\psi^*\right)\psi + \psi^*\frac{\partial}{\partial t}\psi = \frac{\hbar}{2mi}\left((\frac{\partial^2}{\partial x^2}\psi^*)\psi - \psi^*\frac{\partial^2}{\partial x^2}\psi\right) \quad (3.219)$$

The term with the potential is canceled. In contrast to the time derivative of ψ, the time derivative of ρ does not depend on the potential, but only on the current shape of ψ. The expression in brackets on the right hand side is, up to a sign, the x-derivative of

$$j(x, t) = \frac{\hbar}{2mi}\left(\psi^*(x, t)\frac{\partial}{\partial x}\psi(x, t) - \psi(x, t)\frac{\partial}{\partial x}\psi^*(x, t)\right). \quad (3.220)$$

The funtion j is the **probability current density**. In combination with ρ it obeys the **continuity equation**

$$\frac{\partial}{\partial t}\rho + \frac{\partial}{\partial x}j = 0. \quad (3.221)$$

The structure of the equation is analogous to the continuity equation in electrodynamics. We will discuss it further when we generalize to three dimensions.

The continuity equation also holds for pseudo-states—there are no statements about expectation values included.

Exercise 3.13
Show that for the state $|p\rangle$ with the wave function
$\psi_p(x) = \exp(ipx/\hbar)/\sqrt{2\pi\hbar}$ one has $j(x, t) = p/(2\pi m\hbar)$, which is independent of x and t, and the time derivative of ρ vanishes. As expected, the current is proportional to momentum but doesn't lead to a movement of ρ.

For real wave functions $\psi(x)$, already j vanishes (see 3.220), and the time deriv-ative ρ thus all the more. Such a state is therefore "even more stationary" than $|p\rangle$. If, however, the real-valued wave function under consideration is not an eigenfunction of the Hamiltonian operator, this property (being real-valued) holds only for a single moment. The time evolution according to the time-dependent Schrödinger equation changes the state $\psi(x, t)$, a current is created, and hence a change of ρ. This means that the first time derivative of ρ vanishes at this very moment $t = t_0$, but not the second derivative with respect to time. The state is stationary only for this single moment, but not stationary in the usual sense (i.e. permanently, as an eigenfunction of H).

Self-check questions:

1. Can you write down the Schrödinger equation in one-dimensional position space (time dependent/stationary)?
2. Why is it always possible to choose eigenfunctions of a given energy eigenvalue real-valued?
3. What is a probability current? What does the continuity equation say?

3.7 Several Space Dimensions

If the quantum particle moves in d dimensions instead of just one (the standard case is $d = 3$, of course), then the underlying Hilbert space is $L^2(\mathbb{R}^d, \mathbb{C})$, the space of square-integrable functions from $\mathbb{R}^d$ to $\mathbb{C}$. We specify a point $\mathbf{r}$ of a d-dimensional space by d cartesian coordinates $(x_1, \ldots, x_d)$. Associated with each of these coordinates is a separate X- and a separate P-operator:

$$X_i \psi(\mathbf{r}) = x_i \psi(\mathbf{r}), \quad P_i \psi(\mathbf{r}) = -i\hbar \frac{\partial}{\partial x_i} \psi(\mathbf{r}) \tag{3.222}$$

The multiplication of different coordinates commutes, i.e. the order of multiplica-tion is irrelevant. Similarly, partial derivatives with respect to different coordinates commute. The commutators are thus

$$[X_i, X_j] = 0, \quad [P_i, P_j] = 0, \quad [X_i, P_j] = i\hbar \delta_{ij} \mathbf{1}. \tag{3.223}$$

Exercise 3.14
Prove the last one of these three relations.

Nerd's Corner 3.2
The commutator relations (3.223) have the same structure (up to a factor $i\hbar$) as
the classical **Poisson brackets** for the coordinates and momenta. The Poisson
brackets are defined in the following way: Let $f(\mathbf{r}, \mathbf{p}, t)$, $g(\mathbf{r}, \mathbf{p}, t)$ be functions
of the coorinates, the momenta, and of time. Then

$$\{f, g\} := \sum_i \frac{\partial f}{\partial x_i} \frac{\partial g}{\partial p_i} - \frac{\partial f}{\partial p_i} \frac{\partial g}{\partial x_i}. \tag{3.224}$$

Exercise 3.15
Show that

$$\{x_i, x_j\} = \{p_i, p_j\} = 0, \quad \{x_i, p_j\} = \delta_{ij}. \tag{3.225}$$

There are subtle connections between the Poisson brackets of classical mechanics and the commutators of QM. For example, one has

$$\frac{df}{dt} = \{f, h\} + \frac{\partial f}{\partial t}, \tag{3.226}$$

where h is again the Hamiltonian function. Compare this with the Heisenberg
equation (2.231)!

A **canonical transformation** is a transformation

$$x_i \to x_i'(\mathbf{r}, \mathbf{p}, t), \quad p_i \to p_i'(\mathbf{r}, \mathbf{p}, t), \tag{3.227}$$

for which the Poisson brackets (3.225) still hold in the new variables. This
property is inherited to the commutators of the associated operators. One can
then rewrite the wave function as a function of the new variables $\{x_i'\}$. The
operators $\{P_i'\}$ then act again via $\partial/\partial x_i'$.

Exercise 3.16
Show that in one-dimensional space the transformation

$$x \to x' = p, \quad p \to p' = -x \tag{3.228}$$

is canonical. In this way, find a deeper reason for the behavior of X and P in
momentum space, (3.185).

We define the **vector operators** $\mathbf{P}$ and $\mathbf{X}$ as d-tuples $\mathbf{P} = (P_1, \ldots, P_d)$ and
$\mathbf{X} = (X_1, \ldots, X_d)$, respectively. "Scalar products" like $\mathbf{P}^2$ or $\mathbf{P} \cdot \mathbf{X}$ are defined in
analogy to the "normal" scalar product, e.g. $\mathbf{P}^2 = P_1^2 + \cdots + P_d^2$. The result is now
an operator, of course, not a number.

The position eigenstate $|\mathbf{r}'\rangle$ is a simultaneous eigenstate of all X-operators, with eigenvalues $(x_1', \ldots, x_d')$. The corresponding wave function is a product of delta distributions,

$$\psi_{\mathbf{r}'}(\mathbf{r}) = \langle \mathbf{r}|\mathbf{r}'\rangle = \delta(x_1 - x_1')\delta(x_2 - x_2') \cdots \delta(x_d - x_d'). \qquad (3.229)$$

The momentum eigenstate $|\mathbf{p}\rangle$ is a simultaneous eigenstate of all P-operators, with eigenvalues $(p_1, \ldots, p_d)$. The corresponding wave function is, in position space:

$$\psi_{\mathbf{p}}(\mathbf{r}) = \langle \mathbf{r}|\mathbf{p}\rangle = \frac{1}{(2\pi\hbar)^{d/2}} e^{\frac{i}{\hbar}\mathbf{p}\cdot\mathbf{r}} \qquad (3.230)$$

Each dimension has a separate Fourier transformation:

$$\psi(\mathbf{r}) = \int dp_1 \frac{e^{\frac{i}{\hbar}p_1 x_1}}{\sqrt{2\pi\hbar}} \int dp_2 \frac{e^{\frac{i}{\hbar}p_2 x_2}}{\sqrt{2\pi\hbar}} \cdots \int dp_d \frac{e^{\frac{i}{\hbar}p_d x_d}}{\sqrt{2\pi\hbar}} \tilde{\psi}(\mathbf{p}) \qquad (3.231)$$

$$= \frac{1}{(2\pi\hbar)^{d/2}} \int d^d p \, e^{\frac{i}{\hbar}\mathbf{p}\cdot\mathbf{r}} \, \tilde{\psi}(\mathbf{p}) \qquad (3.232)$$

The Hamiltonian operator now reads

$$\langle \mathbf{r}|H|\psi\rangle = (H\psi)(\mathbf{r}) = \left[\left(\frac{\mathbf{p}^2}{2m} + V(\mathbf{X}) \right) \psi \right](\mathbf{r})$$

$$= -\frac{\hbar^2}{2m}\Delta\psi(\mathbf{r}) + V(\mathbf{r})\psi(\mathbf{r}), \qquad (3.233)$$

where Δ is the Laplace operator in d dimensions,

$$\Delta\psi = \frac{\partial^2}{\partial x_1^2}\psi + \cdots + \frac{\partial^2}{\partial x_d^2}\psi. \qquad (3.234)$$

The time-dependent Schrödinger equation reads:

Time-Dependent Schrödinger Equation in Position Space

$$i\hbar\frac{d}{dt}\psi(\mathbf{r}, t) = -\frac{\hbar^2}{2m}\Delta\psi(\mathbf{r}, t) + V(\mathbf{r})\psi(\mathbf{r}, t) \qquad (3.235)$$

The stationary Schrödinger equation reads:

Stationary Schrödinger Equation in Position Space

$$-\frac{\hbar^2}{2m}\Delta\psi(\mathbf{r}) + V(\mathbf{r})\psi(\mathbf{r}) = E\psi(\mathbf{r}) \qquad (3.236)$$

The statements about discrete/continuous spectrum and about real solutions are still valid.

The probability current density (3.220) becomes

$$\mathbf{j}(\mathbf{r}, t) = \frac{\hbar}{2mi}(\psi^*(\mathbf{r}, t)\nabla\psi(\mathbf{r}, t) - \psi(\mathbf{r}, t)\nabla\psi^*(\mathbf{r}, t)) \qquad (3.237)$$

and the continuity equation reads:

Continuity Equation

$$\frac{\partial}{\partial t}\rho(\mathbf{r}, t) + \nabla \cdot \mathbf{j}(\mathbf{r}, t) = 0, \qquad (3.238)$$

where

$$\rho = \psi^*\psi, \qquad \mathbf{j} = \frac{\hbar}{2mi}(\psi^*\nabla\psi - \psi\nabla\psi^*). \qquad (3.239)$$

With the help of Gauss's theorem, the continuity equation can be brought into an integral form. Let V be a finite region of $\mathbb{R}^d$ and S its surface. Then Gauss's theorem says that

$$\int_V d^d x\, \nabla \cdot \mathbf{A} = \oint_S d\mathbf{S} \cdot \mathbf{A} \qquad (3.240)$$

for a vector field $\mathbf{A}$. Applied to the current density and the continuity equation, this leads to

$$\frac{d}{dt}\int_V d^d x\, \rho + \oint_S d\mathbf{S} \cdot \mathbf{j} = 0 \qquad (3.241)$$

with the following interpretation: The probability to find the quantum particle in the region V is reduced by the probability current flowing out of V.

An interesting form of the continuity equation is obtained if ψ is split into absolute value and phase,

$$\psi(\mathbf{r}, t) = A(\mathbf{r}, t)e^{\frac{i}{\hbar}S(\mathbf{r}, t)}, \qquad (3.242)$$

with real functions A and S. The probability density ρ and the current $\mathbf{j}$ then have the form

$$\rho = A^2, \quad \mathbf{j} = A^2 \frac{\nabla S}{m}. \tag{3.243}$$

From this one concludes that

$$\mathbf{u} = \nabla S/m \tag{3.244}$$

is a kind of flow velocity, similarly to electrodynamics: there, the current density is the product of charge density and flow velocity. Here the charge density is replaced by the probability density $\rho = A^2$. For an eigenstate of momentum, $|\mathbf{p}\rangle$, the interpretation of $\nabla S/m$ as a flow velocity is more obvious: now $S = \mathbf{p} \cdot \mathbf{r}$, hence $\nabla S = \mathbf{p}$. If we plug these expressions into (3.238), we obtain the following form of the continuity equation:

$$\frac{\partial \rho}{\partial t} + \nabla \rho \cdot \mathbf{u} + \rho \nabla \mathbf{u} = 0 \tag{3.245}$$

This form is well-known from hydrodynamics: the local change of density is a combined effect of the spatial change of density in the direction of flow (more or less dense amounts of fluid stream into the point $\mathbf{r}$) and the local variation of the flow velocity (compressing or decompressing the fluid). In this sense we can interpret the quantum particle as a streaming fluid—which however collapses into a single point at the moment of a position measurement.

Rewriting ρ and $\mathbf{u}$ in terms of A and S, one gets, after cancellation of a factor A:

$$2m\frac{\partial A}{\partial t} + 2\nabla A \cdot \nabla S + a\Delta S = 0 \tag{3.246}$$

This is just the imaginary part of the Schrödinger equation (3.235), after inserting (3.242). The real part of the Schrödinger equation gives

$$\frac{\partial S}{\partial t} + \frac{(\nabla S)^2}{2m} + V = \frac{\hbar^2}{2m}\frac{\Delta A}{A}. \tag{3.247}$$

In many realistic situations, the expression on the right hand side is much smaller than the expressions on the left hand side. The oscillation of the phase of the wave function in space and time by far outweighs the variation of its amplitude. The **classical approximation** is then defined by ignoring the term on the right hand side:

$$\frac{\partial S}{\partial t} + \frac{(\nabla S)^2}{2m} + V = 0 \tag{3.248}$$

Inserting the expression for the flow velocity $\mathbf{u}$, we obtain

$$\frac{\partial S}{\partial t} + \frac{m\mathbf{u}^2}{2} + V = 0. \tag{3.249}$$

Taking the gradient and again replacing ∇S by $m\mathbf{u}$, one gets

$$m\left(\frac{\partial \mathbf{u}}{\partial t} + \mathbf{u} \cdot \nabla \mathbf{u}\right) + \nabla V = 0. \tag{3.250}$$

The expression in brackets is the total time derivative $d\mathbf{u}/dt$ of the velocity of a particle streaming with the flow (not of the quantum particle, but an element of the fluid, if we imagine the quantum object as as streaming fluid). The change of the velocity of the flowing element of the fluid is a combination of the change of the velocity of the fluid at a given point, $\partial \mathbf{u}/\partial t$, and the spatial variation of the velocity in the direction of motion, at a given time t, $\mathbf{u} \cdot \nabla \mathbf{u}$. This reduces to the classical equation of motion

$$m\frac{d\mathbf{u}}{dt} = -\nabla V. \tag{3.251}$$

In the classical approximation, the Schrödinger equation is therefore nothing else but the description of a streaming fluid moving in the classical potential V. This holds only as long as no measurement takes place. In the moment of a measurement, the fluid instantaneously changes its distribution. During a measurement of position, for instance, it is suddenly compressed into a tiny amount of space. The expression on the right hand side of (3.247) can be considered as a correction term to the classical motion of the fluid.

Nerd's Corner 3.3

There are further connections we want to mention here between (3.248) and classical mechanics as well as geometric optics: Equation (3.248) is nothing but the **Hamilton-Jacobi equation** for a classical particle in the potential V. This equation originates from the Hamilton-Jacobi theory of classical mechanics, a tool used to solve some complicated problems in an elegant way. It replaces momenta with gradients of a so-called **principal function** $S(\mathbf{r}, t)$, $\mathbf{p} = \nabla S$. With (3.248), S becomes the generating function of a canonical transformation, where in the target system the Hamiltonian function is $h = 0$, so that all equations of motion become trivial. The new coordinates Q_i are then arbitrary constant and are identified with the initial conditions of the system at time $t = t_0$, $\mathbf{Q} = \mathbf{r}(t_0) = \mathbf{r}_0$. In the computation of S, the Q_i appear as constants of integration, and one writes $S = S(\mathbf{r}, \mathbf{r}_0, t, t_0)$.

A possible solution S of the Hamilton-Jacobi equation is the classical action,

$$S(\mathbf{r}, \mathbf{r}_0, t, t_0) = \int_{t_0}^{t} L \, dt, \qquad (3.252)$$

where L is the Lagrangian, to be evaluated for the classical trajectory of a particle, located at $\mathbf{r}_0$ for t_0 and at $\mathbf{r}$ for t. This solution can be carried over to the classical approximation of QM: the classical action (with arbitrary t_0, $\mathbf{r}_0$ as parameters) is a solution of equation (3.248), i.e. for the phase of the wave function ψ. We will meet this connection between action and phase again when we discuss the path integral (Chap. 13).

Furthermore, there is a connection to geometric optics: For stationary solutions (energy eigenstates with energy eigenvalue E) one has

$$S(\mathbf{r}, t) = S_0(\mathbf{r}) - Et, \qquad (3.253)$$

hence $\partial S / \partial t = -E$. Equation (3.248) thus becomes

$$(\nabla S_0)^2 = 2m(E - V). \qquad (3.254)$$

In geometric optics with inhomogeneous materials having a position-dependent refraction index $n(\mathbf{r})$, one inserts the ansatz

$$\phi(\mathbf{r}, t) = A(\mathbf{r}) e^{ik_0 L(\mathbf{r}) - \omega t} \qquad (3.255)$$

into the wave equation

$$\Delta \phi - \frac{n^2}{c^2} \frac{\partial^2 \phi}{\partial t^2} = 0, \qquad (3.256)$$

where $k_0 = \omega/c$ is the vacuum wave number corresponding to the frequency ω, and L is the so-called **eikonal**. The real part of the wave equation then reads

$$(\nabla L)^2 - n^2 = \frac{1}{k_0^2} \frac{\Delta A}{A}. \qquad (3.257)$$

As in the classical approximation of QM, one ignores the term on the right hand side (the variation of the amplitude is very small compared to the oscillation of the wave) and obtains the **eikonal equation**

$$(\nabla L)^2 = n^2. \qquad (3.258)$$

Comparison with (3.254) shows: **In analogy to optics, a stationary state in QM can be considered as a radiation field in an inhomogeneous material with refraction index**

$$n(\mathbf{r}) \sim \sqrt{2m(E - V(\mathbf{r}))}. \tag{3.259}$$

Again, the analogy breaks down at the moment of a measurement: The "radiation field" shrinks down (in the case of a position measurement) into a tiny region of space.

The connection between Hamilton-Jacobi theory and geometric optics exists already without quantum mechanics and has been known since 1834. The eikonal equation of optics is formally equivalent to the Hamilton-Jacobi equation. **Already in classical physics there is thus some kind of wave-particle duality.**

In the first half of the 19th century there were two theories that were able to explain the reflection and refraction behavior of radiation: Newton's corpuscular theory, according to which light consists of a ray of particles, and Huygens' wave theory. The analogy between Hamilton-Jacobi theory and geometric optics is the deeper reason why both theories work equally well.

Maxwell's theory of electromagnetism put the wave theory on a broad foundation and led to Newton's approach being considered refuted, until finally quantum mechanics recognized wave and particle as being of equal value again.

With an ingenious twist, quantum mechanics provides a deeper meaning to the wave-particle duality of classical physics, which used to be only a formal analogy between two equations; the ingenious twist being that QM uses the wave as a field which describes the probability for the possible locations of the particle.

Self-check questions:

1. What do position and momentum eigenstates look like in d dimensions?
2. Can you recite the content of the three "boxes" of this section with closed eyes, even under the influences of fatigue and alcohol?
3. What is the classical approximation?

3.8 Several Particles

Finally we consider n quantum particles in d dimensions. Each particle has its own position vector $\mathbf{r}^{(\alpha)}$. The global state $|\psi\rangle$ is described by a wave function $\psi(\mathbf{r}^{(1)}, \mathbf{r}^{(2)}, \ldots, \mathbf{r}^{(n)}, t)$. The corresponding Hilbert space is $L^2(\mathbb{R}^{nd}, \mathbb{C})$, the tensor product of the Hilbert spaces of the individual particles,

$$L^2(\mathbb{R}^{nd}, \mathbb{C}) = \bigotimes_{\alpha=1}^{n} L^2(\mathbb{R}^d, \mathbb{C}). \tag{3.260}$$

In this chapter, we assume that the particles are distinguishable, having different masses, say, so that we can assign them the labels (α) unambiguously and identify them in a measurement. The case of indistinguishable particles will be discussed in Chap. 12.

For each coordinate of each particle there is a separate position and momentum operator $X_i^{(\alpha)}$ and $P_i^{(\alpha)}$ acting as

$$X_i^{(\alpha)} \psi(\mathbf{r}^{(1)}, \ldots, \mathbf{r}^{(n)}, t) = x_i^{(\alpha)} \psi(\mathbf{r}^{(1)}, \ldots, \mathbf{r}^{(n)}, t) \tag{3.261}$$

$$P_i^{(\alpha)} \psi(\mathbf{r}^{(1)}, \ldots, \mathbf{r}^{(n)}, t) = -i\hbar \frac{\partial}{\partial x_i^{(\alpha)}} \psi(\mathbf{r}^{(1)}, \ldots, \mathbf{r}^{(n)}, t) \tag{3.262}$$

and with the commutators

$$[X_i^{(\alpha)}, X_j^{(\beta)}] = 0, \quad [P_i^{(\alpha)}, P_j^{(\beta)}] = 0, \quad [X_i^{(\alpha)}, P_j^{(\beta)}] = i\hbar \delta_{\alpha\beta} \delta_{ij} \mathbf{1}. \tag{3.263}$$

The position eigenstate $|\mathbf{r}'^{(1)}, \ldots, \mathbf{r}'^{(n)}\rangle$ is a simultaneous eigenstate for all X-operators, with eigenvalues $\{x'_i^{(\alpha)} | i = 1, \ldots, d; \ \alpha = 1, \ldots, n\}$. The corresponding wave function is a product of delta distributions,

$$\psi_{\mathbf{r}'^{(1)}, \ldots, \mathbf{r}'^{(n)}}(\mathbf{r}^{(1)}, \ldots, \mathbf{r}^{(n)}) = \prod_{\alpha=1}^{n} \prod_{i=1}^{d} \delta(x'_i^{(\alpha)} - x_i^{(\alpha)}). \tag{3.264}$$

In particular, we have

$$|\mathbf{r}'^{(1)}, \ldots, \mathbf{r}'^{(n)}\rangle = |\mathbf{r}'^{(1)}\rangle \otimes |\mathbf{r}'^{(2)}\rangle \otimes \cdots \otimes |\mathbf{r}'^{(n)}\rangle. \tag{3.265}$$

The momentum eigenstate $|\mathbf{p}^{(1)}, \ldots, \mathbf{p}^{(n)}\rangle$ is a simultaneous eigenstate for all P-operators, with eigenvalues $\{p_i^{(\alpha)} | i = 1, \ldots, d; \ \alpha = 1, \ldots, n\}$. The corresponding wave function reads, in position space:

$$\psi_{\mathbf{p}^{(1)}, \ldots, \mathbf{p}^{(n)}}(\mathbf{r}^{(1)}, \ldots, \mathbf{r}^{(n)}) = \frac{1}{(2\pi\hbar)^{nd/2}} e^{\frac{i}{\hbar} \sum_{\alpha=1}^{n} \mathbf{p}^{()} \cdot \mathbf{r}^{(\alpha)}} \tag{3.266}$$

In particular, we have

$$|\mathbf{p}^{(1)}, \ldots, \mathbf{p}^{(n)}\rangle = |\mathbf{p}^{(1)}\rangle \otimes |\mathbf{p}^{(2)}\rangle \otimes \cdots \otimes |\mathbf{p}^{(n)}\rangle. \tag{3.267}$$

The Hamiltonian operator, for a potential $V(\mathbf{r}^{(1)}, \ldots, \mathbf{r}^{(n)})$ and particle masses $m^{(\alpha)}$, now reads

$$H\psi = \left(\sum_{\alpha=1}^{n} \frac{\mathbf{P}^{(\alpha)2}}{2m^{(\alpha)}} + V(\mathbf{X}^{(1)}, \ldots, \mathbf{X}^{(n)}) \right) \psi \qquad (3.268)$$

$$= -\sum_{\alpha=1}^{n} \frac{\hbar^2}{2m^{(\alpha)}} \Delta^{(\alpha)} \psi + V(\mathbf{r}^{(1)}, \ldots, \mathbf{r}^{(n)})\psi, \qquad (3.269)$$

where $\Delta^{(\alpha)}$ is the Laplace operator with respect to the position vector $\mathbf{r}^{(\alpha)}$. Die Schrödinger equation is

$$i\hbar\frac{d}{dt}\psi = -\sum_{\alpha=1}^{n} \frac{\hbar^2}{2m^{(\alpha)}} \Delta^{(\alpha)} \psi + V\psi. \qquad (3.270)$$

In the simplest case, the global wave function is the tensor product of the wave functions of the individual particles,

$$|\psi(t)\rangle = |\psi^{(1)}(t)\rangle \otimes |\psi^{(2)}(t)\rangle \otimes \cdots \otimes |\psi^{(n)}(t)\rangle, \qquad (3.271)$$

$$\psi(\mathbf{r}^{(1)}, \ldots, \mathbf{r}^{(n)}, t) = \psi^{(1)}(\mathbf{r}^{(1)}, t)\, \psi^{(2)}(\mathbf{r}^{(2)}, t) \,\ldots\, \psi^{(n)}(\mathbf{r}^{(n)}, t). \qquad (3.272)$$

Such solutions of the Schrödinger equation exist if the particles don't interact with each other, each particle being exposed to an external potential which does not depend on the positions of the other particles,

$$V(\mathbf{r}^{(1)}, \ldots, \mathbf{r}^{(n)}) = V^{(1)}(\mathbf{r}^{(1)}) + V^{(1)}(\mathbf{r}^{(2)}) + \cdots + V^{(n)}(\mathbf{r}^{(n)}). \qquad (3.273)$$

For in this cse the Schrödinger equation is separable. Consider, for example, a two-particle system with (3.273) being fulfilled. Then $H = H^{(1)} + H^{(2)}$ with

$$H^{(1)} = \frac{\mathbf{P}^{(1)2}}{2m^{(1)}} + V(\mathbf{X}^{(1)}), \quad H^{(2)} = \frac{\mathbf{P}^{(2)2}}{2m^{(2)}} + V(\mathbf{X}^{(2)}). \qquad (3.274)$$

With the separation ansatz (3.272) each of the two operators acts only on one of the particles,

$$H\psi = (H^{(1)}\psi^{(1)})\psi^{(2)} + \psi^{(1)}H^{(2)}\psi^{(2)}. \qquad (3.275)$$

In particular, $H^{(1)}$ and $H^{(2)}$ commute with each other and also with H, hence we can diagonalize the three operators simultaneously. If we look for stationary solutions, we can consider each particle separately. With

$$H^{(1)}\psi^{(1)} = E^{(1)}\psi^{(1)}, \quad H^{(2)}\psi^{(2)} = E^{(2)}\psi^{(2)} \qquad (3.276)$$

follows

$$H\psi = E\psi, \quad E = E^{(1)} + E^{(2)}. \tag{3.277}$$

The time evolution of the eigenstates is consistent:

$$|\psi(t)\rangle = |\psi(0)\rangle e^{-i\frac{E}{\hbar}t} \tag{3.278}$$

$$= |\psi^{(1)}(0)\rangle e^{-i\frac{E^{(1)}}{\hbar}t} \otimes |\psi^{(2)}(0)\rangle e^{-i\frac{E^{(2)}}{\hbar}t} \tag{3.279}$$

$$= |\psi^{(1)}(t)\rangle \otimes |\psi^{(2)}(t)\rangle, \tag{3.280}$$

compare the discussion in Sect. 2.10.

A different separable case with two particles occurs if the potential is a pure interaction potential, depending only on the relative position of the two particles.

$$V(\mathbf{r}^{(1)}, \mathbf{r}^{(2)}) = V(\mathbf{r}^{(1)} - \mathbf{r}^{(2)}) \tag{3.281}$$

Just as in classical mechanics, such a system can be separated into center-of-gravity and relative motion. However, there are some new aspects we want to explain here. With the new position vectors (CG is for center of gravity, R for relative)

$$\mathbf{r}_{CG} = \frac{m^{(1)}\mathbf{r}^{(1)} + m^{(2)}\mathbf{r}^{(2)}}{m^{(1)} + m^{(2)}}, \quad \mathbf{r}_R = \mathbf{r}^{(2)} - \mathbf{r}^{(1)}, \tag{3.282}$$

the definitions for **total mass** M and **reduced mass** μ,

$$M = m^{(1)} + m^{(2)}, \quad \mu = \frac{m^{(1)}m^{(2)}}{m^{(1)} + m^{(2)}}, \tag{3.283}$$

and the momenta

$$\mathbf{p}_{CG} = M\dot{\mathbf{r}}_{CG} = \mathbf{p}^{(1)} + \mathbf{p}^{(2)} \tag{3.284}$$

$$\mathbf{p}_R = \mu\dot{\mathbf{r}}_R = \frac{m^{(1)}\mathbf{p}^{(2)} - m^{(2)}\mathbf{p}^{(1)}}{m^{(1)} + m^{(2)}} \tag{3.285}$$

(dots are as usual for time derivatives), the Hamiltonian function reads

$$h = h_{CG} + h_R = \frac{\mathbf{p}_{CG}^2}{2M} + \frac{\mathbf{p}_R^2}{2\mu} + V(\mathbf{r}_R). \tag{3.286}$$

Exercise 3.17
Verify this.

What does the associated Hamiltonian operator H look like? We have

$$\mathbf{P}_{CG} = -i\hbar(\nabla^{(1)} + \nabla^{(2)}) \tag{3.287}$$

$$\mathbf{P}_R = -i\hbar\left(\frac{m^{(1)}\nabla^{(2)} - m^{(2)}\nabla^{(1)}}{m^{(1)} + m^{(2)}}\right). \tag{3.288}$$

Exercise 3.18
Show, using the chain rule, that

$$\mathbf{P}_{CG} = -i\hbar\nabla_{CG}, \quad \mathbf{P}_R = -i\hbar\nabla_R, \tag{3.289}$$

where $\nabla_{CG} = (\partial/\partial x_{CG1}, \partial/\partial x_{CG2}, \partial/\partial x_{CG3})$ and similarly for ∇_R (x_{CG1} is the first component of $\mathbf{r}_{CG}$ etc.).

The new momentum operators are, just as the old ones, obtained as the partial derivatives with respect to the coordinates, in particular with the canonical commutation relations

$$[X_{CGi}, P_{CGj}] = i\hbar\delta_{ij}, \quad [X_{Ri}, P_{Rj}] = i\hbar\delta_{ij}, \tag{3.290}$$

$$\left[X_{Ri}, P_{CGj}\right] = 0, \quad [X_{CGi}, P_{Rj}] = 0. \tag{3.291}$$

The new Hamiltonian operator is thus

$$H = H_{CG} + H_R = -\frac{\hbar^2}{2M}\Delta_{CG} - \frac{\hbar^2}{2\mu}\Delta_P + V(\mathbf{X}_R). \tag{3.292}$$

This conservation of the substitution rule $p \to P$ in the new coordinates is not self-evident. It is only valid because we have a **canonical transformation**, see nerd's corner 3.2. With (3.292), the time-independent Schrödinger equation for the wave function, written as a function of the new coordinates, $\hat{\psi}(\mathbf{r}_{CG}, \mathbf{r}_R)$, can now be separated. For stationary states one has

$$\hat{\psi}(\mathbf{r}_{CG}, \mathbf{r}_R) = \psi_{CG}(\mathbf{r}_{CG})\psi_R(\mathbf{r}_R), \tag{3.293}$$

$$H\hat{\psi} = E\hat{\psi}, \quad H\psi_{CG} = E_{CG}\psi_{CG}, \quad H\psi_R = E_R\psi_R, \quad E = E_{CG} + E_R. \tag{3.294}$$

Here ψ_{CG} is the wave function of a free particle with momentum p_S and energy $E_{CG} = P_{CG}^2/(2M)$. ψ_R is the wave function of a particle with mass μ and energy E_R in the external potential V. The original wave function $\psi(\mathbf{r}^{(1)}, \mathbf{r}^{(2)})$ can be retrieved from $\hat{\psi}(\mathbf{r}_{CG}, \mathbf{r}_R)$ by inserting the transformation rule (3.282):

$$\psi(\mathbf{r}^{(1)}, \mathbf{r}^{(2)}) = \hat{\psi}\left(\mathbf{r}_{CG} = \frac{m^{(1)}\mathbf{r}^{(1)} + m^{(2)}\mathbf{r}^{(2)}}{m^{(1)} + m^{(2)}}, \mathbf{r}_R = \mathbf{r}^{(2)} - \mathbf{r}^{(1)}\right) \qquad (3.295)$$

However, this wave function is not a tensor product of the form

$$\hat{\psi}(\mathbf{r}^{(1)}, \mathbf{r}^{(2)}) = \psi^{(1)}(\mathbf{r}^{(1)})\psi^{(2)}(\mathbf{r}^{(2)}). \qquad (3.296)$$

The stationary solutions of the quantum mechanical two-body problem are entangled states of the two particles.

Self-check questions:

1. Is the state of n particles in general described by one or by n wave functions?
2. How do you approach the two-body problem in QM, where the potential depends only on the relative position of the two particles?

Chapter 4
Interpretations

Abstract Several interpretations of the QM formalism are discussed, in particular the Many Worlds Interpretation, the Copenhagen Interpretation, and Bohmian Mechanics.

4.1 The Problem of Interpreting QM

What should we think of the formalism we've defined on the last 120 pp? What does it *mean*? While the formalism of QM was developed in the 1920s within a relatively short time by Schrödinger, Heisenberg, Born, Dirac and others, the progress regarding the interpretation of the theory was distributed over a much larger period of time. At first the Copenhagen Interpretation dominated, going back to Heisenberg and Bohr, but was heavily attacked by Einstein though. The Many Worlds Interpretation was introduced by Everett in 1957 and extended in the subsequent decades. A deterministic variant of QM with hidden variables was first suggested by de Broglie in 1927 and further developed by Bohm in 1952. Bell's inequalities which are so crucial for this problem were formulated not before 1964 by Bell. Due to the progress in Quantum Information Theory and the more and more elaborate experimental construction of entangled states, again some new light was shed on these issues.

The developments regarding the question what QM actually means happened in small steps, but induced some substantial changes in the way how QM is viewed and presented. Today we find ourselves confronted with a broad spectrum of views about what QM exactly says and what a role its state vectors play. You can get an impression of the breadth of this spectrum by reading *Elegance and Enigma* Schlosshauer (2011), a collection of interviews with current representatives of this field of research. How mysterious QM still appears even to experts, can be seen in the following passage:

> It is not at all clear what quantum theory is about. Indeed, it is not at all clear what quantum theory actually says. Is quantum mechanics fundamentally about measurement and observation? Is it about the behavior of macroscopic variables? Or is it about our mental states? Is it about the behavior of wave functions? Or is it about the behavior of suitable fundamental microscopic entities, elementary particles and/or fields? Quantum mechanics provides us with formulas for lots of probabilities. What are these the probabilities of? Of results of measurements? Or are they the probabilities for certain unknown details about the state

© Springer International Publishing Switzerland 2016   127
J.-M. Schwindt, *Conceptual Basis of Quantum Mechanics*,
Undergraduate Lecture Notes in Physics, DOI 10.1007/978-3-319-24526-3_4

of a system, details that exist and are meaningful prior to measurement? (S. Goldstein in Schlosshauer (2011))

We will pick up on some of these questions in the following discussion and will now consider the core problems arising from the QM formalism in detail.

The Measurement Problem: What exactly happens during a measurement? The formalism suggests that a quantum state contains the possible results of a measurement, together with the associated probabilities. The measurement device acts on the state via a projection operator which projects the state into an eigenspace of a hermitian operator associated with the measured variable. The inconsistency between the experienced process of the measurement and its description by the formalism is remarkable. For a measurement device, a box with a pointer, is obviously something completely different than a hermitian operator. So, how does the operator get into the box? And our intuition of a point-like particle is also something very different than a vector in an infinite-dimensional Hilbert space. So, how are these two things related? And why does the state vector collapse into a specific eigenvector at the moment of measurement? What is physically so unique about a measurement that makes it the only process in the world with the power to let a state collapse? Is it something real that collapses there? And when exactly does it collapse? Or is it a continuous, dynamic process? Or is the collapse just an illusion? Are the particle and its state vector one and the same? Does that mean there are in fact no particles at all? Or does the state vector turn into a particle at the moment of measurement? Or does the state vector not really exist "out there", and describes only the subjective information we have about the particle, and it is only this information which collapses? But if, according to Bell's insights, the properties of the particle don't exist prior to the measurement, and the state vector also doesn't exist (because it represents only our subjective information), what does then exist at all prior to a measurement? What should we think of a superposition of different possibilities, what does it mean?

In the context of these questions, the thought experiment with **Schrödinger's cat** became famous, which is of rather metaphorical character though (no physicist would ever support such a cruel animal experiment!): a cat lies in a closed box. Also in the box there is a radioactive substance and a detector which detects the decay of an atom. If the detector reacts within a certain time interval $[t_1, t_2]$, which happens with exactly 50 % likelihood, it triggers a mechanism which kills the cat. Later, at time t_3 an experimenter opens the box and finds a living or a dead cat. In what a condition is the cat during the time interval $[t_2, t_3]$? A superposition of dead and alive, $\frac{1}{\sqrt{2}}(|\text{dead}\rangle + |\text{alive}\rangle)$? And when the experimenter opens the box, this state collapses to either $|\text{dead}\rangle$ or $|\text{alive}\rangle$?

The thought experiment was extended by Wigner with **Wigner's friend**: In a laboratory, Wigner's friend opens the box with the cat. Wigner himself is waiting outside the laboratory for his friend and the information whether the cat is dead or alive. As long as the door is closed, no information leaves the laboratory. Only when Wigner's friend comes out, Wigner is informed about the result. The laboratory thus functions as a second box around the first box. As long as Wigner's friend is in the laboratory, he is in the state $\frac{1}{\sqrt{2}}(|\text{W.F. sees dead cat}\rangle + |\text{W.F. sees living cat}\rangle)$. So, when exactly does the collapse take place? Is there an objective moment when this happens? Or is the collapse subjective, happening only in the observer's mind?

Determinism or Chance: Classical physics is deterministic. If the positions of all particles, the distributions of all fields, and the first time derivatives of these quantities are known for one given moment of time, then one can, in principle, derive the particle positions and field distributions for all other times (forward and backward). In QM this doesn't seem to be the case. But are the measurement results really random? Or are there hidden variables which determine the results in advance? It is similar for a classical die: In fact, the result of a toss is already determined by the momentum and angular momentum of the die at the moment of tossing, combined with its other physical properties (e.g. elasticity) as well as the properties of the table. However, the details of all these properties are hidden from the tosser, so that he assumes all six possible results to have the same probability. Or he deduces the probabilities of $1/6$ statistically, by tossing many times (where each toss slightly differs in momentum and angular momentum, of course). Is it similar in the case of QM? Or maybe, is it even conceivable that **all** possible results actually occur simultaneously, and we just don't notice it?

The Appearance of a Classical World: Until about 100 years ago, classical physics seemed sufficient for describing our world. This is because for most macroscopic phenomena, quantum effects are negligible. Quantum effects here often occur only as small corrections to the classical description. How does it come that classical mechanics is such a good approximation? Why don't we find ourselves surrounded by a fog of fuzzy objects? Why do the planets move along well-defined orbits around the sun, very different from the "electron cloud" around the atomic nucleus? Why does space appear three-dimensional at all, when quantum physics takes place in an infinite-dimensional Hilbert space?

In contrast to most other issues discussed in this chapter, there are some results regarding this topic which are overall agreed upon by most physicists and which solve the problem at least partially: quantum objects take on classical properties by interacting with their environment. This phenomenon is known under the name **decoherence**. Very roughly speaking, one can say: the more macroscopic an object is, the stronger it interacts with its environment (e.g. by collisions or by absorption or emission of radiation), the stronger is the decoherence effect, and the more "classical" it therefore appears to us. How decoherence exactly works is beyond the the scope of this book. We will come back to it in the section on the Many Worlds Interpretation, since decoherence plays an important role there. At this point I want to emphasize that decoherence is not a matter of interpretation, but a direct consequence of the Schrödinger equation. To the interested reader I recommend the book Joos et al. (2003) on the topic.

Information and Existence: We dive a bit deeper into philosophy when we ask about the relation between information and existence in the context of quantum mechanics. For many this is the core problem regarding the interpretation of QM. Wheeler has invented the famous expression "*it from bit*" for this theme, where *it* stands for the reality, *bit* for the information. Is reality in the end nothing else but information? Or is it just described via information? How objective is such a description? Does it maybe just represent a subjective perspective? And what is information after all? Smolin writes:

> The only interpretations of quantum mechanics that make sense to me are those that treat quantum mechanics as a theory of the information that observers in one subsystem of the universe can have about another subsystem. (L. Smolin in Schlosshauer (2011))

So, is the quantum state something purely **epistemic** (i.e. something that has to do with our knowledge) or something **ontic** (something that "really" exists)? And what exactly is the difference? Interestingly, only recently a class of interpretations which consider the quantum state purely epistemic was refuted Pusey et al. (2011).

After we have briefly introduced the central issues, we are now going to meet some interpretations of QM and have a look what answers they have to offer for the questions asked above.

4.2 Many Worlds Interpretation

The Many Worlds Interpretation is in some sense the most conservative, minimal interpretation of QM. That's why I've chosen it to begin with. It is also know as the Everett Interpretation, since Everett was the one who introduced it. David Wallace, one of its present main advocates, writes:

> The Everett interpretation just is ordinary quantum mechanics. (D. Wallace in Schlosshauer (2011))

It is conservative, because it neither gives up the realism nor the determinism of classical physics: quantum states are ontic, i.e. real. Their deterministic evolution is given by the Schrödinger equation. That's it. Everything else follows from that. The measurement process is a quantum physical process like anything else and is described by the time-dependent Schrödinger equation; it doesn't follow any special rules. There are no hidden variables. There is also no collapse of the wave function.

The interpretation considers only the postulates 1 and 4 from Sect. 2.1 as fundamental (Hilbert space, states, Schrödinger equation) and has to solve the problem of deriving the others as consequences of a physical process. This means, it has to explain why results of measurements are eigenvalues of certain hermitian operators and why we subjectively experience the results as random (in the Many Worlds Interpretation, probabilities are only subjective!). This task is difficult, but seems natural. It is reasonable to take a minimal number of postulates as a starting point for a theory and to derive as much as possible from them, hence not adding anything redundant to the theory.

How did the Many World Interpretation get its name? Let's consider the measurement process. It involves a measurement device M and an observed object X. Assume X can have only two linearly independent states, $|X+\rangle$ and $|X-\rangle$. The state

$|X+\rangle$ lets the pointer of M point to the right, $|X-\rangle$ to the left (that's why we have chosen this basis of $\mathcal{H}_X$, the state space of X). Like any other physical object, M obeys the rules of QM, i.e. M is described by a quantum mechanical state. Before the measurement, M is in the state $|M0\rangle$ (pointer is in the middle) after the measurement two states are possible, $|M+\rangle$ (pointer to the right) and $|M-\rangle$ (pointer to the left). In fact, the Hilbert space of an apparatus has a huge number of dimensions, due to its internal degrees of freedom—the states of each atom it consists of. That we restrict ourselves to three dimensions is a huge simplification. The six-dimensional state space $\mathcal{H}_1$ of the quantum system consisting of M and X is the tensor product of the state spaces $\mathcal{H}_M$ (three-dimensional) and $\mathcal{H}_X$ (two-dimensional) of M and X,

$$\mathcal{H}_1 = \mathcal{H}_M \otimes \mathcal{H}_X. \tag{4.1}$$

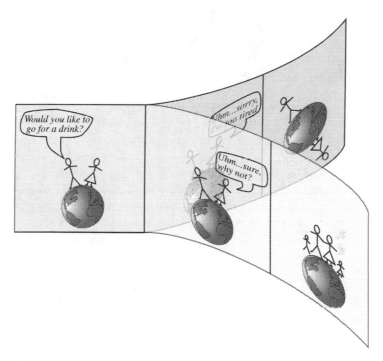

Fig. 4.1 The basic idea of the Many Worlds Interpretation. Cartoon by Max Tegmark, from *Our Mathematical Universe*

The measurement takes place in the time interval $[t_1, t_2]$. At time t_1, X is in the state $|X+\rangle$, say. The total state is

$$|\Psi(t_1)\rangle = |M0\rangle \otimes |X+\rangle. \tag{4.2}$$

Between t_1 and t_2 an interaction takes place between X and M which makes the pointer change its position. At time t_2 the total state is therefore

$$|\Psi(t_2)\rangle = |M+\rangle \otimes |X+\rangle. \tag{4.3}$$

Like any other state, $|\Psi\rangle$ obeys the Schrödinger equation, i.e. $|\Psi(t_2)\rangle$ is obtained from $|\Psi(t_1)\rangle$ via unitary evolution according to some Hamiltonian operator H_1, namely the operator representing the total energy of M and X. What H_1 exactly looks like is not important to us here. In any case, it contains some interaction term which describes the influence of X on M. Equivalent considerations hold if X is in the state $|X-\rangle$.

Now we consider the situation where X is in the state $\alpha|X+\rangle + \beta|X-\rangle$ prior to the measurement, that is, in a superposition of the two basis states. The total state is initially

$$|\Psi(t_1)\rangle = |M0\rangle \otimes (\alpha|X+\rangle + \beta|X-\rangle) \tag{4.4}$$
$$= \alpha|M0\rangle \otimes |X+\rangle + \beta|M0\rangle \otimes |X-\rangle. \tag{4.5}$$

What is the total state at time t_2? The time evolution of the system is linear, implying that the individual terms evolve completely independently (that is the superposition principle!). The first term is, up to a constant factor α, identical to the state (4.2) and therefore evolves just like it, i.e. to the state (4.3). Similar for the second term. After the measurement, at time t_2, the total state is thus

$$|\Psi(t_2)\rangle = \alpha|M+\rangle \otimes |X+\rangle + \beta|M-\rangle \otimes |X-\rangle. \tag{4.6}$$

What does that mean? After the measurement, the state consists of two terms: One term describes an apparatus whose pointer points to the right, and the part of the object which led to this deflection. The other one describes an apparatus whose pointer points to the left, and the part of the object which led to this deflection. Indeed, both possible measurement results are realized. During the measurement, the states of the apparatus and the object are entangled with each other. As a consequence, the total state can no longer be written as a tensor product of the state of the apparatus and the state of the object, but only as a sum of several such products. The state of the apparatus is **relative** to the state of the object.

As a next step, we include an experimenter E who reads off the result from the apparatus in the time interval $[t_2, t_3]$. Within the Many Worlds Interpretation, E is also a quantum mechanical object. Before the measurement, he is in the state $|E0\rangle$ (E without knowledge about the result); after the measurement the states $|E+\rangle$ (E sees pointer to the right) and $|E-\rangle$ (E sees pointer to the left) are possible. Again we have strongly simplified by reducing the Hilbert space $\mathcal{H}_E$ of the experimenter to three dimensions. The total Hilbert space $\mathcal{H}_2$ is now twelve-dimensional:

$$\mathcal{H}_2 = \mathcal{H}_E \otimes \mathcal{H}_M \otimes \mathcal{H}_X \qquad (4.7)$$

The state in this Hilbert space again obeys a Schrödinger equation with a Hamiltonian operator H which is now represented by a 12×12-matrix. H contains an interaction term which leads to an entanglement between M and X in the time interval $[t_1, t_2]$, and another interaction term which leads to an entanglement between E and M in the time interval $[t_2, t_3]$:

$$|\Psi(t_1)\rangle = |E0\rangle \otimes |M0\rangle \otimes (\alpha|X+\rangle + \beta|X-\rangle) \qquad (4.8)$$

$$|\Psi(t_2)\rangle = |E0\rangle \otimes (\alpha|M+\rangle \otimes |X+\rangle + \beta|M-\rangle \otimes |X-\rangle) \qquad (4.9)$$

$$|\Psi(t_3)\rangle = \alpha|E+\rangle \otimes |M+\rangle \otimes |X+\rangle + \beta|E-\rangle \otimes |M-\rangle \otimes |X-\rangle \qquad (4.10)$$

Again, the state consists of two terms in the end: one describes an observer who sees the pointer to the right, together with a apparatus whose pointer indeed points to the right, and the part of the object which lead to this deflection. The other one describes an observer who sees the pointer to the left, together with a apparatus whose pointer indeed points to the left, and the part of the object which lead to this deflection.

The system consisting of E, M and X has been split up, so to speak, into two separate systems which have no knowledge of each other. (In particular, the experimenter has been split up too!) The observer in the state $|E+\rangle$ has no chance to communicate with the experimenter in the state $|E-\rangle$.

This game can be continued. If E is Wigner's friend, then outside the laboratory another experimenter (Wigner) is waiting to receive some information. Wigner is also described by a quantum state, in a further Hilbert space $\mathcal{H}_W$. His state is entangled with E when W hears E's report (another interaction), and this makes W split up too. In this way, the state of the entire world is successively split up: the superposition of the two states of X is spread out further and further by interaction. It is the dynamics of the interaction which determines how fast this split takes place, and along which basis ($|E+\rangle \otimes |M+\rangle \otimes |X+\rangle$ etc.).

Here, spacetime is not *actually* split into two spacetimes. One should rather imagine the whole thing similar to radio stations. There we have the same superposition principle. Signals from all stations are contained within one and the same electromagnetic wave. The different frequencies of the wave propagate independently. Each station is responsible for a specific range of frequencies. If you listen to one station, you don't hear anything from the other stations. In QM, during a measurement one station (state vector) is split into several stations (independent parts of the state vector), and the "range of frequencies" of the original station is thereby split into several smaller intervals. This is the picture of the measurement process drawn by the Many Worlds Interpretation.

Is such a description of the measurement process reasonable? Can we represent the measurement process as a unitary evolution of the quantum state of the observer, the apparatus and the object? We have seen that if we do this, it automatically leads to an entanglement of the participating subsystems and a kind of split of the world. Can we maybe find similar kinds of entanglement in simpler physical processes in nature, which we can take as indicators that our description of the measurement process is correct?

The by now well confirmed theory of **decoherence** says that this is indeed the case, and thereby gives the proponents of the Many Worlds Interpretation an important argument at hand. For decoherence not only leads to the classical appearance of macroscopic objects, but also to the creation of more and more entanglements of the type $(4.4) \rightarrow (4.6)$. The classical appearance holds then only in each term of the state vector separately—for example in one term a classically appearing living cat, in the other a dead one.

Due to the decoherence phenomenon, the description of the measurement process in the Many Worlds Interpretation has a physical foundation. The split into several worlds is accordingly a continuous dynamic process. For an observable with a discrete, finite spectrum, the world is split in as many parts (that is, in a sum of as many tensor product states) as there are possible measurement results. Each part of the world corresponds to one measured value. But what about observables with a continuous or a discrete but infinite spectrum? Here we have understand that each apparatus has a finite resolution and a finite scale. Effectively, it can distinguish between only finitely many values. Decoherence happens through interaction with the environment, and if it doesn't make a difference for the environment whether the object is in the state ψ_1 or ψ_2, there will be no associated entanglement, no split of the world.

So, each apparatus is associated with an observable of finite spectrum, even for a position or momentum measurement. The position or momentum operator has to be modified such that it accounts for the finite resolution of the apparatus. The experimental setup determines which operator is responsible for splitting up the world; it follows from the dynamics of the measurement process, which is given by the Hamiltonian operator H_1, describing the time evolution of the state in $\mathcal{H}_1 = \mathcal{H}_M \otimes \mathcal{H}_X$. More precisely: at first it follows from the setup (the dynamics) which subspace of $\mathcal{H}_X$ will move into which branch of the state vector (which branch of the world). In our example above, the one-dimensional subspace spanned by $|X+\rangle$ ends up in one branch, and the subspace spanned by $|X-\rangle$ in the other one. These subspaces have to be eigenspaces of the sought operator. The eigenvalues, and therefore the operator itself (or the observable), are then defined by the values the experimenter *associates* with the positions of the pointer. If in our example the experimenter E

associates with the pointer positions $|M+\rangle$ and $|M-\rangle$ the measurement values $\pm\hbar/2$ (because he knows that he is doing a spin measurement, say), then he has "measured the state of X with respect to the basis $|X+\rangle$, $|X-\rangle$ with the hermitian operator $\frac{\hbar}{2}\sigma_z$". This implies that the second postulate of QM follows only if the psychology of the observer is taken into account. He has the "impression" that he obtained a unique result for a specific observable that is associated with a specific hermitian operator.

The branching of the world or its state vector happens successively with each decoherence process. After one such process the world is split into n branches. Afterwards, each branch has to be considered separately. If another decoherence happens in the k-th branch, this k-th branch is split again, and so forth.

Problems of the Many Worlds Interpretation:

As already mentioned, the Many Worlds Interpretation is based only on postulates 1 and 4, and has to derive the other two from them, even if only as subjective impressions of an observer. For the second postulate this has been sketched above. Now what about the third postulate? Where do the probabilities come from in a deterministic Many Worlds universe? In the Many Worlds Interpretation, the probability of a measurement result is a purely subjective phenomenon. The big difference to the classical die is that indeed **each** possible result of a quantum measurement is realized (as if in a single toss of a die all six results would come out simultaneously). But the observer doesn't notice this. Each of the two observers $|E+\rangle$ and $|E-\rangle$ in (4.10) finds himself in a world in which only one of the two possible results is realized. Why this makes him assign the probabilities $|\alpha|^2$ and $|\beta|^2$ to the possible results— the squared absolute values of the two terms in (4.10)—is an open question of the Many Worlds Interpretation (at least it is controversial). For many opponents of the interpretation, the fact that the statistical character of the measurement doesn't show up here in a convincing way is already a reason to exclude the Many Worlds Interpretation. Proponents argue though that *just* the Many Worlds Interpretation offers a conceptual framework to study the subjectively experienced probabilities of quantum measurements. For example, one can prove that for repeated measurements the following holds: The length of the part of the state vector corresponding to such branches of the world where the statistical distribution of the measurement results deviates strongly from the probabilities as given by the third postulate, converges to zero. It is controversial whether this is a sufficient explanation of the subjectively experienced probabilities. The question remains open until today.

Another problem is that the already complicated questions regarding human consciousness are getting even worse here. What happens to our consciousness when the world is split up? The most natural answer seems to be that consciousness is split up too. What you consider to be yourself is only one of the countless paths the original person you were born as has taken through the ever branching world (cf. Fig. 4.1). However, there are also advocates of the Many Worlds Interpretation who ask to give up the psycho-physical parallelism and believe that consciousness follows only one of the branches. One can only hope that it's the same branch for everybody, otherwise we are surrounded by mindless zombies within our respective branch.

To me, the most important problem seems to be the decomposition of the universe into subsystems: The state vector of the universe is a vector of norm 1, rotating all by itself in a giant Hilbert space. In a vector space, all vectors of norm 1 look the same. How can the entire rich structure of our world be encoded in such a vector which looks just like any other? Such a structure emerges only when the universe is split into subsystems, i.e. into factors of a tensor product. But what criteria do we have for such a decomposition?

Let's consider a simple example: The space $\mathcal{H}_1$ from (4.1) is, in the first place, just a six-dimensional vector space. Within this space, each state can be expressed as a unit vector. But the unit vectors of a vector space all look the same, as long as no external distinguishing criterion is specified. For example, one can always choose a basis of $\mathcal{H}_1$, such that the transition (4.4)$\rightarrow$(4.6) looks as follows when written out in components:

$$(1, 0, 0, 0, 0, 0) \rightarrow (0, 1, 0, 0, 0, 0) \tag{4.11}$$

Only when $\mathcal{H}_1$ is decomposed into two subsystems in a very special way, $\mathcal{H}_1 = \mathcal{H}_M \otimes \mathcal{H}_X$, this rotation of a unit vector tells us the story of a measurement process. There are infinitely many possibilities to write $\mathcal{H}_1$ as a tensor product of a three- and a two-dimensional vector space! (Choose an arbitrary basis $\{|e_i\rangle\}$ for $\mathcal{H}_1$ and define $|e_i\rangle = |e_j^{(1)}\rangle \otimes |e_k^{(2)}\rangle$ for $i = 1, \ldots, 6, j = 1, 2, 3, k = 1, 2$.) Why should we choose exactly the one decomposition which tells the story of an entanglement in a measurement process? At this point the problem is resolved by the fact that the system itself interacts with its environment. This interaction can single out a preferred basis (it is an external distinguishing criterion): The observer E sees the apparatus as a distinct object, hence the decomposition $\mathcal{H}_1 = \mathcal{H}_M \otimes \mathcal{H}_X$ makes sense to him. But why do we assign a meaning to E at all? For he only appears as a distinct entity if we decompose the twelve-dimensional Hilbert space $\mathcal{H}_2$ in a very specific way, namely $\mathcal{H}_2 = \mathcal{H}_B \otimes \mathcal{H}_1$. What justifies this decomposition, which ascribes an identity to the observer? The justification again happens via external interactions, with Wigner, say, who interacts with E as a separate object. And so forth. In the end, one inevitably gets to the state vector of the entire universe. This is also just a unit vector in a giant Hilbert space, and these unit vectors all look the same, as long as no external distinguishing criterion is given that singles out a preferred basis. But for the entire universe, there is no environment left, no observer outside the system who could provide such a preferred basis. There *is* a tensor decomposition of the universal Hilbert space which tells the story of an entangled universe with galaxies, planets, measurement devices and observers. But this choice is completely arbitrary. With equal right, one could say that the state of the universe is just $(1, 0, 0, 0, \ldots)$.

A state vector contains information only if there is an environment, an external observer looking at it by means of a preferred basis and thereby breathing life into it. For this reason, the Many Worlds Interpretation appears incomplete to me, after all.

To those who want to get a broader impression of the diverse arguments for and against this interpretation of QM, I recommend the book by Saunders et al. (2012).

4.3 Copenhagen Interpretation

The Copenhagen Interpretation is the original interpretation of the QM formalism. Similar to the Many Worlds Interpretation, it assumes that nothing needs to be added to the postulates. In particular there are no hidden variables. But this is as far as the similarity goes. Other than the Many Worlds Interpretation, all postulates are equally fundamental here. One cannot regard some postulates as a subjective experience of consequences of the others. In particular, the statistical character of measurement is fundamental. QM is non-deterministic in principle.

But the central claim of the Copenhagen interpretation is: **The distinction between the model and the experienced reality is fundamental and unavoidable.** Measurement devices, setups of experiments and the registering of their results have to be expressed in the language of classical physics, the language of our experienced reality. In order to predict the results of experiments—at least statistically—one needs, however, a model written in a completely different language, the language of quantum states and hermitian operators. The only purpose and the only meaning of this model, this language, is it to predict measurement results (statistically). Beyond that, they don't have any independent reality. Quantum states are not real. They are epistemic, not ontic.

Within the Copenhagen Interpretation, there were different opinions on what the reason for the uncertainty is. Some of its proponents—among them Heisenberg— assumed that a particle *does* have a position and a momentum simultaneously, but we can always measure only one of them, because of the always occurring interaction between particle an apparatus. The rest of the proponents—among them Bohr— were convinced that the particles don't even have these properties; that the notions of position and momentum don't have a meaning for the particle independent from the context of an associated measurement. Bell's inequalities have decided this question. The violation of Bell's inequalities proves that already the assumption that a particle has both of these properties simultaneously leads to a contradiction, independent of an interaction with the apparatus.

Accordingly, between two measurements we can understand the particle only in terms of tendencies and possibilities. At the moment of measurement, one of the possibilities becomes real. For Schrödinger's cat this means: only at the moment when the observer opens the box, her destiny is decided. In the Many Worlds Interpretation she came off slightly better: in one branch of the world she stayed alive in any case.

The Copenhagen Interpretation speaks firstly of an experienced world which is to be described in the language of classical physics. This experienced world arises from a sequence of observations. Secondly, it speaks of a model world, expressed in the language of quantum physics. However, it does *not at all* speak of a real, objectively existing world that is independent of models and observations. For according to this interpretation, we cannot say anything about such a world in principle. The discrepancy between the experienced world and the model world therefore cannot be resolved by us. The Copenhagen Interpretation takes Wittgenstein's *"Whereof one cannot speak, thereof one must be silent"* serious.

Due to its non-real character, the Copenhagen Interpretation has provoked lots of dissent. Most physicists are realists, i.e. they assume that there is an objective reality and that we can speak about it; that the measurement devices and also the quantities appearing in our models belong to this reality. The interpretation was particularly attacked by Einstein—both its non-real character and its non-determinism disgusted him. In numerous debates with Bohr he tried to disprove it, but he always came out on the short end.

Nevertheless these debates were very fruitful for the understanding of QM. In particular, it was Einstein who found out that a "spooky action at a distance" is inherent to QM. The corresponding EPR thought experiment (Einstein et al. 1935) was a predecessor of the experiments which finally proved the violation of Bell's inequalities.

In any case, the dichotomy between the experienced and the model world, their extremely different nature, their vague connection consisting only of a mysterious predictive power of one of them for the other, leaves us with a bad aftertaste.

A further stumbling block is the role of the measurement process which is very special in the Copenhagen Interpretation, different from all other physical processes: It makes quantum states collapse. However, since quantum states are not real, the same holds for the collapse. Only if the state is considered as real and the collapse as actually happening, the question arises where and when this collapse actually happens. In the apparatus? In the eye of the observer? In his consciousness? We will get back to this in Sect. 4.5. But such considerations do not apply to the Copenhagen Interpretation as I described it here, because here the collapse is not a really happening process, but only a modification in our model world that became necessary by an actualization of the experienced reality.

The question remains if the dichotomy between the experienced and the model world is really necessary. The Many Worlds Interpretation in connection with decoherence theory has shown that the measurement process can in principle be represented as a physical process inside the model world, and that processes of similar kind (as described by the model world) occur all the time in nature. So why should we lift the measurement process and the experienced reality out of the model world and turn it into something else? Why can't we take the model world as the only true, objective reality, just as it is done with all other scientific theories?

I will come back to the discrepancy between Many Worlds and Copenhagen Interpretation in Sect. 4.7.

Self-check question:

4.4 De Broglie–Bohm Theory

The remaining interpretations we are going to introduce here add something to the QM formalism or modify it, mostly in order to make the theory more similar to the more familiar classical physics. More precisely they are no longer interpretations but independent theories, which however have the hitch that they are very difficult to verify experimentally—eventually their predictions must be extremely close to those of "pure" quantum mechanics, such that they are consistent with observations. The most prominent of those theories is the **de Broglie–Bohm theory** (also called **pilot-wave theory** or **Bohmian mechanics**) with its **non-local hidden variables**.

The de Broglie–Bohm theory was introduced 1927 by de Broglie, and refined in the 1950s by Bohm. At first glance it is quite similar to the Many Worlds Interpretation: There is a really existing (i.e. ontic) universal state vector $|\Psi\rangle$ which never collapses and evolves according to the Schrödinger equation with an appropriate Hamiltonian operator:

$$i\hbar\frac{d}{dt}|\Psi\rangle = H|\Psi\rangle \tag{4.12}$$

In particular, $|\Psi\rangle$ branches successively with each decoherence process into several separate terms, analogous to $(4.4)\rightarrow(4.6)$.

The difference to the Many Worlds Interpretation is that in addition the existence of n particles $\{T^{(i)}\}$, $i \in \{1, \ldots, n\}$ with masses $\{m^{(i)}\}$ is postulated, which constitute the actual matter. These particles behave classically in the sense that at any time they have well-defined positions $\{q^{(i)}\}$ in a three-dimensional position space, $q^{(i)} \in \mathbb{R}^3$. These positions $\{q^{(i)}\}$ are the hidden variables of the theory.

In its simplest form the de Broglie–Bohm theory assumes that $|\Psi\rangle$ can be written as a wave function depending on the $3n$ spatial coordinates and time,

$$\Psi = \Psi(\mathbf{x}^{(1)}, \ldots, \mathbf{x}^{(n)}, t). \tag{4.13}$$

The complex-valued function Ψ can be decomposed into absolute value and phase,

$$\Psi(\mathbf{x}^{(1)}, \ldots, \mathbf{x}^{(n)}, t) = A(\mathbf{x}^{(1)}, \ldots, \mathbf{x}^{(n)}, t) \exp\left(\frac{i}{\hbar} S(\mathbf{x}^{(1)}, \ldots, \mathbf{x}^{(n)}, t)\right), \tag{4.14}$$

with real functions R and S. The motion of the particles is determined by the **guiding equation**:

$$\frac{d}{dt} \mathbf{q}^{(i)}(t) = \frac{\nabla^{(i)} S(\mathbf{q}^{(1)}, \ldots, \mathbf{q}^{(n)}, t)}{m^{(i)}}, \tag{4.15}$$

where $\nabla^{(i)}$ is the gradient w.r.t. the three coordinates $\mathbf{x}^{(i)}$. The gradient of S is evaluated with the present positions of the n particles, i.e. for each j, one sets $\mathbf{x}^{(j)} = \mathbf{q}^{(j)}(t)$. The motion of a particle therefore depends on the positions of all other particles in the universe. This interaction is thus non-local, as it has to be the case, due to Bell's theorem, for theories with hidden variables. One says the particles are **guided** by the wave function ("they ride like dust particles on a water surface"), which is therefore called **pilot wave**. On the other hand, the wave function does not depend on the particles, doesn't notice them at all.

In contrast to the Many Worlds Interpretation, in Bohm's theory we consist of particles, not of components of the wave function. The particles move in a deterministic way in three-dimensional position space. On the level of particles, no branching of the world takes place. Instead, the particles in a sense realize one of the possibilities provided by the wave function. Which one of these possibilities is realized depends deterministically on the initial conditions of the particle positions. The fact that we can make only statistical predictions is because we don't know these positions.

The form of (4.15) should not completely surprise you. The decomposition of a single quantum object's wave function (i.e. $n = 1$) into absolute value and phase was already discussed in Sect. 3.8. There we also interpreted $\nabla S/m$ as a velocity, cf. Equation (3.244), namely as the flow speed of a kind of "quantum fluid" with density $\rho = |\Psi|^2$, as suggested by the continuity equation. In Bohmian mechanics, the "real" particle corresponds to a point mass moving along with this fluid.

As in the Many Worlds Interpretation, the postulates 2 and 3 of QM need to be derived from postulates 1 and 4, now including the guiding equation. Why does the wave function appear to collapse for us? Where do the probabilities we ascribe to the measurement results come from? How are the hermitian operators related to the measurement? The de Broglie–Bohm theory was quite successful in achieving these derivations (see Dürr et al. (2003a) for postulate 2 and Dürr et al. (2003b) for postulate 3). It also solves a problem that heavily bothered the Many Worlds Interpretation: The position space basis is automatically preferred due to the particles, and the possible decomposition into subsystems (tensor product decompositions of Hilbert space) are also given by the particles. The interpretation of $|\Psi\rangle$ is therefore no longer arbitrary.

Observables not related to position space can be included into the theory by adding them as components to the wave function. The guiding equation needs to be extended for that. No additional property is assigned to the particles though. Spin belongs to the wave function not to the particle. Due to the modified guiding equation, however, it influences the motion of the particles.

The de Broglie–Bohm theory was criticized for several reasons:

- It is formally inconsistent with special relativity. The guiding equation requires an unambiguous notion of simultaneity, for the coordinates of all particles **at the same time** need to be evaluated. A relativistic generalization of QM works very well for the wave function alone, as we will see in the last chapter of this book. But for the particles of Bohmian mechanics, a distinguished global time coordinate needs to be introduced. However, this has no direct experimental consequences, i.e., even in Bohmian mechanics the theory of relativity is obeyed by experiments. It is only a formal "ugliness".
- The devil of non-reality (Copenhagen Interpretation) or the branching of the world (Many Worlds Interpretation) is cast out by the Beelzebub of non-locality.
- The theory cannot be distinguished experimentally from standard QM, but requires more objects (the particles) and more equations (the guiding equation). By the principle of **Ockham's razor**, for equivalent experimental predictions one should always prefer the theory which makes less assumptions. According to this principle, Bohmian mechanics should be therefore discarded. This objection is made in particular by the proponents of the Many Worlds Interpretation who follow a similar philosophical approach (reality of the wave function, determinism), but believe they can get along without additional particles.

For the interested reader who wants to know more about this topic, I recommend the book by Holland (1995).

4.5 Collapse Models

Now we consider interpretations of QM where the state vector is real, but does not lead to a branching of the world, but rather commits itself to an eigenstate corresponding to one specific measurement value. That is: it collapses. This leads to the questions when and why this collapse takes place. The collapse models can be classified according to how they respond to this question:

Unspecified Collapse in the Apparatus: This point of view assumes that the apparatus behaves classically and therefore cannot be subject to quantum superpositions. When the quantum object hits the apparatus, its state vector is forced to collapse. How exactly that happens remains unspecified. A not very illuminating viewpoint. It also leaves open the question where exactly the border between classical and quantum behavior is localized (**Heisenberg cut**).

Dynamic Collapse: In this category we have models where the Schrödinger equation is modified by an additional term which is supposed to enforce the collapse in a dynamic way. The additional term must be so designed that it has a considerable effect (namely the collapse) only when a quantum object interacts with a macroscopic object like a measurement device, but leaves the probabilities given by the third postulate unchanged. So far this approach hasn't led to any convincing results.

Collapse by Quantum Gravity: In theoretical physics, when facing two different problems that baffle us, it is a popular method to eliminate one of them by claiming that it can be traced back to the other. Quantum gravity is not understood, the collapse of the wave function is not understood, so why not declare the latter as a consequence of the former? And so, in serious research articles the possibility was considered that parts of the wave function are swallowed by tiny baby universes spontaneously born in some spots of spacetime.

The most serious attempt to explain the collapse by gravitation effects was undertaken by Roger Penrose. He postulated a backreaction of spacetime curvature onto the wave function, prohibiting the delocalization of macroscopic objects (like the needle of a measurement device) and thereby enforcing the collapse. He even suggested an experiment to verify his hypothesis, which is however not feasible for practical reasons.

Collapse by Consciousness: These interpretations are based on the same method as the previous ones: Human consciousness is not understood, the collapse of the wave function is not understood, so why not declare the latter as a consequence of the former? The idea that the collapse takes place in the mind or is caused by it was proposed by some prominent physicists, among them Wigner and von Neumann. Some of them saw this even as an expression of free will. The indeterminism of quantum mechanics leaves some room for our freedom. So why not regard the collapse of the wave function as an interface between our mind (or free will) and the material world? If Schrödinger's cat is dead when the box is opened, it was the cruel (yet maybe unconscious) will of the observer. One of the main advocates of this idea (which I have presented here in a somewhat trenchant way) is Henry Stapp. His book *Mind, Matter, and Quantum Mechanics* Stapp (2009) is nevertheless worth reading.

4.6 New Age Interpretation

Fig. 4.3 *Source* http://smbc-comics.com

4.7 Conclusions

What should we think of all this? Each of the presented interpretations has one or another problem, or a bad aftertaste. We may distinguish three different attitudes:

- QM works well for all practical purposes. Philosophy and questions about deeper meaning are not everyone's cup of tea. So, one may simply ignore all the questions asked in this chapter without too much of a bad conscience.

- Or one may follow the position that all these interpretations are so unsatisfying that QM simply cannot be a fundamental theory, and wait for a better, deeper theory to be found, or even participate in this quest.
- Finally, one could accept that QM has proved to be successful on all levels from particle physics to chemistry and solid state physics, and try to understand what it means, and to draw own conclusions from the different possible (or impossible, depending on your point of view) interpretations.

Personally, I follow the third option, and in this section I want to present my own, subjective view.

In my opinion it makes more sense to refer to interpretations which don't add anything to the theory, in particular when additional assumptions and objects (e.g. the particles of Bohmian mechanics) cannot be verified or observed. Therefore, I want to restrict myself to the Copenhagen Interpretation and the Many Worlds Interpretation. The comparison between these two interpretations reminds me of a passage in a philosophical book by Schrödinger, which didn't have anything to do with QM in the first place:

> The thing that bewilders us is the curious double role that the conscious mind acquires. On the one hand it is the stage, and the only stage on which this whole world-process takes place, or the vessel or container that contains it all and outside which there is nothing. On the other hand we gather the impression, maybe the deceptive impression, that within this world-bustle the conscious mind is tied up with certain very particular organs (brains), which while doubtless the most interesting contraption in animal and plant physiology are yet not unique, not sui generis; for like so many others they serve after all only to maintain the lives of their owners, and it is only to this that they owe their having been elaborated in the process of speciation by natural selection.
>
> Sometimes a painter introduces into his large picture, or a poet into his long poem, an unpretending subordinate character who is himself. Thus the poet of the Odyssey has, I suppose, meant himself by the blind bard who in the hall of the Phaeacians sings about the battles of Troy and moves the battered hero to tears. In the same way we meet in the song of the Nibelungs, when they traverse the Austrian lands, with a poet who is suspected to be the author of the whole epic. In Dürer's All-Saints picture two circles of believers are gathered in prayer around the Trinity high up in the skies, a circle of the blessed above, and a circle of humans on the earth. Among the latter are kings and emperors and popes, but also, if I am not mistaken, the portrait of the artist himself, as a humble side-figure that might as well be missing.
>
> To me this seems to be the best simile of the bewildering double role of mind. On the one hand mind is the artist who has produced the whole; in the accomplished work, however, it is but an insignificant accessory that might be absent without detracting from the total effect. (Schrödinger 1958)

The deep philosophical problem discussed above finds a miraculous parallel in quantum mechanics, where the bewildering double role is now played by the measurement process. The Many Worlds Interpretation describes the accomplished work, in which the measurement process is only one of many quantum mechanical processes leading to the entanglement of an object with its environment. The Copenhagen Interpretation, on the other hand, describes the measurement process as the artist, standing outside of the picture, who only turns the abstract quantum picture into something real, by taking one of the possibilities offered by the state over into the realm of classical reality.

Each of the two viewpoints, when standing alone, leaves open some questions, appears unsatisfying or incomplete. Only together they offer a complete view on QM. They are complementary to each other, like wave and particle.

Many people assume that quantum mechanics is not yet the final truth; that it has to be replaced by a deeper theory whose boundary case it is, such as classical mechanics is a boundary case of quantum mechanics.

However, I cannot avoid the impression that the antinomy depicted above, the bewildering double role of mind as well as of the measurement process, is so fundamental that it cannot be resolved within science.

References

D. Dürr, S. Goldstein, N. Zanghi, Quantum equilibrium and the role of operators as observables in quantum theory. J. Stat. Phys. **116**, 959 (2003a), http://arxiv.org/abs/quant-ph/0308038. (Demonstrates how QM postulate 2 is contained in Bohmian mechanics)

D. Dürr, S. Goldstein, N. Zanghi, Quantum equilibrium and the origin of absolute uncertainty. J. Stat. Phys. **67**, 843 (2003b), http://arxiv.org/abs/quant-ph/0308039. (Demonstrates how QM postulate 3 is contained in Bohmian mechanics)

A. Einstein, B. Podolsky, N. Rosen, Can quantum-mechanical description of physical reality be considered complete? Phys. Rev. **47**, 777

P.R. Holland, *The Quantum Theory of Motion* (Cambridge University Press, 1995). (Detailed presentation of Bohmian mechanics)

E. Joos, H.D. Zeh, C. Kiefer, D. Giulini, J. Kupsch, I.O. Stamatescu, *Decoherence and the Appearance of a Classical World in Quantum Theory*, 2nd edn. (Springer, 2003). (Discusses the phenomenon of decoherence and some views on QM, e.g. Many Worlds Interpretation, Consistent Histories, collapse models)

M.F. Pusey, J. Barrett, T. Rudolph, On the reality of the quantum state. Nat. Phys. **8**, 476 (2011), http://arxiv.org/abs/1111.3328. (Relatively new research result which rejects a class of QM interpretations. Shows how active the field of QM interpretations still is)

S. Saunders, J. Barrett, A. Kent, D. Wallace (eds.), *Many Worlds?* (Oxford University Press, 2012). (Discussion of several views on the Many Worlds Interpretation)

M. Schlosshauer (ed.) *Elegance and Enigma* (Springer, 2011). (Interviews regarding the current status of QM interpretations. Demonstrates the diversity of opinions among the leading representatives of this field)

E. Schrödinger, *Mind and Matter* (Cambridge University Press, 1958), http://web.mit.edu/philosophy/religionandscience/mindandmatter.pdf. (Philosophical work by Erwin Schrödinger)

H.P. Stapp, *Mind, Matter and Quantum Mechanics*, 3rd edn. (Springer, 2009). (Stapp explains here his theory about the collapse of the wave function as an interface between mind and matter)

Part II
A Single Scalar Particle
in an External Potential

Chapter 5
One-Dimensional Problems

Abstract Typical features of solutions of the Schrödinger equation in position space are investigated, using the simplest possible potentials in one dimension. As a highlight, we solve the harmonic oscillator with algebraic methods.

In this chapter, we study wave functions in one dimension. The discussed problems are in several ways idealizations: In addition to restricting ourselves to one single quantum object in one dimension, we also assume that the quantum object is **scalar**, i.e. there are no additional **internal degrees of freedom** like spin, for instance, with respect to which the object could assume several states. That is, we assume that the state of the object is given by its wave function alone. In the subsequent chapters, we will increase the number of space dimensions first to two and then to three. Only in Chap. 9, spin is included again, which has served us so faithfully as an example in Chap. 2. Then we will finally see how the finite and infinite-dimensional Hilbert spaces we have considered so far are combined into one global Hilbert space.

The potentials we are going to look at are also quite idealized, but they are sufficient to present some essential phenomena of QM.

At first, the dissolving of a free wave packet is demonstrated, i.e. it is shown how a Gaussian wave packet without an external potential is more and more broadened, and the corresponding quantum object delocalized accordingly. Next, we study the energy eigenstates of piecewise constant potentials. For a **step potential** we introduce **reflection and transmission** of a wave function. For a **potential well** we find **bound and free states**. For a **potential barrier** we meet the **tunnel effect**. As the last problem for this chapter, we are going to face the **harmonic oscillator**. It will turn out to be one of the most beautiful exercises in quantum mechanics, where the entire glory and usefulness of the algebraic approach becomes apparent.

5.1 Dissolving of a Gaussian Wave Packet

The Gaussian wave packet was already studied in Sect. 3.5 for a fixed moment in time, and the associated uncertainty in position and momentum was determined. Now we want to see how the package evolves in time, under the assumption that it describes a **free particle**, i.e. with vanishing potential, $V(x) = 0$.

© Springer International Publishing Switzerland 2016
J.-M. Schwindt, *Conceptual Basis of Quantum Mechanics*,
Undergraduate Lecture Notes in Physics, DOI 10.1007/978-3-319-24526-3_5

The calculations regarding wave packets make strong usage of Gaussian integrals, i.e. integrals where the square of the integration variable is in the exponent. These integrals occur very often and can be looked up. However, it makes sense to derive them at least once. In the following exercise, we therefore want to generalize the integrals used in Sect. 3.5 in a systematic way.

Exercise 5.1

(a) The basic Gaussian integral is (3.189). Prove it. Begin with

$$\left[\int_{-\infty}^{\infty} dy\, e^{-y^2} \right]^2 = \int_{-\infty}^{\infty} dx \int_{-\infty}^{\infty} dy\, e^{-(x^2+y^2)}, \qquad (5.1)$$

rewrite this expression in polar coordinates, $\int dx \int dy \to \int d\rho\, \rho \int d\phi$, and substitute $u = \rho^2$.

(b) Show

$$\int_{-\infty}^{\infty} dy\, e^{-\frac{(y-y_0)^2}{a}} = \sqrt{\pi a}, \qquad (5.2)$$

using the linear substitution $u = (y - y_0)/\sqrt{a}$.

Remark Equation (5.2) also holds when y_0 or a is complex valued. The only precondition is that a has a positive real part, so that the function vanishes at infinity. After the substitution, the path of integration is tilted in the complex plane. With the help of the residual theorem one can show that the result is the same as when the path is along the real axis. You should do that only if you feel destined to.

(c) Show that

$$\int_{-\infty}^{\infty} dy\, y\, e^{-\frac{(y-y_0)^2}{a}} = \sqrt{\pi a}\, y_0. \qquad (5.3)$$

Use the fact that the integral $\int_{-\infty}^{\infty}$ of an odd function (i.e. a function f with $f(-x) = -f(x)$) vanishes.

(d) Show that

$$\int_{-\infty}^{\infty} dy\, y^2\, e^{-\frac{(y-y_0)^2}{a}} = \sqrt{\pi a}\left(\frac{a}{2} + y_0^2 \right). \qquad (5.4)$$

Substitute as in (b) and perform a partial integration for the quadratic term $u^2 e^{-u^2}$, setting $v(u) = u$, $w'(u) = u\, e^{-u^2}$.

Now we proceed with the wave packet. Due to $V(x) = 0$, the Hamiltonian operator contains only the momentum operator,

$$H = \frac{P^2}{2m},$$ (5.5)

and therefore momentum eigenstates are also energy eigenstates. The time-dependent Schrödinger equation is most easily expressed in momentum space,

$$i\hbar \frac{d}{dt}\tilde{\psi}(p, t) = \frac{p^2}{2m}\tilde{\psi}(p, t),$$ (5.6)

with solutions

$$\tilde{\psi}(p, t) = \tilde{\psi}(p, 0)e^{-i\frac{p^2}{2m\hbar}t}.$$ (5.7)

Let's assume that the initial state at time $t = 0$ is just the Gaussian package (3.187). The wave function in momentum space at time t is then given by

$$\tilde{\psi}(p, t) = \frac{\sqrt{\sigma}}{\sqrt{\hbar}\pi^{1/4}}e^{-\frac{\sigma^2(p-p_0)^2}{2\hbar^2} - i\frac{p^2}{2m\hbar}t}.$$ (5.8)

The absolute value $|\tilde{\psi}(p, t)|$ does not depend on t, since the time evolution consists of pure phase rotations. When calculating expectation values of the form

$$\langle P^n \rangle_\psi = \int_{-\infty}^{\infty} dp\, p^n \tilde{\psi}^*(p, t)\tilde{\psi}(p, t),$$ (5.9)

the phase cancels with its complex conjugate, i.e. the expectation values of powers of momentum don't change. At all times one has the values determined in Sect. 3.5,

$$\langle P \rangle_\psi = p_0, \quad (\Delta P)_\psi = \frac{\hbar}{\sigma\sqrt{2}}.$$ (5.10)

Now, the simplest way to compute the expectation value and uncertainty of X is to represent the X-operator as a derivative in momentum space (see 3.185).

Exercise 5.2
Perform this calculation. Compare with the values below (5.18) and (5.20).

However, we are also interested how the wave function in position space $\psi(x, t)$ exactly looks. Hence we take the effort to calculate it, making use of the Gaussian integrals derived in the exercise above:

$$\psi(x, t) = \frac{1}{\sqrt{2\pi\hbar}} \int_{-\infty}^{\infty} dp \, e^{i\frac{p}{\hbar}x} \tilde{\psi}(p, t) \tag{5.11}$$

$$= \frac{\sqrt{\sigma}}{\sqrt{2}\,\hbar\pi^{3/4}} \int_{-\infty}^{\infty} dp \, \exp\left(i\frac{p}{\hbar}x - \frac{\sigma^2(p - p_0)^2}{2\hbar^2} - i\frac{p^2}{2m\hbar}t\right) \tag{5.12}$$

$$= \frac{\sqrt{\sigma}}{\pi^{1/4}} \frac{1}{\sqrt{\alpha(t)}} \exp\left[-\frac{\left(x - \frac{p_0 t}{m}\right)^2}{2\alpha(t)} + i\frac{p_0}{\hbar}\left(x - \frac{p_0 t}{2m}\right)\right] \tag{5.13}$$

with

$$\alpha(t) = \sigma^2 + i\frac{\hbar t}{m}. \tag{5.14}$$

Exercise 5.3
Several intermediate steps were skipped between the second and third line, please reproduce them. At first, the exponent has to be brought into the form of (5.2). Constant factors can be pulled out of the integral. Then the integral can be evaluated according to (5.2). The result needs to be rearranged somewhat to take on the form of (5.13). Don't worry if you flounder on the way, that's normal.

The wave function looks quite complicated. What interests us most though is the probability density

$$|\psi(x, t)|^2 = \psi^*(x, t)\psi(x, t) = \frac{1}{\sqrt{\pi}\,\beta(t)} \exp\left(-\frac{\left(x - \frac{p_0 t}{m}\right)^2}{\beta^2(t)}\right) \tag{5.15}$$

with

$$\beta(t) = \sqrt{\sigma^2 + \frac{\hbar^2 t^2}{\sigma^2 m^2}}. \tag{5.16}$$

Here we used for the exponent that

$$\frac{1}{2\alpha(t)} + \frac{1}{2\alpha^*(t)} = \frac{\alpha^*(t) + \alpha(t)}{2\alpha(t)\alpha^*(t)} = \frac{\text{Re}(\alpha(t))}{|\alpha(t)|^2} = \frac{1}{\beta^2(t)}. \tag{5.17}$$

The function $\beta(t)$ replaced the constant σ and represents the width of the Gaussian distribution (the distribution of the probability density is still Gaussian). For large t, β grows linearly with time, and so does the position uncertainty. From the numerator of the exponent one can see that the peak of the Gaussian curve moves with the speed p_0/m, as expected. Of course, one can also calculate this explicitly:

$$\langle X \rangle_\psi = \int_{-\infty}^{\infty} dx \, x \, |\psi(x, t)|^2 = \frac{p_0 t}{m} \tag{5.18}$$

$$\langle X^2 \rangle_\psi = \int_{-\infty}^{\infty} dx \, x^2 \, |\psi(x,t)|^2 = \frac{\beta^2(t)}{2} + \frac{p_0^2 t^2}{m^2} \tag{5.19}$$

$$(\Delta X)_\psi = \sqrt{\langle X^2 \rangle_\psi - \langle X \rangle_\psi^2} = \frac{\beta(t)}{\sqrt{2}} \tag{5.20}$$

The combined uncertainty of position and momentum is

$$(\Delta X)_\psi (\Delta P)_\psi = \frac{\hbar}{2} \frac{\beta(t)}{\sigma}. \tag{5.21}$$

For all times $t \neq 0$ this value is larger than the one required by the Uncertainty Principle, $\hbar/2$. The pure Gaussian package at time $t = 0$ was thus an exception.

The wave packet therefore dissolves with time, which is due to the fact that contributions of different momenta move with different velocities through space. This result remains unchanged in three dimensions (except that the package then dissolves in all three directions). In practice, however, particles are never free for a long time. Even in outer space, particles are exposed to magnetic fields and the interaction with radiation. Such interactions lead to decoherence, as we briefly discussed in Chap. 4. The result is in general similar to that of a position measurement: the particle gets localized with respect to its environment; a kind of effective collapse of the wave function has taken place.

Self-check questions:

1. Do the momentum and position uncertainties change for a free wave packet, and why?
2. What stops the dissolving of wave packets in practice?

5.2 Piecewise Constant Potentials

5.2.1 General Remarks

The simplest potentials you can look at are those which are constant almost everywhere, only jumping between one value and another in one or two places. It is no problem at all that these jumps are discontinuities. In contrast, it simplifies the calculations. One only has to figure out once what this implies for the continuity of the wave function and its derivatives. The very simplest potential (right after the free particle with $V(x) = 0$) contains only one jump, the **step potential**. For two jumps there are three possibilities:

- Two steps in the same direction. But this gives us no new insights compared to the single step and is therefore not discussed here.
- the **potential well**, where the potential between the two steps is lower than in the outer region. This leads to the existence of bound states.

- the **potential barrier**, where the potential between the steps is higher than in the outer region. Here the tunnel effect occurs, where a wave whose energy is lower than the potential in the middle section can "tunnel" through this potential.

We consider only energy eigenstates here. General states can always be combined out of energy eigenstates. For the time evolution one then has to follow a similar approach as for the wave packet in the previous section:

1. Decompose the initial wave function $\psi(x)$ into energy eigenstates. For the free wave packet this simply meant to operate in momentum space, since there momentum eigenstates were also energy eigenstates.
2. For each eigenstate apply the time evolution factor $\exp(-i(e/\hbar)t)$. In the case of the free wave packet, this was the establishment of $\tilde{\psi}(p, t)$, (5.8).
3. From there reconstruct the wave function $\psi(x, t)$. In the case of the free wave packet this was the Fourier back transformation leading to (5.13).

The wave function $\psi(x)$ of an energy eigenstate $|E\alpha\rangle$ is subject to the stationary Schrödinger equation

$$-\frac{\hbar^2}{2m}\frac{d^2}{dx^2}\psi(x) = (E - V(x))\psi(x). \tag{5.22}$$

We assume that there can be several linearly independent eigenstates with the same energy eigenvalue E. One then calls the eigenvalue E **degenerate**. We thus add an additional subscript α to $|E\rangle$ to enumerate states with the same eigenvalue. One can always normalize eigenstates such that

$$\langle E\alpha|E'\beta\rangle = \delta(E - E')\delta_{\alpha\beta}, \quad \langle E_i\alpha|E_j\beta\rangle = \delta_{ij}\delta_{\alpha\beta}, \tag{5.23}$$

where the first equation holds for the continuous, the second for the discrete part of the spectrum. In the first case the eigenstates are actually pseudo-states, in the second case actual states, i.e. elements of the Hilbert space.

The normalization becomes important when one wants to decompose a generic wave function into its eigenstate components, or when one wants to compute expectation values. But this is not what we are going to do in this section, and therefore we keep the states unnormalized, saving the extra computational effort involved in normalization.

In piecewise constant potentials, there are one or several discontinuities, i.e. at some position x_0 the potential jumps from one value V_0 to a different value V_1. What does this imply for the wave function $\psi(x)$ and its derivatives? The Schrödinger equation connects the potential with the second derivative of ψ. Hence the second derivative of ψ will also have a discontinuity at x_0. The integral over V is continuous though, even at x_0, and thus the first derivative of ψ is also continuous, and all the more ψ itself. We will see that solutions for ψ are determined by these continuity conditions. (Some textbooks also discuss "delta-like" potentials, $V(x) = V_0\delta(x - x_0)$. In this case also the integral over V has a discontinuity, and thus the first derivative of ψ. Only ψ itself is then continuous.)

For piecewise constant potentials, we can separate the x-axis into regions $J =$ I, II, . . . , such that the potential is constant in each region, $V(x) = V_J$ for $x \in J$. In each region, the solutions of the Schrödinger equation have the form

$$\psi(x) = a_J e^{ik_J(E)x} + b_J e^{-ik_J(E)x} \tag{5.24}$$

with

$$k_J(E) = \sqrt{\frac{2m}{\hbar^2}(E - V_J)}, \tag{5.25}$$

if $E > V_J$, or

$$\psi(x) = c_J e^{\kappa_J(E)x} + d_J e^{-\kappa_J(E)x} \tag{5.26}$$

with

$$\kappa_J(E) = \sqrt{\frac{2m}{\hbar^2}(V_J - E)}, \tag{5.27}$$

if $V_J > E$. The second case is not allowed by classical physics. The potential energy of a particle cannot be larger than its total energy, since this would imply negative kinetic energy. A classical particle cannot be located in such a region, its too small energy prevents it from entering them. A region J with $V_J > E$ is therefore called **classically forbidden region**. In QM, however, the wave function does not vanish in these regions. If J extends to $+\infty$ $(-\infty)$, the c_J (d_J) have to vanish for the sake of normalizability. How the solutions have to be put together at the borders of the regions is determined by the continuity conditions for ψ and its first derivative. They result in relations between the coefficients a_J, b_J, c_J, d_J.

Exercise 5.4
Clarify to yourself that the first term in (5.24) describes a wave moving from left to right, the second term a wave running from right to left. Make use of the corresponding solution of the time-dependent Schrödinger equation, and analyze the motion of points with constant phase. Alternatively, you can calculate the currents associated with the two terms.

5.2.2 Potential Step

We analyze a potential step at $x = 0$,

$$V(x) = V_{\mathrm{I}} \quad \text{f\"ur } x < 0, \quad V(x) = V_{\mathrm{II}} \quad \text{f\"ur } x > 0. \tag{5.28}$$

Without loss of generality we assume that $V_{\mathrm{II}} > V_{\mathrm{I}}$ (Fig. 5.1).

Fig. 5.1 One-dimensional
potential step

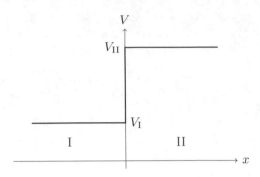

(1) **E < V$_I$**

Solutions with $E < V_I$ don't exist. For then we would need in both regions solutions of type (5.26), where in regions I the parameter d_I, in region II the parameter c_{II} would have to vanish. The continuity conditions for $\psi(x)$ and $\psi'(x)$ at $x = 0$ are them

$$c_I = d_{II}, \quad \kappa_I c_I = -\kappa_{II} d_{II}, \tag{5.29}$$

which obviously don't have a solution.

(2) **V$_I$ < E < V$_{II}$**

In I we now need a solution of type (5.24), in II one of type (5.26) with $c_{II} = 0$. The continuity conditions are

$$a_I + b_I = d_{II} \tag{5.30}$$
$$i(a_I - b_I)k_I = -\kappa_{II} d_{II}, \tag{5.31}$$

which leads to

$$\frac{b_I}{a_I} = \frac{k_I - i\kappa_{II}}{k_I + i\kappa_{II}} \tag{5.32}$$
$$\frac{d_{II}}{a_I} = \frac{2k_I}{k_I + i\kappa_{II}} \tag{5.33}$$

The expression on the right hand side of the first equation has norm 1, i.e. it is a pure phase, and so $|b_I| = |a_I|$. The solution has thus the following meaning: A wave arriving from the left with amplitude $|a_I|$ is completely reflected at the step (outgoing wave to the left with amplitude $|b_I| = |a_I|$) thereby receiving a phase shift. The phase shift is the smaller, the closer E is to V_{II}, see (5.27). For $E = V_I$, hence $k_I = 0$, the phase shift is π; for $E = V_{II}$, hence $\kappa_{II} = 0$, it vanishes. During the reflection, the wave penetrates the classically forbidden region II, but decays there exponentially. The depth of penetration is the larger (i.e. the exponential decay slower), the closer E is to V_{II}.

(3) $E > V_{II}$

Now we need in both regions solutions of type (5.24). This is the most interesting case. To solve it, we choose a specific approach: We set initially $b_{II} = 0$, i.e. region II contains only an outgoing wave to the right, not an incoming wave from the right. The interpretation is that the incoming wave (from the left) in I (amplitude a_I) is the *cause* of the entire spectacle. A part of this wave is let through by the potential step, this is the part outgoing to the right (amplitude a_{II}). The rest of the wave is reflected. This is the part outgoing to the left (amplitude b_I). The continuity conditions are

$$a_I + b_I = a_{II} \tag{5.34}$$
$$i(a_I - b_I)k_I = ia_{II}k_{II}, \tag{5.35}$$

leading to

$$\frac{b_I}{a_I} = \frac{k_I - k_{II}}{k_I + k_{II}} \tag{5.36}$$
$$\frac{a_{II}}{a_I} = \frac{2k_I}{k_I + k_{II}} \tag{5.37}$$

Comparing this with case (2), we see that only d_{II} was replaced by a_{II} and $i\kappa_{II}$ by k_{II}.

To interpret the result, we define the **transmission coefficient** T and the **reflection coefficient** R. The former is defined as the ratio between the outgoing current to the right (the part of the wave that was let through by the potential step) j_d and the incoming current j_0; similarly, the latter is defined as the ratio between the reflected current (outgoing to the left) j_r and the incoming current j_0:

$$T = \frac{|j_d|}{|j_0|}, \quad R = \frac{|j_r|}{|j_0|} \tag{5.38}$$

The calculation of the currents yields (cf. Exercise 3.13)

$$j_0 = \frac{\hbar k_I |a_I|^2}{m}, \quad j_d = \frac{\hbar k_{II} |a_{II}|^2}{m}, \quad j_r = -\frac{\hbar k_I |b_I|^2}{m}. \tag{5.39}$$

From the solution for the coefficients (5.36), (5.37) we get

$$T = \frac{4k_I k_{II}}{(k_I + k_{II})^2}, \quad R = \left(\frac{k_I - k_{II}}{k_I + k_{II}}\right)^2. \tag{5.40}$$

One has $R + T = 1$. It has to be like this, for in a stationary solution the probability density is time-independent, $\dot{\rho} = 0$, and the continuity equation gives $\frac{d}{dx} j = 0$. All

of j_0 thus has to move on without loss, either in the reflected or in the transmitted
wave.

For $E = V_{II}$ the wave gets completely reflected, since $k_{II} = 0$. For higher values
of E, T increases and converges for $E \to \infty$ to 1, because k_I/k_{II} approaches 1 there.
By the way, the values in (5.40) don't depend on whether or not $V_{II} > V_I$, as long as
$E > \max(V_I, V_{II})$. So, a part of the wave is also reflected if the potential step goes
down instead of up.

We have set $b_{II} = 0$ to describe an incoming wave from the left, partially reflected
and partially transmitted. From this solution, further solutions can be constructed.
Let's denote the solution above as solution 1, and the corresponding coefficients as
$a_I^{(1)}, b_I^{(1)}$ and $a_{II}^{(1)}$. One of the coefficients remains undetermined—it makes sense to
choose $a_I^{(1)}$ for that, as the incoming, "causing" part—and the others were derived
from that one. The free coefficient $a_I^{(1)}$ can be fixed when dealing with normalization,
which we don't want to do here though. A second solution can be obtained by taking
the complex conjugate of the first one. This new solution 2 has then the coefficients

$$a_I^{(2)} = (b_I^{(1)})^*, \quad b_I^{(2)} = (a_I^{(1)})^*, \quad a_{II}^{(2)} = 0, \quad b_{II}^{(2)} = (a_{II}^{(1)})^*. \tag{5.41}$$

In this solution, all currents run in the opposite direction as compared to solution
1. The previously outgoing parts are now incoming and vice versa. Two waves,
incoming from the left and from the right, respectively, merge at $x = 0$ in such a way
that the transmitted part of the wave coming from the left and the reflected part of
the wave coming from the right interfere destructively, giving $a_{II}^{(2)} = 0$. One might
call solution 1 **causal** and solution 2 **final**. Solution 1 describes an incoming wave
causing two outgoing waves. Solution 2 describes two incoming waves finetuned
with each other in such a way as to lead to only *one* outgoing wave. Another solution
can be constructed by exchanging left and right in solution 1, i.e. we start with a wave
incoming from the right, causing two outgoing waves. The coefficients are obtained
by exchanging I and II as well as a and b in (5.36) and (5.37). If we then identify the
coefficient b_{II} of this solution (corresponding to the wave incoming from the right)
with the coefficient a_I of the first solution (wave incoming from the left), we get:

$$a_I^{(3)} = 0, \quad b_I^{(3)} = \frac{k_{II}}{k_I} a_{II}^{(1)}, \quad a_{II}^{(3)} = -b_I^{(1)}, \quad b_{II}^{(3)} = a_I^{(1)} \tag{5.42}$$

Complex conjugation of this solution yields another *final* solution,

$$a_I^{(4)} = \frac{k_{II}}{k_I} (a_{II}^{(1)})^*, \quad b_I^{(4)} = 0, \quad a_{II}^{(4)} = (a_I^{(1)})^*, \quad b_{II}^{(4)} = -(b_I^{(1)})^*. \tag{5.43}$$

This time the two incoming waves merge in such a way that they produce only one
outgoing wave to the right.

Of the four solutions, only two are linearly independent. The other two can be
written as linear combinations of those.

Exercise 5.5
Verify this.

Usually one chooses the two causal solutions 1 and 3 as the starting point, for they are more intuitive than the two "conspirative" solutions 2 and 4.

In summary we can say: For $E < V_I$ there is no solution; for any E with $V_I < E < V_{II}$ there is one independent solution; and for any $E > V_{II}$ there are two independent solutions.

5.2.3 Potential Well

Given a potential

$$V(x) = 0 \quad \text{f''ur } |x| > x_0, \quad V(x) = V_0 < 0 \quad \text{f''ur } |x| < x_0. \quad (5.44)$$

There are three regions:

$$\text{I} = (-\infty, -x_0], \quad \text{II} = [-x_0, +x_0], \quad \text{III} = [x_0, \infty) \quad (5.45)$$

with potentials $V_I = V_{III} = 0$ and $V_{II} = V_0$ (Fig. 5.2). This time there are two discontinuities, hence four continuity conditions. We take a similar approach as for the potential step. First we classify the solutions by the value of E, determining which type of solution is given in which region. The possible cases are $V_0 < E < 0$ and $E > 0$.

(1) $V_0 < E < 0$, bound states

In regions I and III solutions of type (5.26) are needed, where for the sake of normalizability d_I and c_{III} need to vanish. So, in the external regions the wave function decays exponentially, i.e. we have **bound states** whose main contribution is localized

Fig. 5.2 One-dimensional potential well

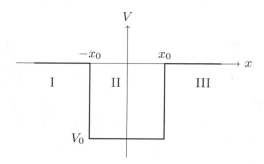

inside the potential well. In this case, the wave function can be normalized to 1, it is a real element of the Hilbert space. In region II the solution has to be of type (5.24), which we are now going to rewrite somewhat though (for reasons that will become apparent):

$$\psi_{\mathrm{II}}(x) = \tilde{a}_{\mathrm{II}} \cos k_{\mathrm{II}} x + \tilde{b}_{\mathrm{II}} \sin k_{\mathrm{II}} x \tag{5.46}$$

For simplicity we denote $\kappa = \kappa_{\mathrm{I}} = \kappa_{\mathrm{III}}$, $k = k_{\mathrm{II}}$, $c = c_{\mathrm{I}}$, $d = d_{\mathrm{III}}$, $a = \tilde{a}_{\mathrm{II}}$ and $b = \tilde{b}_{\mathrm{II}}$. The four continuity conditions are then:

$$c\, e^{-\kappa x_0} = a \cos k x_0 - b \sin k x_0 \tag{5.47}$$

$$\kappa c\, e^{-\kappa x_0} = k(a \sin k x_0 + b \cos k x_0) \tag{5.48}$$

$$d\, e^{-\kappa x_0} = a \cos k x_0 + b \sin k x_0 \tag{5.49}$$

$$-\kappa d\, e^{-\kappa x_0} = k(-a \sin k x_0 + b \cos k x_0) \tag{5.50}$$

We have four coefficients a, b, c, d, of which we can choose one freely (or fix it later for normalization). So, we have four equations for three unknowns. This system of equations will no longer have a solution for any energy eigenvalue E. This suggests the presence of a discrete spectrum, as it is to be expected for bound states. We are in the first place interested in the allowed energy values. For that, we search for an expression for k or κ in the continuity conditions, since E is contained in these.

The second equation divided by the first gives

$$\kappa = k \frac{a \sin k x_0 + b \cos k x_0}{a \cos k x_0 - b \sin k x_0}. \tag{5.51}$$

The fourth equation divided by the third yields

$$\kappa = k \frac{a \sin k x_0 - b \cos k x_0}{a \cos k x_0 + b \sin k x_0}. \tag{5.52}$$

these two expressions for κ are consistent with each other only if a or b vanishes.

Exercise 5.6
Verify this; equate the two right hand sides and multiply with the denominators.

We consider the cases separately. If $b = 0$, one gets

$$\tan k x_0 = \frac{\kappa}{k} = \frac{\sqrt{|U_0| - k^2}}{k} \tag{5.53}$$

with the definition

$$U_0 = \frac{2m}{\hbar^2} V_0. \tag{5.54}$$

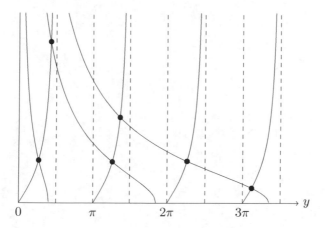

Fig. 5.3 Graphical determination of the energy eigenvalues of a potential well, part 1

The second equation in (5.53) follows from

$$\kappa^2 = -\frac{2m}{\hbar^2}E, \quad k^2 = \frac{2m}{\hbar^2}(E - V_0). \tag{5.55}$$

With $y = kx_0$ we can rewrite (5.53) to

$$\tan y = \frac{\sqrt{x_0^2|U_0| - y^2}}{y}. \tag{5.56}$$

This equation is best visualized in a diagram where the left and right hand sides are plotted as functions of y and the intersections can be looked up, cf. Fig. 5.3.

The ascending curves are the branches of the tangent function, the descending ones correspond to the right hand side of (5.56) for several values of $x_0^2|U_0|$. Each of these curves meets the y-axis (the horizontal axis of the diagram) at $y = x_0\sqrt{|U_0|}$. One can see that, depending on the value of $x_0^2|U_0|$, there are one or several intersections. The larger $x_0^2|U_0|$, the more intersections. The intersections are located in the intervals $[n\pi, (n + \frac{1}{2})\pi]$. To each intersection belongs a value of $k = y/x_0$, and to each k an energy eigenvalue

$$E = V_0 + \frac{\hbar^2}{2m}k^2. \tag{5.57}$$

If $a = 0$, one follows the same procedure, obtaining

$$-\cot y = \tan(y + \frac{\pi}{2}) = \frac{\sqrt{x_0^2|U_0| - y^2}}{y}. \tag{5.58}$$

Again, a diagram helps, cf. Fig. 5.4.

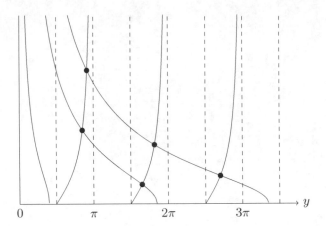

Fig. 5.4 Graphical determination of the energy eigenvalues of a potential well, part 2

This time there is no solution if $x_0\sqrt{|U_0|} < \pi/2$. The solutions for $x_0\sqrt{|U_0|} > \pi/2$ are located in the intervals $[(n - \frac{1}{2})\pi, n\pi]$ and thus *between* the intervals with the solutions for $b = 0$. The corresponding energy eigenvalues are determined as in the case $b = 0$. Together they form the discrete spectrum of the bound states of the potential well.

The corresponding wave functions are *even* in the case $b = 0$, i.e. $\psi(-x) = \psi(x)$, in particular $d = c$, *odd* in the case $a = 0$, i.e. $\psi(-x) = -\psi(x)$, in particular $d = -c$. This can be directly inferred from the continuity conditions.

(2) E > 0, free states

Now we need solutions of type (5.24) in all three regions, i.e. the corresponding wave functions are distributed over the entire one-dimensional space, in contrast to the bound states, where the main part of the wave function was located inside region II. We again analyze waves incoming from the left, setting $b_{III} = 0$. Other solutions can be obtained again by right–left exchange and by complex conjugation, as for the potential step. The incoming wave is now reflected in two places, namely each of the two discontinuities of the potential. The two reflected waves overlap in I and interfere with each other. We expect that under certain conditions the interference can be fully destructive where the two reflected parts of the wave cancel each other completely in region I.

The continuity conditions are now

$$a_I e^{-ik_I x_0} + b_I e^{ik_I x_0} = a_{II} e^{-ik_{II} x_0} + b_{II} e^{ik_{II} x_0} \tag{5.59}$$

$$k_I \left(a_I e^{-ik_I x_0} - b_I e^{ik_I x_0}\right) = k_{II} \left(a_{II} e^{-ik_{II} x_0} - b_{II} e^{ik_{II} x_0}\right) \tag{5.60}$$

$$a_{II} e^{ik_{II} x_0} + b_{II} e^{-ik_{II} x_0} = a_{III} e^{ik_I x_0} \tag{5.61}$$

$$k_{II} \left(a_{II} e^{ik_{II} x_0} - b_{II} e^{-ik_{II} x_0}\right) = k_I a_{III} e^{ik_I x_0}, \tag{5.62}$$

where we have used $k_{III} = k_I$.

Exercise 5.7

Use your personal favorite method for the solution of systems of linear equation to solve this system of linear equations for b_I, a_{II}, b_{II} and a_{III}. (The parameter a_I is again treated as the input, as the strength of the incoming wave, which "causes" the other parts.) Defining

$$z = k_{II}/k_I, \tag{5.63}$$

the solution reads:

$$\frac{b_I}{a_I} = \frac{2i(z^2-1)\sin(2k_{II}x_0)e^{-2ik_Ix_0}}{(z+1)^2e^{-2ik_{II}x_0} - (z-1)^2e^{2ik_{II}x_0}} \tag{5.64}$$

$$\frac{a_{II}}{a_I} = \frac{2(z+1)e^{-i(k_I+k_{II})x_0}}{(z+1)^2e^{-2ik_{II}x_0} - (z-1)^2e^{2ik_{II}x_0}} \tag{5.65}$$

$$\frac{b_{II}}{a_I} = \frac{2(z-1)e^{i(k_{II}-k_I)x_0}}{(z+1)^2e^{-2ik_{II}x_0} - (z-1)^2e^{2ik_{II}x_0}} \tag{5.66}$$

$$\frac{a_{III}}{a_I} = \frac{4ze^{-2ik_Ix_0}}{(z+1)^2e^{-2ik_{II}x_0} - (z-1)^2e^{2ik_{II}x_0}} \tag{5.67}$$

This looks quite complicated, but in fact we are only interested in the reflection and transmission coefficients.

Exercise 5.8

Calculate R and T. The solution is:

$$R = \frac{(z^2-1)^2\sin^2(2k_{II}x_0)}{4z^2 + (z^2-1)^2\sin^2(2k_{II}x_0)} \tag{5.68}$$

$$T = \frac{4z^2}{4z^2 + (z^2-1)^2\sin^2(2k_{II}x_0)} \tag{5.69}$$

From that, we infer two things:

1. If $E \to \infty$ and thus $z \to 1$, the reflection goes to zero.
2. If $2k_{II}x_0$ is an integer multiple of π, the reflection vanishes, due to destructive interference of the two reflected parts.

Altogether we have seen that for a potential well the spectrum of the Hamiltonian operator consists of a discrete part with certain energy values $E < 0$ and a continuous part for $E > 0$. the energy eigenstates for $E < 0$ are bound states whose wave functions decay exponentially outside of region II. The (pseudo-) eigenstates for $E > 0$ are free states whose wave functions are distributed over the entire one-dimensional space. A generic state can be composed of both free and bound parts.

5.2.4 Potential Barrier

For a potential barrier we have the same potential as for the potential well, the only difference being that the value V_0 of the potential in the middle region is now larger than 0 (Fig. 5.5). An incoming wave from the left has now to overcome a "hurdle". The case $E > V_0$ has no new features to offer, the behavior is just as for the free states of the potential well. The ansatz, the continuity conditions and therefore even the solutions are identical. Only the case $0 < E < V_0$ is new. Here we need solutions of type (5.24) in I and III, and solutions of type (5.26) in II. Again we set $b_{III} = 0$, considering incoming waves from the left.

In the calculation, the changes compared to the free states of the potential well are minimal. The coefficients in region II are now called c_{II} and d_{II} instead of a_{II} and b_{II}, and in the exponents one has to replace ik_{II} by κ_{II}. Defining

$$w = \kappa_{II}/k_I \tag{5.70}$$

and using $\sin i\alpha = i \sinh \alpha$, the solution for the transmission coefficient can be taken over from (5.69), with only small changes:

$$T = \frac{4w^2}{4w^2 + (w^2 + 1)^2 \sinh^2(2\kappa_{II}x_0)} \tag{5.71}$$

Fig. 5.5 One-dimensional potential barrier

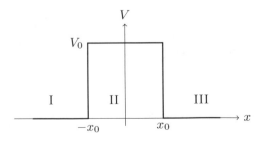

> **Exercise 5.10**
> Verify this.

This equation expresses the famous **tunnel effect**: Although region II is classically forbidden—the potential has there a higher value than the energy of the quantum object—a part of the wave can "tunnel" through the barrier. For large α, one has $\sinh \alpha \approx e^{\alpha}/2$. With increasing $\kappa_{II} x_0$, the transmission decreases exponentially. The broader and higher the barrier (remember that $\kappa_{II} \sim \sqrt{V_0 - E}$), the smaller the portion that reaches the right side, as one would have expected intuitively.

Self-check questions:

1. What is the procedure to solve the stationary Schrödinger equation for piecewise constant potentials?
2. How are reflection and transmission coefficients defined?
3. What is the tunnel effect, and which parameters determine its strength?

5.3 Harmonic Oscillator

As a harmonic oscillator we denote a potential of the form

$$V(x) = V_0 + \frac{m\omega^2}{2}(x - x_0)^2. \tag{5.72}$$

This potential is of great importance, since *any* potential can, in the proximity of a minimum, be approximated by a harmonic oscillator, if only its second derivative with respect to x does not vanish at the minimum. For close to the position x_0 where a potential $U(x)$ acquires a minimum, one has

$$U(x) = U(x_0) + U''(x_0)(x - x_0)^2 + O((x - x_0)^3), \tag{5.73}$$

which corresponds to a harmonic oscillator with $m\omega^2/2 = U''(x_0)$. It is therefore to be expected that for low energy eigenvalues, where a large part of the wave function is localized in the proximity of x_0, eigenvalues as well as eigenfunctions can be approximated by those of the corresponding harmonic oscillator.

Without loss of generality, we may set $V_0 = 0$ and $x_0 = 0$ in (5.72): V_0 represents only an irrelevant shift of the minimal energy, and x_0 can be simply removed by a coordinate shift. The potential to be studied is thus

$$V(x) = \frac{m\omega^2}{2}x^2. \tag{5.74}$$

We want to determine the associated energy eigenvalues and eigenstates, solving the stationary Schrödinger equation. As a differential equation in position space it reads

$$-\frac{\hbar^2}{2m}\psi''(x) + \frac{m\omega^2}{2}x^2\psi(x) = E\psi(x). \tag{5.75}$$

However, solving such a differential equation is quite cumbersome, and it turns out there is a simpler way which makes use of the abstract formulation of QM in an ingenious way. We therefore forget about the representation of the state $|\psi\rangle$ as a wave function for a moment, and only use the abstract operators X and P, for which we assume nothing but the fundamental commutator, $[X, P] = i\hbar\mathbf{1}$. The Hamiltonian operator is

$$H = \frac{P^2}{2m} + \frac{m\omega^2}{2}X^2. \tag{5.76}$$

The stationary Schrödinger equation reads then

$$\left(\frac{P^2}{2m} + \frac{m\omega^2}{2}X^2\right)|\psi\rangle = E|\psi\rangle. \tag{5.77}$$

We will now reformulate this in a tricky way, by introducing a new operator A,

$$A = \frac{1}{\sqrt{2\hbar}}\left(\sqrt{m\omega}X + \frac{i}{\sqrt{m\omega}}P\right). \tag{5.78}$$

Its hermitian conjugate operator is

$$A^\dagger = \frac{1}{\sqrt{2\hbar}}\left(\sqrt{m\omega}X - \frac{i}{\sqrt{m\omega}}P\right). \tag{5.79}$$

The combination $A^\dagger A$ is called N, for reasons that will soon become apparent. Now (5.76) can be rewritten as

$$H = \hbar\omega\left(A^\dagger A + \frac{1}{2}\right) = \hbar\omega\left(N + \frac{1}{2}\right). \tag{5.80}$$

The additional $+\frac{1}{2}$ occurs, because when multiplying out $A^\dagger A$ we have to take into account the generalization of the third binomial formula to operators:

$$(B + C)(B - C) = B^2 - C^2 - [B, C], \tag{5.81}$$

here applied to

$$B = \frac{1}{\sqrt{2\hbar}}\sqrt{m\omega}X, \quad C = \frac{1}{\sqrt{2\hbar}}\frac{(-i)}{\sqrt{m\omega}}P. \tag{5.82}$$

From (5.80) we read off that the energy eigenvalues E_n can be expressed through the eigenvalues n of N:

$$E_n = \hbar\omega \left(n + \frac{1}{2} \right) \tag{5.83}$$

The eigenstates of H are thus simultaneously eigenstates of N. We denote the eigenstate of N with eigenvalue n by $|n\rangle$. To be precise, it could be that the eigenvalue n is degenerate, i.e. that there are several states with this eigenvalue. Then we would need further information to specify the state. However, we will soon catch up on proving that n is not degenerate. But first we show that n cannot be negative: Let $|v\rangle = A|n\rangle$. Then

$$n = \langle n|N|n\rangle = \langle n|A^\dagger A|n\rangle = \langle v|v\rangle \geq 0. \tag{5.84}$$

Exercise 5.11
Show, using not more than $[X, P] = i\hbar$, the following commutator relation:

$$[A, A^\dagger] = 1 \tag{5.85}$$

Conclude

$$[N, A] = -A, \quad [N, A^\dagger] = A^\dagger. \tag{5.86}$$

From (5.86) follows

$$N A|n\rangle = A(N-1)|n\rangle = (n-1)A|n\rangle \tag{5.87}$$
$$N A^\dagger|n\rangle = A^\dagger(N+1)|n\rangle = (n+1)A^\dagger|n\rangle. \tag{5.88}$$

$A|n\rangle$ is therefore an eigenstate of N with eigenvalue $n-1$, and $A^\dagger|n\rangle$ is an eigenstate of N with eigenvalue $n+1$. By repeated application of $A^\dagger$ one "climbs" up the "ladder" of N-eigenvalues in steps of 1, and by repeated application of A one descends. $A^\dagger$ is therefore called **raising operator**, A correspondingly **lowering operator**.

But since we know that n cannot become negative, the descending must come to an end at some point. There must be a **ground state** $|n_0\rangle$, from where the application of A does not lead any further, which is only possible if $A|n_0\rangle$ is not a state at all, i.e. $A|n_0\rangle = 0$. As a consequence,

$$0 = ||A|n_0\rangle||^2 = \langle n_0|A^\dagger A|n_0\rangle = \langle n_0|N|n_0\rangle = n_0 \tag{5.89}$$

and so $n_0 = 0$. So, $|0\rangle$ is the only possible ground state, And we conclude that the **ground state energy** of the harmonic oscillator is

$$E_0 = \frac{\hbar\omega}{2}. \tag{5.90}$$

The values n in (5.83) therefore have to be non-negative integers.

It remains to show that there is only one state $|0\rangle$, and not several with the same eigenvalue, and therefore we go back to position space. The equation $A|0\rangle = 0$ there reads

$$\frac{1}{\sqrt{2\hbar}} \left(\sqrt{m\omega}x + \frac{\hbar}{\sqrt{m\omega}} \frac{d}{dx} \right) \psi_0(x) = 0. \tag{5.91}$$

This is a linear differential equation of first order and has thus only one independent solution. With some intuition, one can see that this solution is a Gaussian function. Properly normalized it reads

$$\psi_0(x) = \frac{\sqrt{\sigma}}{\pi^{1/4}} e^{-\frac{\sigma^2}{2}x^2}, \quad \sigma^2 = \frac{m\omega}{\hbar}. \tag{5.92}$$

Hence, the ground state is not degenerate, and since all higher states can be obtained from the ground state by repeated application of $A^\dagger$, these higher states are also non-degenerate. Let's normalize these states:

$$||A^\dagger|n\rangle||^2 = \langle n|AA^\dagger|n\rangle = \langle n|A^\dagger A + 1|n\rangle = n + 1 \tag{5.93}$$

The normalized state $|n + 1\rangle$ is therefore

$$|n + 1\rangle = \frac{1}{\sqrt{n + 1}} A^\dagger |n\rangle \tag{5.94}$$

or, if we start from the ground state and apply $A^\dagger$ repeatedly,

$$|n\rangle = \frac{1}{\sqrt{n!}} (A^\dagger)^n |0\rangle. \tag{5.95}$$

Expressing this relation again in position space, we get

$$\psi_n(x) = \frac{1}{\sqrt{2^n n!}} \left(\frac{m\omega}{\hbar}x - \frac{\hbar}{m\omega} \frac{d}{dx} \right)^n \psi_0(x). \tag{5.96}$$

The result is—after some short consideration—a polynomial of degree n times a Gaussian function. Since the eigenfunctions have to be orthogonal to each other, the polynomials *have to be* the Hermite polynomials H_n mentioned in Sect. 3.2. Using the definition of the Hermite polynomials,

$$H_n(x) = e^{x^2/2} \left(x - \frac{d}{dx} \right)^n e^{-x^2/2}, \tag{5.97}$$

one immediately gets

$$\psi_n(x) = \frac{1}{\sqrt{2^n n!}} H_n\left(\sqrt{\frac{m\omega}{\hbar}}x\right)\psi_0(x). \qquad (5.98)$$

The first few Hermite polynomials are

$$H_0(x) = 1, \quad H_1(x) = 2x, \quad H_2(x) = 4x^2 - 2. \qquad (5.99)$$

The expectation values of position and momentum can be determined through A and $A^\dagger$ without referring to the wave function. With

$$X = \sqrt{\frac{\hbar}{2m\omega}}(A + A^\dagger) \qquad (5.100)$$

$$P = -i\sqrt{\frac{\hbar m\omega}{2}}(A - A^\dagger) \qquad (5.101)$$

one obtains

$$\langle X\rangle_n = \langle n|X|n\rangle = \sqrt{\frac{\hbar}{2m\omega}}(\langle n|A|n\rangle + \langle n|A^\dagger|n\rangle) = 0 \qquad (5.102)$$

$$\langle P\rangle_n = \langle n|X|n\rangle = -i\sqrt{\frac{\hbar m\omega}{2}}(\langle n|A|n\rangle - \langle n|A^\dagger|n\rangle) = 0. \qquad (5.103)$$

Here we have used that $A|n\rangle$ lies in a different eigenspace and therefore has no overlap with $|n\rangle$, and the same holds for $A^\dagger|n\rangle$. So, the expectation values vanish. This is how it has to be: The expectation values obey the Ehrenfest equations and therefore the classical equations of motion for an oscillation around $x = 0$. Since energy eigenstates are stationary, there can be no motion of $\langle X\rangle_n$ though; $\langle X\rangle_n$ thus has to be at zero all the time. This holds for arbitrarily large energies, where in classical physics the amplitude gets higher and higher. However, a wave packet composed of several energy eigenstates will in general show the oscillating behavior: the expectation values of position and momentum will swing forth and back with the angular frequency ω, as required by the Ehrenfest equations.

For energy eigenstates though, the energy dependent "deflection" shows itself only in the form of increased uncertainties:

$$(\Delta X)_n^2 = \langle n|X^2|n\rangle = \frac{\hbar}{2m\omega}\langle n|A^2 + AA^\dagger + A^\dagger A + (A^\dagger)^2|n\rangle \qquad (5.104)$$

$$= \frac{\hbar}{2m\omega}\langle n|AA^\dagger + A^\dagger A|n\rangle \qquad (5.105)$$

$$= \frac{\hbar}{2m\omega}\langle n|2N + 1|n\rangle = \frac{\hbar}{m\omega}\left(n + \frac{1}{2}\right) \qquad (5.106)$$

The operators A^2 and $(A^\dagger)^2$ were dropped in the second row, because their contributions vanish due to the orthogonal eigenspaces: $\langle n|A^2|n\rangle = 0$, and similarly for $(A^\dagger)^2$. In the same way one gets

$$(\Delta P)_n^2 = \hbar m\omega \left(n + \frac{1}{2}\right). \qquad (5.107)$$

The combination of the two uncertainties yields

$$(\Delta X)_n(\Delta P)_n = \hbar \left(n + \frac{1}{2}\right). \qquad (5.108)$$

So, only for $n = 0$ the minimal value required by the Uncertainty Relation is realized. For all higher energy eigenstates the uncertainty is higher too.

Self-check questions:

1. Why do the raising and lowering operators have these names?
2. What are the energy eigenvalues of the harmonic oscillator?
3. Which differential equation remains to be solved when the algebraic method is used (only in words, not the exact expression)?

Chapter 6
Two-Dimensional Systems

Abstract This is just a stopover between one and three dimensions. It allows for a simplified introduction into rotation symmetric potential, angular momentum, and separation of variables.

6.1 Cartesian Coordinates

The stationary Schrödinger equation in one dimension is an ordinary differential equation: there is only one variable, x. In two dimensions, it becomes a partial differential equation

$$-\frac{\hbar^2}{2m}\left(\frac{\partial^2}{\partial x^2}\psi(x, y) + \frac{\partial^2}{\partial y^2}\psi(x, y)\right) = (E - V(x, y))\psi(x, y) \qquad (6.1)$$

and is therefore in general more difficult to solve.

There are two situations where the problem is simplified significantly and can be reduced to one-dimensional, i.e. ordinary differential equations:

- The potential consists of two separate potentials for x and y, $V(x, y) = V_1(x) + V_2(y)$.
- The potential is that of a **central force**, i.e. it depends only on the distance $r = \sqrt{x^2 + y^2}$ from the origin.

The second case requires polar coordinates and will be studied in the next section. The **isotropic harmonic oscillator**,

$$V(x, y) = \frac{m\omega^2}{2}(x^2 + y^2) = \frac{m\omega^2}{2}r^2, \qquad (6.2)$$

fortunately fulfills both conditions and can therefore be used as an example in both cases, so we can compare the results. "Isotropic" here means that the parameter ω is the same in both directions, whereas the potential of a generic harmonic oscillator in two dimensions reads

© Springer International Publishing Switzerland 2016
J.-M. Schwindt, *Conceptual Basis of Quantum Mechanics*,
Undergraduate Lecture Notes in Physics, DOI 10.1007/978-3-319-24526-3_6

$$V(x, y) = \frac{m}{2}(\omega_x^2 x^2 + \omega_y^2 y^2).$$ (6.3)

In the first case, $V(x, y) = V_1(x) + V_2(y)$, the Schrödinger equation can be simplified via the product ansatz

$$\psi(x, y) = u(x)v(y).$$ (6.4)

This is called a **separation of variables**. Inserting this ansatz into the Schrödinger equation and then dividing by u and v, one obtains

$$\left(-\frac{\hbar^2}{2m}\frac{u''(x)}{u(x)} + V_1(x)\right) + \left(-\frac{\hbar^2}{2m}\frac{v''(y)}{v(y)} + V_2(y)\right) = E.$$ (6.5)

The content of the first bracket does not depend on y. However, it can also not depend on x, since the remaining part of the equation (the second bracket and the right hand side) doesn't. So, it has to be a constant, which we call E_1. Similarly, the content of the second bracket has to be another constant, E_2. We have thus reduced the two-dimensional Schrödinger equation to two one-dimensional ones:

$$-\frac{\hbar^2}{2m}u''(x) + V_1(x)u(x) = E_1 u(x)$$ (6.6)

$$-\frac{\hbar^2}{2m}v''(y) + V_2(y)v(y) = E_2 v(y)$$ (6.7)

and we have $E = E_1 + E_2$.

Let's consider the isotropic harmonic oscillator (6.2) as an example. Because of

$$V_1(x) = \frac{m\omega^2}{2}x^2, \quad V_2(y) = \frac{m\omega^2}{2}y^2,$$ (6.8)

the one-dimensional equations are just those of the one-dimensional harmonic oscillator, whose solutions ψ_n and energy eigenvalues E_n we already know. The energy eigenstates of the two-dimensional oscillator can be represented in the form $|n_x n_y\rangle$ with the associated wave functions

$$\psi_{n_x n_y}(x, y) = \psi_{n_x}(x)\psi_{n_y}(y).$$ (6.9)

The energy eigenvalues $E_{n_x n_y}$ are

$$E_{n_x n_y} = E_{n_x} + E_{n_y} = \hbar\omega(n_x + n_y + 1).$$ (6.10)

We can see that the energy eigenvalue $\hbar\omega(n + 1)$ can be obtained in $n + 1$ different ways, implying that it is $(n + 1)$-fold degenerate: For fixed $n = n_x + n_y$, n_x can assume any integer value from 0 to n, and the corresponding value for n_y is $n - n_x$ in each case.

One arrives at the same result if one starts from scratch with the algebraic procedure that we have used for the one-dimensional oscillator. The Hamiltonian operator is

$$H = \left(\frac{P_x^2}{2m} + \frac{m\omega^2}{2} X^2 \right) + \left(\frac{P_y^2}{2m} + \frac{m\omega^2}{2} Y^2 \right), \tag{6.11}$$

and again we can introduce lowering operators, one for the x- and one for the y-direction,

$$A_x = \frac{1}{\sqrt{2\hbar}} \left(\sqrt{m\omega} X + \frac{i}{\sqrt{m\omega}} P_x \right), \tag{6.12}$$

$$A_y = \frac{1}{\sqrt{2\hbar}} \left(\sqrt{m\omega} Y + \frac{i}{\sqrt{m\omega}} P_y \right), \tag{6.13}$$

so that, after a short calculation, one obtains the following form of H:

$$H = \hbar\omega \left(A_x^\dagger A_x + A_y^\dagger A_y + 1 \right), \tag{6.14}$$

similar to the one-dimensional case. The raising operator $A_x^\dagger$ is responsible for the excitation of the oscillator in x-direction, $A_y^\dagger$ for the excitation in y-direction. The amount of excitation is expressed in terms of the operators

$$N_x = A_x^\dagger A_x, \quad N_y = A_y^\dagger A_y, \tag{6.15}$$

which just as in the one-dimensional case have non-negative integer eigenvalues, which can be identified with the values n_x and n_y in (6.10). The ground state can be determined as in the one-dimensional case, see (5.91), but this time with the differential equation holding separately in the x- and the y-direction, with the solution

$$\psi_{0,0}(x, y) = \psi_0(x)\psi_0(y). \tag{6.16}$$

In one dimension, the ground state energy $\frac{1}{2}\hbar\omega$ was due to a term $+1/2$ in the Hamiltonian operator (5.80) which appeared in the process of rewriting expressions of X^2- and P^2 into expressions of $A^\dagger A$. In d dimensions this extra term occurs d times, since d position and momentum operators are rewritten into d distinct $A^\dagger A$-expressions. The ground state energy of the isotropic oscillator in d dimensions is therefore $E_0 = \frac{d}{2}\hbar\omega$. In particular, it is simply $\hbar\omega$ in two dimensions.

Beginning with the ground state, one can get the higher states by applying the raising operators:

$$|n_x n_y\rangle = \frac{1}{\sqrt{n_x! n_y!}} (A_x^\dagger)^{n_x} (A_y^\dagger)^{n_y} |0\rangle \tag{6.17}$$

To get a bit more concrete, we want to determine the wave functions of the first excited energy eigenvalue $E = 2\hbar\omega$, $\psi_{n_x=1,n_y=0}$ and $\psi_{n_x=0,n_y=1}$, which we will later compare with the result obtained in polar coordinates. In position space we have

$$A_x^\dagger = \sqrt{\frac{m\omega}{2\hbar}} \left(x - \frac{\hbar}{m\omega} \frac{\partial}{\partial x} \right) \tag{6.18}$$

$$A_y^\dagger = \sqrt{\frac{m\omega}{2\hbar}} \left(y - \frac{\hbar}{m\omega} \frac{\partial}{\partial y} \right). \tag{6.19}$$

The wave function of the ground state is

$$\psi_{n_x=0,n_y=0}(x, y) = \sqrt{\frac{m\omega}{\pi\hbar}} \exp\left(-\frac{m\omega}{2\hbar}(x^2 + y^2) \right). \tag{6.20}$$

Application of the raising operators yields

$$\psi_{n_x=1,n_y=0}(x, y) = A_x^\dagger \psi_{n_x=0,n_y=0}(x, y) \tag{6.21}$$

$$= \sqrt{\frac{2}{\pi} \frac{m\omega}{\hbar}} \, x \, \exp\left(-\frac{m\omega}{2\hbar}(x^2 + y^2) \right) \tag{6.22}$$

$$\psi_{n_x=0,n_y=1}(x, y) = A_y^\dagger \psi_{n_x=0,n_y=0}(x, y) \tag{6.23}$$

$$= \sqrt{\frac{2}{\pi} \frac{m\omega}{\hbar}} \, y \, \exp\left(-\frac{m\omega}{2\hbar}(x^2 + y^2) \right). \tag{6.24}$$

Alternatively, one could have inferred this from (6.9) and (5.98).

Self-check question:

1. How does the separation of variables work in the case $V(x, y) = V_1(x) + V_2(y)$?

6.2 Polar Coordinates

Polar coordinates (r, ϕ) are defined via the coordinate transformation

$$x = r \cos\phi, \quad y = r \sin\phi, \tag{6.25}$$

or vice versa

$$r = \sqrt{x^2 + y^2}, \quad \phi = \arctan\frac{y}{x}. \tag{6.26}$$

We were a bit sloppy is the last equation, because the arc tangent usually maps into the interval $[-\pi/2, +\pi/2]$, but we want ϕ to run from 0 to 2π. It therefore has to be understood with an additional $+\pi$ in the second and third quadrant, and an additional $+2\pi$ in the fourth.

In order to express the derivative operators, and in particular the Laplacian, in the new coordinates, the chain rule has to be applied. (You have certainly done this already in a course on classical mechanics or electrodynamics. However, we want to repeat this here, for the sake of completeness.) A function $f(x, y)$ is rewritten in polar coordinates by replacing each x and y in the function expression via (6.25),

$$f(x, y) \to f(r, \phi) := f(x(r, \phi), y(r, \phi)), \tag{6.27}$$

for instance

$$V(x, y) = \frac{m\omega^2}{2}(x^2 + y^2) \tag{6.28}$$

$$\to V(r, \phi) = \frac{m\omega^2}{2}(r^2 \cos^2 \phi + r^2 \sin^2 \phi) = \frac{m\omega^2}{2} r^2. \tag{6.29}$$

The inverse transformation works in the same way

$$f(r, \phi) \to f(x, y) := f(r(x, y), \phi(x, y)). \tag{6.30}$$

The derivatives are then obtained via the chain rule, applied to (6.30), for instance

$$\frac{\partial}{\partial x} f(x, y) = \left(\frac{\partial r}{\partial x} \frac{\partial}{\partial r} + \frac{\partial \phi}{\partial x} \frac{\partial}{\partial \phi} \right) f(r, \phi). \tag{6.31}$$

The partial derivatives $\partial r / \partial x$ and $\partial \phi / \partial x$ are given by (6.26), at first in terms of x and y,

$$\frac{\partial r}{\partial x} = \frac{x}{\sqrt{x^2 + y^2}}, \quad \frac{\partial \phi}{\partial x} = -\frac{y}{x^2 + y^2}, \tag{6.32}$$

which then again has to be rewritten in terms of r and ϕ:

$$\frac{\partial r}{\partial x} = \cos \phi, \quad \frac{\partial \phi}{\partial x} = -\frac{\sin \phi}{r} \tag{6.33}$$

Altogether, (6.31) yields the replacement

$$\frac{\partial}{\partial x} \to \cos \phi \frac{\partial}{\partial r} - \frac{1}{r} \sin \phi \frac{\partial}{\partial \phi}. \tag{6.34}$$

Similarly we get the transformation of the y-derivative:

$$\frac{\partial}{\partial y} \to \sin \phi \frac{\partial}{\partial r} + \frac{1}{r} \cos \phi \frac{\partial}{\partial \phi} \tag{6.35}$$

For higher derivatives, the product rule has to be taken into account. For instance, in

$$\frac{\partial^2}{\partial x^2} f(x, y) \rightarrow \left(\cos\phi\frac{\partial}{\partial r} - \frac{1}{r}\sin\phi\frac{\partial}{\partial\phi}\right)\left[\left(\cos\phi\frac{\partial}{\partial r} - \frac{1}{r}\sin\phi\frac{\partial}{\partial\phi}\right) f(r, \phi)\right]$$
(6.36)

the r-derivative in the left brackets acts once on $1/r$ in the right brackets, and once on $f(r, \phi)$.

Exercise 6.1
Show that

$$\Delta = \frac{\partial^2}{\partial x^2} + \frac{\partial^2}{\partial y^2} \rightarrow \frac{\partial^2}{\partial r^2} + \frac{1}{r}\frac{\partial}{\partial r} + \frac{1}{r^2}\frac{\partial^2}{\partial\phi^2}.$$
(6.37)

The Schrödinger equation in polar coordinates thus reads

$$\left[-\frac{\hbar^2}{2m}\left(\frac{\partial^2}{\partial r^2} + \frac{1}{r}\frac{\partial}{\partial r} + \frac{1}{r^2}\frac{\partial^2}{\partial\phi^2}\right) + V(r, \phi)\right]\psi(r, \phi) = E\psi(r, \phi).$$
(6.38)

Angular momentum:

An important physical quantity is the angular momentum l, which is a scalar quantity in two dimensions, $l = xp_y - yp_x$, and equals the z-component l_z of the angular momentum vector $\mathbf{l}$ in three dimensions. The associated operator is

$$L = XP_y - YP_x = -i\hbar(x\frac{\partial}{\partial y} - y\frac{\partial}{\partial x}).$$
(6.39)

In polar coordinates, L looks simple. Transforming the content of the brackets via (6.25), (6.34), (6.35) into polar coordinates, a short calculation yields

$$L = -i\hbar\frac{\partial}{\partial\phi}.$$
(6.40)

Just as the ordinary momentum, angular momentum also equals $-i\hbar$ times a partial derivative. The eigenvalue equation

$$L\psi(r, \phi) = \hbar l\,\psi(r, \phi)$$
(6.41)

is therefore just as simple to solve: the eigenfunctions have the form

$$\psi_l(r, \phi) = f(r)e^{il\phi},$$
(6.42)

where $f(r)$ is an arbitrary differentiable function of r. Since ψ_l is supposed to be continuous, $e^{il\phi} = e^{il(\phi+2\pi)}$ and hence l being integer is required. Angular momentum is therefore **"quantized"** in QM, with possible eigenvalues $\hbar l$. Note that we used the letter l first to denote the classical angular momentum, but then the **angular momentum quantum number**. Please don't confuse these two things! The physical value of an angular momentum with quantum number l is $\hbar l$.

Central force:

The potential V of a central force depends only on r, by definition. The Hamiltonian operator is then

$$H = -\frac{\hbar^2}{2m}\left(\frac{\partial^2}{\partial r^2} + \frac{1}{r}\frac{\partial}{\partial r}\right) + \frac{L^2}{2mr^2} + V(r) \tag{6.43}$$

(note that we expressed the second derivatives w.r.t. ϕ in terms of L^2), which commutes with L,

$$[L, H] = 0. \tag{6.44}$$

For all terms in H depend only on r, and therefore

$$\frac{\partial}{\partial \phi}(H\psi) = H\frac{\partial}{\partial \phi}\psi. \tag{6.45}$$

This has two consequences:

- Angular momentum is a conserved quantity, its expectation value does not change with time. So, the classical result that angular momentum is conserved for a central force is reproduced in QM.
- H and L can be diagonalized simultaneously; we can thus choose the energy eigenstates to be also eigenstates of angular momentum.

This simplifies the Schrödinger equation substantially. For an energy eigenstate, we can set

$$\psi(r, \phi) = f(r)e^{il\phi} \tag{6.46}$$

and thereby turn the Schrödinger equation into an ordinary differential equation

$$-\frac{\hbar^2}{2m}\left(\frac{d^2}{dr^2} + \frac{1}{r}\frac{d}{dr}\right)f(r) + V_{\text{eff}}(r)f(r) = Ef(r). \tag{6.47}$$

Here V_{eff} is the **effective potential**

$$V_{\text{eff}}(r) = V(r) + \frac{\hbar^2 l^2}{2mr^2}, \tag{6.48}$$

which should look familiar from classical mechanics. Equation (6.47) is the so-called **radial equation**. Angular momentum leads to a **centrifugal term** in the effective potential: the effective potential diverges for $r \to 0$ to $+\infty$. In classical physics, this is associated with a pseudo force, the centrifugal force which drives objects outwards. The analogy in QM is that the centrifugal term in the radial equation for $l > 0$ makes the wave function vanish at $r \to 0$. We save the proof of that for the three-dimensional case.

It remains the task to solve the radial equation—or to avoid this work by smart algebraic considerations. As an example we again consider the isotropic harmonic oscillator.

Isotropic harmonic oscillator

We look for eigenstates of the Hamiltonian operator which are simultaneously eigenstates of angular momentum. The radial equation reads

$$-\frac{\hbar^2}{2m}\left(\frac{d^2}{dr^2} + \frac{1}{r}\frac{d}{dr}\right)f(r) + \left(\frac{m\omega^2}{2}r^2 + \frac{\hbar^2 l^2}{2mr^2}\right)f(r) = Ef(r). \qquad (6.49)$$

This equation can be solved with some effort and some good choices for the ansatz. For example we may guess (since we already know the solutions in cartesian coordinates) that $f(r)$ is of the form

$$f(r) = g(r)\exp\left(-\frac{m\omega}{2\hbar}r^2\right) \qquad (6.50)$$

with some polynomial $g(r)$. However, it is much more elegant to take an algebraic approach again. And since this is such a nice exercise, I leave it to you.

Exercise 6.2

(a) Show that with the definitions (6.12), (6.13) one has:

$$L = i\hbar(A_x A_y^\dagger - A_x^\dagger A_y) \qquad (6.51)$$

(b) We define new lowering operators A_L and A_R:

$$A_L = \frac{1}{\sqrt{2}}(A_x + iA_y), \quad A_R = \frac{1}{\sqrt{2}}(A_x - iA_y) \qquad (6.52)$$

Show that

$$[A_R, A_R^\dagger] = [A_L, A_L^\dagger] = 1, \qquad (6.53)$$

$$[A_R, A_L^\dagger] = [A_L, A_R^\dagger] = [A_R, A_L] = [A_R^\dagger, A_L^\dagger] = 0. \qquad (6.54)$$

The new operators A_R, $A_R^\dagger$, A_L, $A_L^\dagger$ therefore obey the same commutation relations as A_x, $A_x^\dagger$, A_y, $A_y^\dagger$.

(c) Show that

$$A_R^\dagger A_R + A_L^\dagger A_L = A_x^\dagger A_x + A_y^\dagger A_y. \tag{6.55}$$

In particular, the Hamiltonian operator (6.14) can be written as

$$H = \hbar\omega \left(A_R^\dagger A_R + A_L^\dagger A_L + 1 \right). \tag{6.56}$$

As a consequence, the entire procedure to derive the energy eigenstates can be performed with A_R, $A_R^\dagger$, A_L, $A_L^\dagger$ just as well as with A_x, $A_x^\dagger$, A_y, $A_y^\dagger$. We define

$$N_R = A_R^\dagger A_R, \quad N_L = A_L^\dagger A_L \tag{6.57}$$

with integer eigenvalues n_R and n_L. Applying $(A_R^\dagger)^{n_R}(A_L^\dagger)^{n_L}$ to the ground state one gets the state $|n_R n_L\rangle$ with the energy $E = \hbar\omega(n_R + n_L + 1)$.

(d) The big advantage of the new raising and lowering operators becomes apparent when we consider angular momentum. Show that

$$L = \hbar \left(A_R^\dagger A_R - A_L^\dagger A_L \right). \tag{6.58}$$

In contrast to $|n_x n_y\rangle$, $|n_R n_L\rangle$ is thus also an eigenstate of angular momentum,

$$L|n_R n_L\rangle = \hbar(n_R - n_L)|n_R n_L\rangle. \tag{6.59}$$

The angular momentum quantum number is therefore $l = n_R - n_L$. Clarify for yourself that for fixed $n = n_R + n_L$ (i.e. for fixed energy $\hbar\omega(n + 1)$), l can take on the values $n, n - 2, n - 4, \cdots, -n$.

(e) Show, starting from (6.18) and (6.19), that

$$A_R^\dagger = \sqrt{\frac{m\omega}{4\hbar}} e^{i\phi} \left[r - \frac{\hbar}{m\omega} \left(\frac{\partial}{\partial r} + \frac{i}{r} \frac{\partial}{\partial \phi} \right) \right], \tag{6.60}$$

$$A_L^\dagger = \sqrt{\frac{m\omega}{4\hbar}} e^{-i\phi} \left[r - \frac{\hbar}{m\omega} \left(\frac{\partial}{\partial r} - \frac{i}{r} \frac{\partial}{\partial \phi} \right) \right]. \tag{6.61}$$

(f) Use this to determine the two states $\psi_{n_R=1, n_L=0}$ and $\psi_{n_R=0, n_L=1}$ from the ground state

$$\psi_{n_R=0, n_L=0}(r, \phi) = \sqrt{\frac{m\omega}{\pi\hbar}} \exp\left(-\frac{m\omega}{2\hbar} r^2 \right) \tag{6.62}$$

The solution is:

$$\psi_{n_R=1,n_L=0}(r,\phi) = \frac{1}{\sqrt{\pi}}\frac{m\omega}{\hbar}e^{i\phi}r\,\exp\left(-\frac{m\omega}{2\hbar}r^2\right) \tag{6.63}$$

$$\psi_{n_R=0,n_L=1}(r,\phi) = \frac{1}{\sqrt{\pi}}\frac{m\omega}{\hbar}e^{-i\phi}r\,\exp\left(-\frac{m\omega}{2\hbar}r^2\right) \tag{6.64}$$

(g) The states $|n_R n_L\rangle$ differ from the states $|n_x n_x\rangle$. Nevertheless, the eigenspaces for the same energy eigenvalue $E_n = \hbar\omega(n+1)$ must be identical, i.e. each state $|n_R n_L\rangle$ must be a linear combination of states $|n_x n_y\rangle$ with $n_x + n_y = n_R + n_L$. Show that

$$|n_R = 1, n_L = 0\rangle = \frac{1}{\sqrt{2}}\left(|n_x = 1, n_y = 0\rangle + i|n_x = 0, n_y = 1\rangle\right)$$

$$|n_R = 0, n_L = 1\rangle = \frac{1}{\sqrt{2}}\left(|n_x = 1, n_y = 0\rangle - i|n_x = 0, n_y = 1\rangle\right).$$

Self-check questions:

1. What does the angular momentum operator in polar coordinates look like, and what are its eigenvalues?
2. Under what condition is it a conserved quantity?
3. What is the effective potential?

Chapter 7
Three-Dimensional Systems

Abstract The behavior of wave functions in three dimensions is investigated, with a focus on angular momentum and spherically symmetric potentials. As a highlight, we determine the energy levels of the hydrogen atom. Again, algebraic methods turn out to be very useful and elegant.

Many considerations of the previous chapter can be taken over to three dimensions. For example, for a potential of the form

$$V(x, y, z) = V_1(x) + V_2(y) + V_3(z) \tag{7.1}$$

a separation of variables can be performed with the product ansatz

$$\psi(x, y, z) = u(x)v(y)w(z), \tag{7.2}$$

yielding three one-dimensional Schrödinger equations for u, v and w, completely analogous to the two-dimensional case. For the isotropic harmonic oscillator this leads to the energy eigenvalues

$$E_n = \hbar\omega(n + \frac{3}{2}), \tag{7.3}$$

where $n = n_x + n_y + n_z$ is the sum of the N-eigenvalues of the three one-dimensional oscillators in x-, y- and z-direction.

Exercise 7.1
Show that the eigenvalue E_n is g_n-fold degenerate with

$$g_n = \frac{1}{2}(n + 1)(n + 2), \tag{7.4}$$

i.e. there are g_n ways to represent n as a sum of three non-negative integers.

© Springer International Publishing Switzerland 2016   181
J.-M. Schwindt, *Conceptual Basis of Quantum Mechanics*,
Undergraduate Lecture Notes in Physics, DOI 10.1007/978-3-319-24526-3_7

For potentials with a cylinder symmetry,

$$V(x, y, z) = V_1(\rho) + V_2(z), \tag{7.5}$$

with $\rho = \sqrt{x^2 + y^2}$, one chooses a product ansatz in cylindrical coordinates

$$\psi(\rho, \phi, z) = u(\rho)v(\phi)w(z), \tag{7.6}$$

where for u and v everything from the previous section on central forces can be applied. In particular $v(\phi) = e^{im\phi}$ is an eigenfunction of the z-component of angular momentum, with integer m; and u obeys the radial equation (6.47), now with r replaced by ρ.

A complete novelty in three dimensions is the **angular momentum algebra**: In contrast to the two-dimensional case, angular momentum is a vector in three dimensions. Associated with it are three operators (one for each component), L_x, L_y, L_z, forming a **vector operator L**. The spectrum of eigenvalues can be deduced from the commutator relations of the three components of **L** and the operator $\mathbf{L}^2$. The corresponding eigenfunctions Y_{lm} (the meaning of l and m will be explained) depend only on the angles θ and ϕ in spherical coordinates, and are called **spherical harmonics**.

For a spherically symmetric potential (central force)

$$V(x, y, z) = V(r), \tag{7.7}$$

with $r = \sqrt{x^2 + y^2 + z^2}$, angular momentum is, just as in two dimensions, a conserved quantity. As the angular part of the energy eigenstates of eigenvalue E_n, we can again choose the eigenfunctions of angular momentum,

$$\psi_{nlm}(r, \theta, \phi) = R_{nl}(r)Y_{lm}(\theta, \phi). \tag{7.8}$$

The function $R_{nl}(r)$ again obeys a radial equation with an effective potential. To solve the radial equation is the remaining task for a given spherically symmetric potential. We will study two examples for that: the free particle and the Coulomb potential. The latter is of special importance, since it explains the basic properties of the hydrogen atom.

7.1 Angular Momentum Algebra

The angular momentum **l** is defined as

$$\mathbf{l} = \mathbf{r} \times \mathbf{p}. \tag{7.9}$$

The associated operators are

$$L_x = Y P_z - Z P_y = -i\hbar \left(y \frac{\partial}{\partial z} - z \frac{\partial}{\partial y} \right), \tag{7.10}$$

$$L_y = Z P_x - X P_z = -i\hbar \left(z \frac{\partial}{\partial x} - x \frac{\partial}{\partial z} \right), \tag{7.11}$$

$$L_z = X P_y - Y P_x = -i\hbar \left(x \frac{\partial}{\partial y} - y \frac{\partial}{\partial x} \right). \tag{7.12}$$

Note that a component of the position operator commutes with a *different* component of the momentum operator, e.g. $[Y, P_z] = 0$, so that the order of the operators in each term above is irrelevant. The operator associated with the norm squared of angular momentum is

$$\mathbf{L}^2 = L_x^2 + L_y^2 + L_z^2. \tag{7.13}$$

It won't be necessary to write out $\mathbf{L}^2$ in terms of position and momentum operators, thanks to some clever methods we are going to apply. We now want to compute the commutators between these four operators. First

$$[L_x, L_y] = [Y P_z, Z P_x] - [Y P_z, X P_z] - [Z P_y, Z P_x] + [Z P_y, X P_z] \tag{7.14}$$

$$= Y P_x [P_z, Z] - 0 - 0 + P_y X [Z, P_z] \tag{7.15}$$

$$= -i\hbar Y P_x + i\hbar X P_y \tag{7.16}$$

$$= i\hbar L_z. \tag{7.17}$$

Similarly we get

$$[L_y, L_z] = i\hbar L_x, \quad [L_z, L_x] = i\hbar L_y. \tag{7.18}$$

If we replace the indices x, y, z with $1, 2, 3$, this can be written as

$$[L_i, L_j] = i\hbar \sum_{k=1}^{3} \epsilon_{ijk} L_k. \tag{7.19}$$

Here ϵ_{ijk} are the components of the **epsilon tensor** (also called totally antisymmetric tensor or Levi-Civita tensor),

$$\epsilon_{123} = \epsilon_{231} = \epsilon_{312} = -\epsilon_{213} = -\epsilon_{321} = -\epsilon_{132} = 1, \tag{7.20}$$

$\epsilon_{ijk} = 0$ for all other combinations of (ijk), i.e. all combinations where an index value occurs at least twice.

Nerd's Corner 7.1
More precisely, ϵ is a **tensor density**, not a tensor. A tensor is defined through its behavior under coordinate transformations. For a linear transformation

$$r_i' = \sum_{j=1}^{3} A_{ij} r_j \tag{7.21}$$

with transformation matrix A, a tensor T transforms with respect to each index via A or A^{-1}, for example

$$T_{ijk}' = \sum_{l=1}^{3} \sum_{m=1}^{3} \sum_{n=1}^{3} A_{il} A_{jm} A_{kn} T_{lmn}. \tag{7.22}$$

In the case of ϵ this leads to

$$\epsilon_{123}' = \sum_{l=1}^{3} \sum_{m=1}^{3} \sum_{n=1}^{3} A_{1l} A_{2m} A_{3n} \epsilon_{lmn} = \det A. \tag{7.23}$$

The definition of ϵ requires though that ϵ_{123} equals 1 also in the new coordinate system. Therefore one has to demand that

$$\epsilon_{ijk}' = (\det A)^{-1} \sum_{l=1}^{3} \sum_{m=1}^{3} \sum_{n=1}^{3} A_{il} A_{jm} A_{kn} \epsilon_{lmn} \tag{7.24}$$

which is just what characterizes a tensor density: a tensor density of weight w is defined such that the transformation rule (7.22) gets an additional factor of $(\det A)^w$ on the right hand side.

$\mathbf{L}^2$ commutes with all components of angular momentum,

$$[\mathbf{L}^2, L_i] = 0, \tag{7.25}$$

since

$$[\mathbf{L}^2, L_x] = [L_x^2, L_x] + [L_y^2, L_x] + [L_z^2, L_x] \tag{7.26}$$

$$= 0 + L_y[L_y, L_x] + [L_y, L_x]L_y + L_z[L_z, L_x] + [L_z, L_x]L_z \tag{7.27}$$

$$= i\hbar(-L_y L_z - L_z L_y + L_z L_y + L_y L_z) \tag{7.28}$$

$$= 0 \tag{7.29}$$

and similarly

$$[\mathbf{L}^2, L_y] = 0, \quad [\mathbf{L}^2, L_z] = 0. \tag{7.30}$$

In summary:

Angular Momentum Algebra

$$[L_i, L_j] = i\hbar \sum_{k=1}^{3} \epsilon_{ijk} L_k, \quad [\mathbf{L}^2, L_i] = 0 \tag{7.31}$$

As a consequence, $\mathbf{L}^2$ and one arbitrary component of angular momentum are simultaneously diagonalizable, i.e. have common eigenstates. One usually chooses the component L_z for that. Further components of L cannot be included, for they don't commute with L_z. We write down the eigenvalue equations,

$$\mathbf{L}^2 |\lambda m \alpha\rangle = \hbar^2 \lambda |\lambda m \alpha\rangle, \quad L_z |\lambda m \alpha\rangle = \hbar m |\lambda m \alpha\rangle \tag{7.32}$$

Here α stands for another quantum number (eigenvalue of another operator A), necessary in addition to λ and m in order to specify a state uniquely. In a spherically symmetric potential this could be for example the energy (the eigenvalue of the Hamiltonian operator), as we will see. The unspecified operator A thus forms with $\mathbf{L}^2$ and L_z a complete set of commuting observables. It commutes with $\mathbf{L}^2$ and L_z. For simplicity we want to assume that A also commutes with L_x and L_y. This is the case for the Hamiltonian operator in a spherically symmetric potential. Then follows that the values of α are unchanged under the action of L_i:

$$A(L_i |\lambda m \alpha\rangle) = L_i (A|\lambda m \alpha\rangle) = \alpha (L_i |\lambda m \alpha\rangle), \tag{7.33}$$

i.e. $L_i |\lambda m \alpha\rangle$ belongs to the same A-eigenvalue as $|\lambda m \alpha\rangle$.

Exercise 7.2
Let A be the operator of kinetic energy, $A = \mathbf{P}^2/(2m)$. Show that $[A, L_i] = 0$. In Sect. 7.4 we will show that this A forms a *complete* set of commuting observables with $\mathbf{L}^2$ and L_z.

We can already guess the possible values for the so-called **magnetic quantum number** m, since it is analogous to the quantum number l from the previous chapter: it will be integer numbers, for the operator L_z corresponds to the scalar angular momentum in two dimensions. Additionally, we know that $\lambda \geq 0$, because the L_i are hermitian (why?), and therefore

$$\hbar^2 \lambda = \langle \lambda m\alpha | \mathbf{L}^2 | \lambda m\alpha \rangle \tag{7.34}$$

$$= \langle \lambda m\alpha | \sum_{i=1}^{3} L_i^\dagger L_i | \lambda m\alpha \rangle \tag{7.35}$$

$$= \sum_{i=1}^{3} ||L_i | \lambda m\alpha \rangle||^2 \geq 0. \tag{7.36}$$

In the following we will derive the spectrum of possible (λ, m)-combinations, without referring to the form of the operators in position space, only using the commutator relations (7.31). We apply a method that has already proven successful for the harmonic oscillator: we define raising and lowering operators, which in this case raise or lower the value of m by 1,

$$L_\pm = L_x \pm i L_y. \tag{7.37}$$

One has

$$[L_z, L_+] = [L_z, L_x] + i[L_z, L_y] = i\hbar(L_y - iL_x) \tag{7.38}$$
$$= \hbar(L_x + iL_y) = \hbar L_+, \tag{7.39}$$

and similarly

$$[L_z, L_-] = -\hbar L_-. \tag{7.40}$$

It follows

$$L_z(L_+|\lambda m\alpha\rangle) = ([L_z, L_+] + L_+ L_z)|\lambda m\alpha\rangle \tag{7.41}$$
$$= \hbar L_+|\lambda m\alpha\rangle + L_+(L_z|\lambda m\alpha\rangle) \tag{7.42}$$
$$= \hbar L_+|\lambda m\alpha\rangle + \hbar m L_+|\lambda m\alpha\rangle \tag{7.43}$$
$$= \hbar(m+1)(L_+|\lambda m\alpha\rangle). \tag{7.44}$$

In the same way one obtains

$$L_z(L_-|\lambda m\alpha\rangle) = \hbar(m-1)(L_-|\lambda m\alpha\rangle). \tag{7.45}$$

Hence, $L_\pm|\lambda m\alpha\rangle$ are eigenstates of L_z with eigenvalue $\hbar(m\pm1)$; so, $L_\pm$ raises/lowers the quantum number m in steps of 1. The other quantum numbers are not affected, since the corresponding operators commute with L_x and L_y, and therefore also with $L_\pm$, see (7.33). Now we conclude

$$L_\pm|\lambda m\alpha\rangle = c_{\lambda m\pm}|\lambda, m \pm 1, \alpha\rangle, \tag{7.46}$$

where the $c_{\lambda m\pm}$ are normalization constant we are going to determine. We observe that

$$L_+L_- = (L_x + iL_y)(L_x - iL_y) = L_x^2 + L_y^2 - i[L_x, L_y] \qquad (7.47)$$
$$= \mathbf{L}^2 - L_z^2 + \hbar L_z \qquad (7.48)$$

and similarly

$$L_-L_+ = \mathbf{L}^2 - L_z^2 - \hbar L_z. \qquad (7.49)$$

Because of $L_+^\dagger = L_-$, this yields

$$||L_+|\lambda m\alpha\rangle||^2 = \langle\lambda m\alpha|L_-L_+|\lambda m\alpha\rangle \qquad (7.50)$$
$$= \langle\lambda m\alpha|\mathbf{L}^2 - L_z^2 - \hbar L_z|\lambda m\alpha\rangle \qquad (7.51)$$
$$= \hbar^2(\lambda - m^2 - m)\langle\lambda m\alpha|\lambda m\alpha\rangle \qquad (7.52)$$
$$= \hbar^2(\lambda - m(m + 1)). \qquad (7.53)$$

In the same way we get

$$||L_-|\lambda m\alpha\rangle||^2 = \hbar^2(\lambda - m(m - 1)) \qquad (7.54)$$

and therefore

$$c_{\lambda m\pm} = \hbar\sqrt{\lambda - m(m \pm 1)}. \qquad (7.55)$$

Then we find that for given λ the values of m have to be bounded from above and below. This is a consequence of

$$0 \le ||L_x|\lambda m\alpha\rangle||^2 + ||L_y|\lambda m\alpha\rangle||^2 = \langle\lambda m\alpha|L_x^2 + L_y^2|\lambda m\alpha\rangle \qquad (7.56)$$
$$= \langle\lambda m\alpha|\mathbf{L}^2 - L_z^2|\lambda m\alpha\rangle = \hbar^2(\lambda - m^2). \qquad (7.57)$$

So, it is required that $|m| \le \sqrt{\lambda}$. Since L_+ and L_- raise/lower the value of m further and further by repeated application, this is only possible if a state $|\lambda m_{max}\alpha\rangle$ is annihilated by L_+, and a state $|\lambda m_{min}\alpha\rangle$ by L_-. This implies

$$0 = ||L_+|\lambda m_{max}\alpha\rangle||^2 = \hbar^2(\lambda - m_{max}(m_{max} + 1)), \qquad (7.58)$$

$$0 = ||L_-|\lambda m_{min}\alpha\rangle||^2 = \hbar^2(\lambda - m_{min}(m_{min} - 1)). \qquad (7.59)$$

One denotes m_{max} with the letter l. Equation (7.58) is then, for given λ, solved by

$$\lambda = l(l + 1). \qquad (7.60)$$

On the other hand, one can write λ in the form (7.60) already from the start (any non-negative number λ can be written in this way, with a unique non-negative l), then getting $m_{max} = l$ as a consequence. Equation (7.59) has the solutions $m_{min} = l + 1$ and $m_{min} = -l$. But since m_{min} has to be smaller than m_{max}, the first solution can be excluded, leaving us with the important result: **For given $\lambda = l(l + 1)$, the possible values of m go from $-l$ to $+l$.**

Since m changes by application of $L_\pm$ in unit steps only, l must be integer or half-integer, $l = 0, \frac{1}{2}, 1, \frac{3}{2}, 2, \ldots$. This is as much as we can derive from the angular momentum algebra (7.31) alone. But since we know from the representation of L_z as an angular derivative that m even has to be integer, l must be integer too; l is the so-called **orbital quantum number**.

In general one uses l instead of λ to characterize a state, i.e. one writes $|lm\alpha\rangle$. The normalization constants (7.55) are also rewritten in terms of l, resulting in

$$L_+|l, m, \alpha\rangle = \hbar\sqrt{(l - m)(l + m + 1)}\,|l, m + 1, \alpha\rangle \qquad (7.61)$$

$$L_-|l, m, \alpha\rangle = \hbar\sqrt{(l + m)(l - m + 1)}\,|l, m - 1, \alpha\rangle. \qquad (7.62)$$

Exercise 7.3

One gets $|l, m, \alpha\rangle$ by applying L_+ $(l + m)$ times to $|l, -l, \alpha\rangle$, or L_- $(l - m)$ times to $|l, l, \alpha\rangle$. Show that

$$|l, m, \alpha\rangle = \hbar^{-l-m}\sqrt{\frac{(l - m)!}{(2l)!(l + m)!}}\,L_+^{l+m}|l, -l, \alpha\rangle \qquad (7.63)$$

$$= \hbar^{m-l}\sqrt{\frac{(l + m)!}{(2l)!(l - m)!}}\,L_-^{l-m}|l, l, \alpha\rangle. \qquad (7.64)$$

So far we have (except for the remark that m and therefore l need to be integer) only used the commutators of L_i and $\mathbf{L}^2$ for our derivation. In Chap. 2 we have seen that the components S_i of the spin and $\mathbf{S}^2$ obey exactly the same algebraic relations (see Exercises 2.19 and 2.25),

$$[S_i, S_j] = i\hbar \sum_{k=1}^{3} \epsilon_{ijk} S_k, \quad [\mathbf{S}^2, S_i] = 0. \qquad (7.65)$$

And so, the relations between the eigenvalues derived above must also hold. Indeed, we already found that S^2 is a multiple of the unit operator, the only eigenvalue being

$$\frac{3}{4}\hbar^2 = \hbar^2 \frac{1}{2}\left(\frac{1}{2} + 1\right), \qquad (7.66)$$

whereas S_z has the eigenvalues $\pm\frac{1}{2}\hbar$. The value of the quantum number l is obviously $\frac{1}{2}$ in this case. In Chap. 9 we will see how spin and angular momentum can be combined.

Exercise 7.4
Verify explicitly, using the Pauli matrices, that (7.61) and (7.62) are valid in the case of spin, with $S_\pm = S_x \pm i S_y$.

Self-check questions:
1. Which subsets of $\{L_x, L_y, L_z, \mathbf{L}^2, \mathbf{P}^2\}$ can be diagonalized simultaneously?
2. What are the raising and lowering operators in the case of angular momentum, and which quantum number do they raise/lower?

7.2 Spherical Harmonics

After we've determined the eigenvalues of $\mathbf{L}^2$ and L_z, we now turn to the eigenfunctions. For this, we have to grapple a bit with spherical coordinates first. They are defined by

$$x = r \sin\theta \cos\phi \tag{7.67}$$
$$y = r \sin\theta \sin\phi \tag{7.68}$$
$$z = r \cos\theta \tag{7.69}$$

or vice versa

$$r = \sqrt{x^2 + y^2 + z^2} \tag{7.70}$$

$$\theta = \arctan \frac{\sqrt{x^2 + y^2}}{z} \tag{7.71}$$

$$\phi = \arctan \frac{y}{x}. \tag{7.72}$$

Here r runs from 0 to ∞, θ from 0 ("north pole") to π ("south pole") and ϕ from 0 to 2π. Regarding the usage of the arc tangent, see the comment below (6.26).

Just as with polar coordinates in two dimensions we start with transforming the partial derivatives:

$$\frac{\partial}{\partial x} = \frac{\partial r}{\partial x}\frac{\partial}{\partial r} + \frac{\partial\theta}{\partial x}\frac{\partial}{\partial\theta} + \frac{\partial\phi}{\partial x}\frac{\partial}{\partial\phi} \tag{7.73}$$

$$= \frac{x}{r}\frac{\partial}{\partial r} + \frac{1}{1 + \frac{x^2+y^2}{z^2}}\frac{x}{z\sqrt{x^2+y^2}}\frac{\partial}{\partial\theta} + \frac{1}{1 + \frac{y^2}{x^2}}\left(\frac{-y}{x^2}\right)\frac{\partial}{\partial\phi} \tag{7.74}$$

$$= \frac{x}{r} \frac{\partial}{\partial r} + \frac{xz}{r^2 \sqrt{x^2 + y^2}} \frac{\partial}{\partial \theta} - \frac{y}{x^2 + y^2} \frac{\partial}{\partial \phi} \qquad (7.75)$$

$$= \sin \theta \cos \phi \frac{\partial}{\partial r} + \frac{\cos \theta \cos \phi}{r} \frac{\partial}{\partial \theta} - \frac{\sin \phi}{r \sin \theta} \frac{\partial}{\partial \phi} \qquad (7.76)$$

In a similar way we get

$$\frac{\partial}{\partial y} = \sin \theta \sin \phi \frac{\partial}{\partial r} + \frac{\cos \theta \sin \phi}{r} \frac{\partial}{\partial \theta} + \frac{\cos \phi}{r \sin \theta} \frac{\partial}{\partial \phi} \qquad (7.77)$$

$$\frac{\partial}{\partial z} = \cos \theta \frac{\partial}{\partial r} - \frac{\sin \theta}{r} \frac{\partial}{\partial \theta} \qquad (7.78)$$

und from that after a lengthy calculation (product rule!)

$$\Delta = \frac{\partial^2}{\partial x^2} + \frac{\partial^2}{\partial y^2} + \frac{\partial^2}{\partial z^2} \qquad (7.79)$$

$$= \frac{\partial^2}{\partial r^2} + \frac{2}{r} \frac{\partial}{\partial r} + \frac{1}{r^2} \frac{\partial^2}{\partial \theta^2} + \frac{1}{r^2} \cot \theta \frac{\partial}{\partial \theta} + \frac{1}{r^2 \sin^2 \theta} \frac{\partial^2}{\partial \phi^2} \qquad (7.80)$$

$$= \frac{1}{r} \frac{\partial^2}{\partial r^2} r + \frac{1}{r^2 \sin \theta} \frac{\partial}{\partial \theta} \sin \theta \frac{\partial}{\partial \theta} + \frac{1}{r^2 \sin^2 \theta} \frac{\partial^2}{\partial \phi^2}. \qquad (7.81)$$

The expressions in the last row are to be understood through their action on a function $\psi(r, \theta, \phi)$, "from right to left", for example

$$\left(\frac{\partial^2}{\partial r^2} r \right) \psi := \frac{\partial^2}{\partial r^2} (r \psi) \qquad (7.82)$$

and

$$\left(\frac{\partial}{\partial \theta} \sin \theta \frac{\partial}{\partial \theta} \right) \psi := \frac{\partial}{\partial \theta} \left(\sin \theta \frac{\partial}{\partial \theta} \psi \right). \qquad (7.83)$$

Next, we turn our attention to the angular momentum and obtain, again after some calculation

$$L_x = -i\hbar(y \frac{\partial}{\partial z} - z \frac{\partial}{\partial y}) \qquad (7.84)$$

$$= i\hbar \left(\sin \phi \frac{\partial}{\partial \theta} + \cot \theta \cos \phi \frac{\partial}{\partial \phi} \right) \qquad (7.85)$$

$$L_y = -i\hbar(z \frac{\partial}{\partial x} - x \frac{\partial}{\partial z}) \qquad (7.86)$$

$$= -i\hbar \left(\cos \phi \frac{\partial}{\partial \theta} - \cot \theta \sin \phi \frac{\partial}{\partial \phi} \right) \qquad (7.87)$$

$$L_z = -i\hbar(x\frac{\partial}{\partial y} - y\frac{\partial}{\partial x}) \tag{7.88}$$

$$= -i\hbar\frac{\partial}{\partial \phi} \tag{7.89}$$

$$\mathbf{L}^2 = L_x^2 + L_y^2 + L_z^2 \tag{7.90}$$

$$= -\hbar^2 \left(\frac{\partial^2}{\partial\theta^2} + \cot\theta\frac{\partial}{\partial\theta} + \frac{1}{\sin^2\theta}\frac{\partial^2}{\partial\phi^2}\right). \tag{7.91}$$

The expression for L_z was expected, after our results in two dimensions. The expression for $\mathbf{L}^2$ is contained in the Laplace operator (7.80):

$$\Delta = \frac{\partial^2}{\partial r^2} + \frac{2}{r}\frac{\partial}{\partial r} - \frac{1}{\hbar^2 r^2}\mathbf{L}^2 \tag{7.92}$$

Exercise 7.5
Use a rainy Sunday afternoon to verify the equations (7.77)–(7.91), one after another. In particular, enjoy the long calculations for Δ and $\mathbf{L}^2$, where countless terms cancel each other or can be summarized in a miraculous way.

Phew, that was a hard piece of work! Now we only need $L_\pm$, then we have all needed operators at hand:

$$L_+ = L_x + iL_y \tag{7.93}$$

$$= i\hbar\left[(\sin\phi - i\cos\phi)\frac{\partial}{\partial\theta} + \cot\theta(\cos\phi + i\sin\phi)\frac{\partial}{\partial\phi}\right] \tag{7.94}$$

$$= \hbar e^{i\phi}\left(\frac{\partial}{\partial\theta} + i\cot\theta\frac{\partial}{\partial\phi}\right) \tag{7.95}$$

$$L_- = L_x - iL_y \tag{7.96}$$

$$= i\hbar\left[(\sin\phi + i\cos\phi)\frac{\partial}{\partial\theta} + \cot\theta(\cos\phi - i\sin\phi)\frac{\partial}{\partial\phi}\right] \tag{7.97}$$

$$= \hbar e^{-i\phi}\left(-\frac{\partial}{\partial\theta} + i\cot\theta\frac{\partial}{\partial\phi}\right) \tag{7.98}$$

We recognize that all angular momentum operators depend only on θ and ϕ; r does not show up in them, neither in a derivative nor in a factor. The eigenfunctions of $\mathbf{L}^2$ and L_z can therefore be written in the form

$$\psi_{lm\alpha}(r, \theta, \phi) = f(r)Y_{lm\alpha}(\theta, \phi) \tag{7.99}$$

with an (as long as α is not further specified) *arbitrary* function $f(r)$. At the word "arbitrary", the mathematician may knit his brow once again. Yes, there again criteria regarding continuity and differentiability, stubbornly ignored by us physicists. In addition, the wave function should be normalizable. The norm of $\psi_{lm\alpha}$ is given by

$$\|\psi_{lm\alpha}\|^2 = \int_0^\infty dr \int_0^\pi d\theta \int_0^{2\pi} d\phi\, r^2 \sin\theta \times \qquad (7.100)$$

$$f^*(r)f(r)\, Y_{lm\alpha}^*(\theta, \phi) Y_{lm\alpha}(\theta, \phi) \qquad (7.101)$$

$$= \left[\int_0^\infty dr\, r^2 f^*(r) f(r)\right] \times \qquad (7.102)$$

$$\left[\int_0^\pi d\theta \int_0^{2\pi} d\phi\, \sin\theta\, Y_{lm\alpha}^*(\theta, \phi) Y_{lm\alpha}(\theta, \phi)\right]. \qquad (7.103)$$

In order to normalize $\psi_{lm\alpha}$ to 1, we can normalize f and $Y_{lm\alpha}$ separately to 1, i.e. set the contents of each of the two square brackets to 1. Do you remember why for integrals in spherical coordinates there is always this factor $r^2 \sin\theta$? It is the inverse Jacobi determinant of the transformation from cartesian to spherical coordinates.

Exercise 7.6
Verify that

$$\det \begin{pmatrix} \frac{\partial x}{\partial r} & \frac{\partial y}{\partial r} & \frac{\partial z}{\partial r} \\ \frac{\partial x}{\partial \theta} & \frac{\partial y}{\partial \theta} & \frac{\partial z}{\partial \theta} \\ \frac{\partial x}{\partial \phi} & \frac{\partial y}{\partial \phi} & \frac{\partial z}{\partial \phi} \end{pmatrix} = \left[\det \begin{pmatrix} \frac{\partial r}{\partial x} & \frac{\partial \theta}{\partial x} & \frac{\partial \phi}{\partial x} \\ \frac{\partial r}{\partial y} & \frac{\partial \theta}{\partial y} & \frac{\partial \phi}{\partial y} \\ \frac{\partial r}{\partial z} & \frac{\partial \theta}{\partial z} & \frac{\partial \phi}{\partial z} \end{pmatrix}\right]^{-1} = r^2 \sin\theta. \qquad (7.104)$$

In many cases calculations are simplified when functions $f(\theta)$ are rewritten as functions of $v = \cos\theta$, for instance

$$f(\theta) = \sin^2\theta \to f(v) = 1 - v^2. \qquad (7.105)$$

One then substitutes in the integrals

$$\int_0^\pi d\theta \sin\theta \to \int_{-1}^1 dv \qquad (7.106)$$

(following the rules for variable substitution in integrals). Instead of giving the new variable a name (v in our case), it is common to just denote it as $\cos\theta$, writing $\int_{-1}^1 d\cos\theta$.

Let's have a closer look at $Y_{lm\alpha}$. Since $\mathbf{L}^2$ and L_z have no effect on $f(r)$, the eigenvalue equations have to be valid for $Y_{lm\alpha}$ alone:

$$\mathbf{L}^2 Y_{lm\alpha}(\theta, \phi) = \hbar^2 l(l+1) Y_{lm\alpha}(\theta, \phi) \qquad (7.107)$$

$$L_z Y_{lm\alpha}(\theta, \phi) = \hbar m Y_{lm\alpha}(\theta, \phi) \qquad (7.108)$$

Due to (7.89), $Y_{lm\alpha}$ as an eigenfunction of L_z must have the form

$$Y_{lm\alpha}(\theta, \phi) = u_{lm\alpha}(\theta)e^{im\phi}. \tag{7.109}$$

This confirms once again that m and thus l have to be integers.

We have now two options: (a) We can determine Y_{ll} and derive the other Y_{lm} from that using L_-, or (b) we can determine $Y_{l,-l}$ and derive the other Y_{lm} from that using L_+. We choose the second variant. Now, $L_- Y_{l,-l}$ has to vanish, and with (7.95) and (7.109) this yields the differential equation

$$\frac{\partial}{\partial\theta} u_{l,-l,\alpha}(\theta) = l\cot\theta\, u_{l,-l,\alpha}(\theta) \tag{7.110}$$

with the solution

$$u_{l,-l,\alpha}(\theta) = c_l \sin^l \theta, \tag{7.111}$$

where c_l is a constant which we want to choose such that $Y_{l,-l,\alpha}$ is normalized to 1. Since the differential equation (7.110) is of first order, the solution (7.111) is unique. Hence there is only one function $Y_{l,-l,\alpha}$ for the quantum numbers l and $m = -l$. Since the other $Y_{lm\alpha}$ are obtained from $Y_{l,-l,\alpha}$ via the action of L_+, these are also unique. The index α is therefore unnecessary, the **spherical harmonics** $Y_{lm}(\theta, \phi)$ are uniquely determined by l and m. Each Y_{lm} is normalized to 1, and different Y_{lm} are orthogonal to each other, since they belong to different eigenspaces of $\mathbf{L}^2$ or L_z. The normalization condition is

$$\int_0^\pi d\theta \int_0^{2\pi} d\phi \sin\theta\, Y_{lm}^*(\theta, \phi) Y_{l'm'}(\theta, \phi) = \delta_{ll'}\delta_{mm'}. \tag{7.112}$$

The normalization constant c_l in (7.111) can be derived from that. After inserting the known expressions for Y_{lm}, the ϕ-integration is simple. The integral of $\sin^l \theta$ can be looked up. The result for c_l is

$$c_l = \frac{1}{2^l l!}\sqrt{\frac{(2l+1)!}{4\pi}}. \tag{7.113}$$

So, we found

$$Y_{l,-l}(\theta, \phi) = \frac{1}{2^l l!}\sqrt{\frac{(2l+1)!}{4\pi}} \sin^l \theta\, e^{-il\phi}. \tag{7.114}$$

Exercise 7.7
Verify with the help of (7.91) that $Y_{l,-l}$ is an eigenfunction of $\mathbf{L}^2$ with eigenvalue $\hbar^2 l(l+1)$.

Now we apply L_+ n times to $Y_{l,-l}$ and show by induction that

$$(L_+)^n Y_{l,-l}(\theta, \phi) = c_l(-\hbar)^n e^{i(n-l)\phi} \sin^{n-l}\theta \frac{d^n}{d\cos\theta^n}[(1-\cos^2\theta)^l]. \quad (7.115)$$

The statement is obviously true for $n = 0$, for then $\sin^{-l}\theta$ and $(1-\cos^2\theta)^l = \sin^{2l}\theta$ combine to $\sin^l\theta$ and yield (7.114). For the induction step we assume the statement to hold for n, and show that it then also holds for $n+1$. We apply

$$L_+ = \hbar e^{i\phi}\left(\frac{\partial}{\partial\theta} + i\cot\theta\frac{\partial}{\partial\phi}\right) \quad (7.116)$$

to the right hand side of (7.115). This gives three terms:

- The ϕ-derivatve $i\cot\theta\frac{\partial}{\partial\phi}$ acts on $e^{i(n-l)\phi}$, creating a factor $(l-n)\cot\theta$.
- The θ-derivative, applied to $\sin^{n-l}\theta$, creates a term with the factor

$$\frac{\partial}{\partial\theta}\sin^{n-l}\theta = (n-l)\cos\theta\sin^{n-l-1}\theta = (n-l)\cot\theta\sin^{n-l}\theta, \quad (7.117)$$

which just cancels the first term.
- The θ-derivative, applied to $\frac{d^n}{d\cos\theta^n}[(1-\cos^2\theta)^l]$, gives, due to

$$\frac{d}{d\theta} = \frac{d\cos\theta}{d\theta}\frac{d}{d\cos\theta} = -\sin\theta\frac{d}{d\cos\theta}, \quad (7.118)$$

just the right hand side of (7.115) with the replacement $n \to n+1$. This completes the induction step, and (7.115) is proven.

Exercise 7.8
Reproduce the steps sketched in words above in all details.

In order to reach Y_{lm}, we have to apply L_+ $(l+m)$ times. We thus set $n = l+m$ in (7.115):

$$(L_+)^{l+m} Y_{l,-l}(\theta, \phi) = c_l(-\hbar)^{l+m} e^{im\phi}\sin^m\theta\frac{d^{l+m}}{d\cos\theta^{l+m}}[(1-\cos^2\theta)^l] \quad (7.119)$$

The **associated Legendre functions** $P_{lm}(x)$ are defined by

$$P_{lm}(x) = (-1)^{l+m}\frac{1}{2^l l!}(1-x^2)^{m/2}\frac{d^{l+m}}{dx^{l+m}}(1-x^2)^l. \quad (7.120)$$

One immediately recognizes that these functions appear in (7.119) in the form of $P_{lm}(\cos\theta)$. In order to get the final expression for Y_{lm}, we can replace $|l, m, \alpha\rangle$ by

Y_{lm} in (7.63); for α plays no role for the action of $L_\pm$, and the Y_{lm} are by themselves (i.e. without regarding $f(r)$ in 7.99) normalized. One can now combine (7.119), (7.120), (7.63) and (7.113) to obtain:

$$Y_{lm}(\theta, \phi) = \sqrt{\frac{(2l+1)}{4\pi} \frac{(l-m)!}{(l+m)!}} e^{im\phi} P_{lm}(\cos\theta) \tag{7.121}$$

Exercise 7.9
Verify this.

The subset $P_l(x) := P_{l0}(x)$ of the functions P_{lm} with $m = 0$ are the **Legendre polynomials** which were already mentioned in Sect. 3.2. The other P_{lm} are "associated" with them. (The name came about in the following way: The P_l are solutions to a differential equation D1; the P_{lm} are solutions to a slightly more complicated differential equation D2, which can however be derived from the solutions of D1 via a variable transformation; the solutions are thus "associated" with each other.) One can show that the $P_l(x)$ are polynomials of degree l, and that they are orthogonal in the interval $[-1, 1]$,

$$\int_{-1}^{1} dx \, P_l(x) P_{l'}(x) = \frac{2}{2l+1} \delta_{ll'}, \tag{7.122}$$

but due to the factor $\frac{2}{2l+1}$ they are not orthonormal. The Legendre polynomials P_l defined here differ therefore from those in Sect. 3.2 by a factor $\sqrt{\frac{2}{2l+1}}$. Note that the P_{lm} for odd m are *not* polynomials, due to the occurring square root $(1 - x^2)^{m/2}$. For $m = 0$, the Y_{lm} are independent ϕ:

$$Y_{l0}(\theta) = \sqrt{\frac{2l+1}{4\pi}} P_l(\cos\theta) \tag{7.123}$$

The normalization condition (7.112) then reads

$$2\pi \int_{-1}^{1} d\cos\theta \, Y_{l0}^* Y_{l'0} \tag{7.124}$$

$$= \frac{\sqrt{(2l+1)(2l'+1)}}{2} \int_{-1}^{1} d\cos\theta \, P_l(\cos\theta) P_{l'}(\cos\theta) \tag{7.125}$$

$$= \delta_{ll'}, \tag{7.126}$$

consistent with equation (7.122).

With (7.121) and (7.120) one can easily calculate the spherical harmonics for $l = 0, 1, 2$ (but since this is neither fun nor very instructive, and you've already done so much calculation in this section, we don't turn this into an exercise):

$$Y_{00}(\theta, \phi) = \frac{1}{\sqrt{4\pi}} \tag{7.127}$$

$$Y_{10}(\theta, \phi) = \sqrt{\frac{3}{4\pi}} \cos\theta \tag{7.128}$$

$$Y_{1\pm1}(\theta, \phi) = \mp\sqrt{\frac{3}{8\pi}} \sin\theta\, e^{\pm i\phi} \tag{7.129}$$

$$Y_{20}(\theta, \phi) = \sqrt{\frac{5}{16\pi}} (3\cos^2\theta - 1) \tag{7.130}$$

$$Y_{2\pm1}(\theta, \phi) = \mp\sqrt{\frac{15}{8\pi}} \sin\theta\cos\theta\, e^{\pm i\phi} \tag{7.131}$$

$$Y_{2\pm2}(\theta, \phi) = \sqrt{\frac{15}{32\pi}} \sin^2\theta\, e^{\pm 2i\phi} \tag{7.132}$$

Let $\mathcal{H}_\Omega$ be the Hilbert space of square-integrable functions on a sphere, i.e. of the functions $f(\theta, \phi)$ with the property

$$\int d\cos\theta\, d\phi\, f^*(\theta, \phi) f(\theta, \phi) < \infty. \tag{7.133}$$

One can then show that the spherical harmonics Y_{lm} form a Schauder basis of $\mathcal{H}_\Omega$. This means that any such function can be written as an (infinite) linear combination of the Y_{lm}; one says $f(\theta, \phi)$ can be *expanded* in terms of the Y_{lm}. In particular, in $\mathcal{H}_\Omega$ the completeness relation holds,

$$\mathbf{1} = \sum_{l=0}^{\infty} \sum_{m=-l}^{l} |lm\rangle\langle lm|. \tag{7.134}$$

If we introduce with $\{|\theta, \phi\rangle\}$ a pseudo-basis similar to $\{|x\rangle\}$,

$$\langle\theta, \phi|\mathbf{1}|\theta', \phi'\rangle = \delta(\cos\theta - \cos\theta')\delta(\phi - \phi') \tag{7.135}$$

(the delta distribution contains $\cos\theta$, so that it is correctly defined in the integral $\int d\cos\theta$),

$$\langle\theta, \phi|lm\rangle = Y_{lm}(\theta, \phi), \tag{7.136}$$

then we can rewrite the completeness relation to

$$\sum_{l=0}^{\infty} \sum_{m=-l}^{l} Y_{lm}^*(\theta, \phi) Y_{lm}(\theta', \phi') = \delta(\cos\theta - \cos\theta')\delta(\phi - \phi'). \tag{7.137}$$

Self-check questions:

1. The spherical harmonics are orthonormal. Can you express this sentence in the form of an integral?

7.3 Spherically Symmetric Potential

We want to solve the stationary Schrödinger equation for a spherically symmetric potential $V(r)$. With the expression (7.92) for the Laplace operator it reads

$$\left[-\frac{\hbar^2}{2m} \left(\frac{\partial^2}{\partial r^2} + \frac{2}{r}\frac{\partial}{\partial r} - \frac{1}{\hbar^2 r^2}\mathbf{L}^2 \right) + V(r) \right] \psi(r, \theta, \phi) = E\psi(r, \theta, \phi). \quad (7.138)$$

The expression in square brackets on the left hand side is the Hamiltonian operator H. It obviously commutes with all components L_i of the angular momentum operator, since

$$[\mathbf{L}^2, L_i] = [r, L_i] = [\frac{\partial}{\partial r}, L_i] = 0. \quad (7.139)$$

Therefore, just as in classical mechanics, angular momentum is conserved for a spherically symmetric potential; its expectation value does not change with time. From $[H, L_i] = 0$ and $[H, \mathbf{L}^2] = 0$ also follows that we can choose the energy eigenstates such that they are simultaneously eigenstates of L_z and $\mathbf{L}^2$. The ansatz

$$\psi_{nlm}(r, \theta, \phi) = R_{nl}(r)Y_{lm}(\theta, \phi) \quad (7.140)$$

yields, inserted into (7.138), the radial equation

$$\left[-\frac{\hbar^2}{2m} \left(\frac{d^2}{dr^2} + \frac{2}{r}\frac{d}{dr} \right) + \frac{\hbar^2 l(l+1)}{2mr^2} + V(r) \right] R_{nl}(r) = E_n R_{nl}(r). \quad (7.141)$$

Here n is a quantum number we use to enumerate energy eigenvalues. For bound states the spectrum is discrete, so n can be positive integers. For free states the possible energies are continuous, and instead of n one uses the letter k, suggesting something like a wave number. We will find examples for both cases. Some care is needed regarding the letter m which is used for two different things here: mass and magnetic quantum number. The risk of a confusion is hopefully low, for the magnetic quantum number m does not appear in the radial equation; the Laplace operator contains only $\mathbf{L}^2$ and thus the orbital quantum number l. For this reason, m is not contained in the index of R_{nl}. The radial function depends only on the eigenvalues of energy and $\mathbf{L}^2$ ab.

The angular momentum term and V can again (as in two dimensions) be combined into an effective potential,

$$V_{\text{eff}}(r) = V(r) + \frac{\hbar^2 l(l+1)}{2mr^2}. \tag{7.142}$$

The radial equation is further simplified by substituting

$$U_{nl}(r) = r\, R_{nl}(r). \tag{7.143}$$

It then reads

$$-\frac{\hbar^2}{2m} U_{nl}''(r) + V_{\text{eff}}(r) U_{nl}(r) = E_n U_{nl}(r). \tag{7.144}$$

For bound states, we are interested in the behavior of U close to the origin. We make the assumption that if $V(r)$ diverges for $r \to 0$, it does so slower than r^{-2}, i.e.

$$\lim_{r \to 0} V(r) r^2 = 0. \tag{7.145}$$

The radial equation is then for $l > 0$ in the proximity of $r = 0$ dominated by the angular momentum term,

$$U_{nl}''(r) \approx \frac{l(l+1)}{r^2} U_{nl}(r). \tag{7.146}$$

For the possible behavior of solutions in the limit $r \to 0$ this implies:

$$U_{nl} \sim r^{l+1} \quad \text{bzw.} \quad U_{nl} \sim r^{-l} \tag{7.147}$$

But the second solution is not normalizable:

$$\int_0^\varepsilon dr\, r^2\, R^*(r) R(r) = \int_0^\varepsilon dr\, U^*(r) U(r) \sim \int_0^\varepsilon dr\, r^{-2l} = \infty \tag{7.148}$$

Therefore, we must have

$$U_{nl} \sim r^{l+1} \tag{7.149}$$

and so

$$R_{nl} \sim r^l \tag{7.150}$$

(for $l > 0$; for $l = 0$ we cannot say anything yet, in this case it depends on the potential). In particular, for $l > 0$ the probability density of a quantum object vanishes at the origin. That makes sense, because at the origin an infinite momentum would be required to produce a finite angular momentum.

Nerd's Corner 7.2

We now want to derive a statement about the behavior of U_{nl} at the origin for $l = 0$. It will be shown that $U_{n0} \to 0$ also for $r \to 0$, under the assumption that $V(r)$ does not contain a delta function at $r = 0$ and that again $r \to 0$ diverges slower than r^{-2}.

Why are these two preconditions necessary? You probably remember from electrodynamics the important relation

$$\Delta \frac{1}{r} = -4\pi \delta^3(\mathbf{r}), \tag{7.151}$$

where $\delta^3(\mathbf{r}) = \delta(x)\delta(y)\delta(z)$.

Exercise 7.10

Prove (7.151) using the Gauss's theorem

$$\int_S d\mathbf{S} \cdot \left(\nabla \frac{1}{r} \right) = \int_V d^3x \, \Delta \frac{1}{r}. \tag{7.152}$$

Choose the unit sphere as the volume of integration. Verify $\Delta(1/r) = 0$ for $r > 0$. Evaluate the left hand side of (7.152) and conclude that the corresponding contribution on the right hand side must be localized in the origin $r = 0$ alone.

Now let's assume that U_{n0} converges for $r \to 0$ to a non-zero constant c. Then $R_{n0}(r) = U_{n0}(r)/r$ behaves for $r > 0$ like c/r. The second derivative of R_{n0} in (7.141) turns this into a delta function, which must be accompanied by a corresponding delta function in the potential for the radial equation (7.141) to hold. But this is forbidden by assumption, and so $U_{n0} \to c$ is not allowed.

What about the possibility $U_{n0} \to \pm\infty$ for $r \to 0$? Indeed functions of the form

$$U_{n0}(r) = c\, r^{-\alpha}, \quad 0 < \alpha < \frac{1}{2} \quad \text{or} \quad U_{n0}(r) = c \log r \tag{7.153}$$

are square-integrable at the origin (verify this!). But they require a potential which diverges at least like r^{-2}: In the first case (7.144) yields

$$U_{n0}''(r) = c\,\alpha(\alpha + 1)r^{-\alpha-2} = \alpha(\alpha+1)r^{-2}U_{n0}(r) \tag{7.154}$$

$$\Rightarrow \quad V(r) = \frac{\hbar^2}{2m}\alpha(\alpha+1)r^{-2} \tag{7.155}$$

(all statements hold in the limit $r \to 0$). In the second case we have

$$U_{n0}''(r) = -c\,r^{-2} = -\frac{1}{r^2 \log r} U_{n0}(r) \qquad (7.156)$$

$$\Rightarrow \quad V(r) = -\frac{\hbar^2}{2m} \frac{1}{r^2 \log r}. \qquad (7.157)$$

Under the mentioned assumptions for the allowed potentials we therefore conclude $U_{n0}(r) \to 0$ for $r \to 0$.

This is connected with the claim that the operator

$$D_r := \frac{\partial}{\partial r} + \frac{1}{r}, \qquad (7.158)$$

which will soon play a role when we study the free particle, is antihermitian. We remind ourselves of a similar discussion regarding the operator $D = \frac{d}{dx}$ in Sect. 3.2. The result crucially depended on the function space under consideration. Let's refresh our memory: we had

$$\langle f|D|g\rangle = \langle f|Dg\rangle = \int_{-\infty}^{\infty} dx\, f^*(x) g'(x) \qquad (7.159)$$

and

$$\langle f|D^\dagger|g\rangle = \langle Df|g\rangle = \int_{-\infty}^{\infty} dx\, f^{*\prime}(x) g(x) \qquad (7.160)$$

$$= -\int_{-\infty}^{\infty} dx\, f^*(x) g'(x) + \left[f^*(x) g(x)\right]_{-\infty}^{\infty}. \qquad (7.161)$$

By definition, D is antihermitian if and only if

$$\langle f|D|g\rangle = -\langle f|D^\dagger|g\rangle \qquad (7.162)$$

holds for all functions f and g in the function space under consideration, which is true if the boundary term (7.161) vanishes for all f and g. For square-integrable functions this is the case, since their limits for $x \to \pm\infty$ have to vanish. Now we turn to D_r:

$$\langle nlm|D_r|n'l'm'\rangle = \int dr\, d\cos\theta\, d\phi\, r^2 R_{nl}^* Y_{lm}^* D_r R_{n'l'} Y_{l'm'}$$

$$= \left[\int dr\, r^2 R_{nl}^* D_r R_{n'l'}\right]\left[\int d\cos\theta\, d\phi\, Y_{lm}^* Y_{l'm'}\right]$$

$$= \int dr\, r^2 R_{nl}^* D_r R_{n'l'} \delta_{ll'} \delta_{mm'}$$

$$= \delta_{ll'} \delta_{mm'} \int dr\, r\, R_{nl}^* \left(1 + r\frac{d}{dr}\right) R_{n'l'}$$

$$= \delta_{ll'} \delta_{mm'} \int dr\, r\, R_{nl}^* \frac{d}{dr}(r R_{n'l'})$$

$$= \delta_{ll'} \delta_{mm'} \int_0^\infty dr\, U_{nl}^* \frac{d}{dr} U_{n'l'}$$

The third equation follows from the normalization condition (7.112). The last equation shows that while D_r acts in the form $\frac{d}{dr} + \frac{1}{r}$ on R_{nl}, it acts simply as $\frac{d}{dr}$ on U_{nl}. Similarly one gets

$$\langle nlm|D_r^\dagger|n'l'm'\rangle = \delta_{ll'} \delta_{mm'} \int_0^\infty dr \left(\frac{d}{dr} U_{nl}^*\right) U_{n'l'} \tag{7.163}$$

and for $l = l'$, $m = m'$ partial integration again yields

$$\langle nlm|D_r|n'lm\rangle = -\langle nlm|D_r^\dagger|n'lm\rangle + \left[U_{nl}^* U_{n'l}\right]_0^\infty. \tag{7.164}$$

At infinity the U_{nl} vanish due to normalizability (we assume bound states), and at 0 they vanish due to the considerations above (for appropriate potentials). D_r is therefore antihermitian in the function space spanned by the $|nlm\rangle$.

All considerations above hold also for two particle systems, as long as the interaction potential depends only on the distance between the particles, $V = V(|\mathbf{r}_2 - \mathbf{r}_1|)$. For then one can, as was shown in Sect. 3.8, reformulate the two particle problem into a one particle problem. The mass m has to be replaced by the reduced mass μ, and the wave function $\psi_{nlm}(r, \theta, \phi)$ represents the probability distribution of the distance vector $\mathbf{r}_2 - \mathbf{r}_1$, in spherical coordinates. We will make use of that in the treatment of the hydrogen atom.

Self-check questions:

1. How is the radial function $U_{nl}(r)$ defined?
2. What is the effective potential for a spherically symmetric potential?
3. How does the wave function behave for $l > 0$ in the proximity of the origin?

7.4 Free Particle

The simplest of all spherically symmetric potentials is $V(r) = 0$. The free particle in three dimensions is, in cartesian coordinates, just a generalization of the free particle in one dimension. The energy eigenfunctions and eigenvalues are

$$\psi_{\mathbf{k}}(\mathbf{r}) = e^{i\mathbf{k}\cdot\mathbf{r}}, \quad E_k = \frac{\hbar^2 k^2}{2m}. \tag{7.165}$$

Here $\mathbf{k}$ is a **wave number vector** of arbitrary direction and norm k. The energy value E_k is infinitely degenerate, since for each k there are infinitely many possible directions of $\mathbf{k}$. The $\psi_{\mathbf{k}}(\mathbf{r})$ are pseudo-vectors, which can only be normalized via delta functions. We will ignore this kind of normalization in this section to avoid any discussion of normalization constants. The wave function $\psi_{\mathbf{k}}(\mathbf{r})$ describes a plane wave moving in $\mathbf{k}$-direction. The solution of the time-dependent Schrödinger equation is

$$\psi_{\mathbf{k}}(\mathbf{r}, t) = e^{i\mathbf{k}\cdot\mathbf{r} - i\frac{E_k}{\hbar}t}. \tag{7.166}$$

We say that the wave "moves", although the norm of the wave function is constant in time and space, so that there are no moving "wave peaks" in the strict sense. But if one combines several $\psi_{\mathbf{k}}(\mathbf{r})$ with average value $\mathbf{k} = \bar{\mathbf{k}}$ into a square-integrable wave packet, then this wave packet indeed moves with velocity $\mathbf{v} = \hbar\bar{\mathbf{k}}/m$ through space, similar to the Gaussian packet we have studied in one dimension.

Since $V(r) = 0$ is a (vanishing) spherically symmetric potential, we can use spherical coordinates to find energy eigenstates which are also eigenstates of angular momentum,

$$\psi_{klm}(r, \theta, \phi) = \frac{U_{kl}(r)}{r} Y_{lm}(\theta, \phi). \tag{7.167}$$

These states are then **spherical waves**, propagating radially in all directions, thereby containing angular momentum in an abstract way, encoded in Y_{lm}. A combination (superposition) of such eigenfunctions can, however, again be a wave packet whose center moves linearly in one direction, since the classical equations of motion have to be obeyed by expectation values.

We have to solve the radial equation

$$\frac{\hbar^2}{2m}\left(-U_{kl}''(r) + \frac{l(l+1)}{r^2}U_{kl}(r)\right) = \frac{\hbar^2 k^2}{2m}U_{kl}(r). \tag{7.168}$$

Once again we help ourselves with raising and lowering operators. The idea is always the same: There is a differential equation of second order to be solved. One finds an operator (and its adjoint) which connects different solutions with each other and thereby increases some quantum number by 1, whereas the adjoint operator lowers it by 1. For the harmonic oscillator this was the energy quantum number n, for the spherical harmonics the magnetic quantum number m. For the free particle it is going to be the orbital quantum number l. Each of these quantum numbers has a minimal value whose associated eigenfunction can be found relatively easily. All other eigenfunctions can then be obtained via the raising operator. For the free particle there are several options how to proceed. We follow here mainly the approach presented in Shankar (2011).

We divide (7.168) on both sides by $\hbar^2 k^2/(2m)$, substitute the variable r by $\rho = kr$ and get

$$\left(-\frac{d}{d\rho^2} + \frac{l(l+1)}{\rho^2}\right)U_l(\rho) = U_l(\rho). \tag{7.169}$$

One sees that k no longer appears in the equation. The solutions, written as functions of ρ, are independent of the wave number k. Hence we have dropped the index k from U. The k-dependence comes back only when the solutions are rewritten as functions of $r = \rho/k$. The solutions for $l = 0$ are obviously

$$U_0^{(1)}(\rho) = \sin \rho, \quad U_0^{(2)}(\rho) = \cos \rho. \tag{7.170}$$

But the second solution converges to 1 for $\rho \to 0$, and is therefore not allowed, following the considerations of the previous nerd's corner. Hence we can restrict ourselves to $U_0(\rho) = \sin \rho$.

Now we define the operators

$$B_l = \frac{d}{d\rho} + \frac{l+1}{\rho} \tag{7.171}$$

on the space of possible U-functions. Here $d/d\rho$ is, as an operator acting on U, up to a factor k identical to the antihermitian operator D_r from the previous nerd's corner. Hence $B_l^\dagger$ is given by

$$B_l^\dagger = -\frac{d}{d\rho} + \frac{l+1}{\rho}. \tag{7.172}$$

The product of the two operators is, applied to a function $f(\rho)$,

$$B_l B_l^\dagger f = \left(\frac{d}{d\rho} + \frac{l+1}{\rho}\right)\left(-\frac{d}{d\rho} + \frac{l+1}{\rho}\right) f \tag{7.173}$$

$$= -\frac{d^2}{d\rho^2} f + \frac{d}{d\rho}\left(\frac{l+1}{\rho} f\right) - \frac{l+1}{\rho}\frac{d}{d\rho} f + \frac{(l+1)^2}{\rho^2} f \tag{7.174}$$

$$= -\frac{d^2}{d\rho^2} f + \left(\frac{d}{d\rho}\frac{l+1}{\rho}\right) f + \frac{(l+1)^2}{\rho^2} f \tag{7.175}$$

$$= -\frac{d^2}{d\rho^2} f + \frac{l(l+1)}{\rho^2} f \tag{7.176}$$

and so

$$B_l B_l^\dagger = -\frac{d^2}{d\rho^2} + \frac{l(l+1)}{\rho^2}. \tag{7.177}$$

Similarly one gets

$$B_l^\dagger B_l = -\frac{d^2}{d\rho^2} + \frac{(l+1)(l+2)}{\rho^2} = B_{l+1} B_{l+1}^\dagger. \tag{7.178}$$

With (7.169) follows

$$B_l B_l^\dagger U_l = U_l. \tag{7.179}$$

So, U_l is an eigenfunction of $B_l B_l^\dagger$ with eigenvalue 1. Furthermore one gets

$$B_{l+1} B_{l+1}^\dagger B_l^\dagger U_l = B_l^\dagger B_l B_l^\dagger U_l = B_l^\dagger U_l. \tag{7.180}$$

The first equation follows from (7.178), the second from (7.179). Hence $B_l^\dagger U_l$ is an eigenfunction of $B_{l+1} B_{l+1}^\dagger$ with eigenvalue 1. This implies $B_l^\dagger U_l \sim U_{l+1}$, and since we want to ignore normalization in this section, we simply set

$$U_{l+1} = B_l^\dagger U_l. \tag{7.181}$$

$B_l^\dagger$ turned out to be a raising operator increasing the quantum number l to $l + 1$. B_l is the associated lowering operator which reduces the quantum number from $l + 1$ to l. This is a bit different from our previous cases, as now each value of the running quantum number has its own raising and lowering operator. To $l = 0$ corresponds the raising operator $B_0^\dagger$, to $l = 1$ $B_1^\dagger$ etc. A further peculiarity is that U_0 is not annihilated by its lowering operator $B_{-1} = d/d\rho$. (Instead, it leads to $U_{-1}(\rho) = \cos \rho$ which is, as discussed, not an allowed function. But we already know from the angular momentum algebra that $l = 0$ is the minimal possible value.) Therefore, we cannot obtain U_0 from an equation $B_{-1} U_0 = 0$. Good that we could already infer U_0 from (7.168)!

The other U_l can now be derived from U_0 recursively:

$$U_{l+1} = B_l^\dagger U_l = \left(-\frac{d}{d\rho} + \frac{l+1}{\rho} \right) U_l = \rho^{l+1} \left(-\frac{d}{d\rho} \right) \frac{U_l}{\rho^{l+1}} \tag{7.182}$$

$$\Rightarrow \quad \frac{U_{l+1}}{\rho^{l+2}} = -\frac{1}{\rho} \frac{d}{d\rho} \frac{U_l}{\rho^{l+1}} \tag{7.183}$$

The recursion steps can be combined into the formula:

$$\frac{U_{l+1}}{\rho^{l+2}} = -\frac{1}{\rho} \frac{d}{d\rho} \frac{U_l}{\rho^{l+1}} = \left(-\frac{1}{\rho} \frac{d}{d\rho} \right)^2 \frac{U_{l-1}}{\rho^l} = \cdots = \left(-\frac{1}{\rho} \frac{d}{d\rho} \right)^{l+1} \frac{U_0}{\rho} \tag{7.184}$$

or, with $R_l(\rho) = U_l(\rho)/r = k U_l(\rho)/\rho$,

$$R_l(\rho) = (-\rho^l) \left(\frac{1}{\rho} \frac{d}{d\rho} \right)^l R_0(\rho). \tag{7.185}$$

For $U_0(\rho) = \sin \rho$ we would have $R_0(\rho) = k \sin \rho/\rho$. But since we don't care about normalization, we simply divide R_0 by k and get

$$R_0(\rho) = \frac{\sin \rho}{\rho}. \tag{7.186}$$

With this, equation (7.185) is just the definition of the so-called **spherical Bessel functions** j_l,

$$R_l(\rho) = j_l(\rho) := (-\rho^l) \left(\frac{1}{\rho}\frac{d}{d\rho}\right)^l \frac{\sin\rho}{\rho}. \tag{7.187}$$

This completes the solution for the free particle in spherical coordinates; the common eigenstates of H, $\mathbf{L}^2$ and L_z are

$$\psi_{klm}(r, \theta, \phi) = j_l(kr)Y_{lm}(\theta, \phi). \tag{7.188}$$

In the special case $l = 0$ there is a superposition of an incoming and an outgoing spherical wave:

$$\psi_{k00}(r, \theta, \phi) = j_0(kr)Y_{00}(\theta, \phi) = \frac{1}{\sqrt{4\pi}}\frac{\sin kr}{kr} \tag{7.189}$$

$$= \frac{1}{4i\sqrt{\pi}}\frac{e^{ikr} - e^{-ikr}}{kr} \tag{7.190}$$

How do you recognize that the part with e^{ikr} is outgoing and the one with e^{-ikr} incoming? One can infer that from the motion of a constant phase in the time-dependent solution:

$$\psi_{k00}(r, \theta, \phi, t) = \frac{1}{4i\sqrt{\pi}}\frac{e^{i(kr - E_kt/\hbar)} - e^{i(-kr - E_kt/\hbar)}}{kr} \tag{7.191}$$

If t is increased in the first term, r has to be increased too for the phase $(kr - E_kt/\hbar)$ to remain unchanged. The wave thus moves outward. For the second term it is the other way round, cf. Exercise 5.4.

Remark If a free particle solution is constrained to a region of space which does not include the origin, the other solution $U_0(\rho) = \cos\rho$ has to be taken into account too, as a starting point of the raising procedure. The result are the **spherical Neumann functions**

$$n_l(\rho) := (-\rho^l) \left(\frac{1}{\rho}\frac{d}{d\rho}\right)^l \frac{\cos\rho}{\rho}. \tag{7.192}$$

They play a role for the spherical potential well, for instance, when the solution of the outer region is determined.

How are the ψ_{klm} related to the plane waves $e^{i\mathbf{k}\cdot\mathbf{r}}$? It must be possible to represent a plane wave as a superposition of several ψ_{klm} with the same k. Choosing $\mathbf{k}$ in z-direction (or choosing the z-direction such that it is the direction of $\mathbf{k}$), the exponential is

$$e^{i\mathbf{k}\cdot\mathbf{r}} = e^{ikr\cos\theta}. \tag{7.193}$$

Then there is no ϕ-dependency, and therefore all contributions ψ_{klm} with $m \neq 0$ have to vanish. Using some properties of the spherical Bessel functions and the Legendre polynomials, one can show:

$$e^{ikr\cos\theta} = \sum_{l=0}^{\infty} i^l (2l+1) j_l(kr) P_l(\cos\theta) = \sum_{l=0}^{\infty} i^l \sqrt{4\pi(2l+1)}\, \psi_{kl0} \qquad (7.194)$$

Self-check questions:

1. What is a spherical wave?
2. Under what conditions are only the spherical Bessel functions allowed as solutions for the free particle, and when are the Neumann functions also needed?

7.5 Coulomb Potential and Hydrogen Atom

In QM there are only few problems that can be solved exactly, without the help of approximation schemes. Apart from the free particle and the stepwise constant potentials, there are more or less only the harmonic oscillator and the hydrogen atom. Now we want to analyze the latter. This time we don't have raising and lowering operators at hand; we will have to swallow the pill and solve a second order differential equation.

Regarding the Coulomb potential

$$V(r) = -\frac{\alpha}{r} \qquad (7.195)$$

there are two possible points of view:

- We consider the potential as fixed in space. A single particle moves in this external potential.
- The potential energy is due to the interaction between two particles. We assume that both particles have the same charge e, one of them positive, the other one negative. The interaction potential is then

$$V(r) = -\frac{e^2}{r}, \qquad (7.196)$$

where r is the distance between the particles. If one of the particles (e.g. a nucleus with several protons) has the charge Ze instead, e^2 has to be replaced by Ze^2.

The two-body problem is then reduced to a one-body problem by introducing center of mass and relative coordinates, as well as the reduced mass μ.

We look for bound states and their binding energies, that is, for the negative energy eigenvalues. The most prominent example is the hydrogen atom, a class of bound states of proton and electron. Other examples are positronium (electron and positron), muonium (anti-muon and electron or vice versa), muonic hydrogen (proton and muon) and hydrogen-like ions (higher order nucleus and one electron). All of them are treated in the same way, the difference is only in the masses. In fact, what is exactly solvable is only the **naive hydrogen atom**, the naive positronium etc. "Naive" is that we ignore a number of effects causing a modification of the Coulomb potential, for example relativistic effects and the interaction between spin and angular momentum. The relative strength of the modification depends very much on the system under consideration. For the hydrogen atom, the "naive" calculation is delightfully accurate. We will briefly pick up on the modifications occurring in the real hydrogen atom in Sect. 11.1.3.

We start with the form (7.196) of the potential. The radial equation for $U_{nl}(r)$ reads

$$\left[-\frac{\hbar^2}{2m} \frac{d}{dr^2} - \frac{e^2}{r} + \frac{\hbar^2 l(l+1)}{2mr^2} - E_n \right] U_{nl}(r) = 0. \qquad (7.197)$$

For the hydrogen atom $m = m_e$ is the electron mass if we consider the proton as fixed, or the reduced mass $m = \mu_H$ if we consider the setup as a two-body problem (which makes more sense, as it is difficult to staple a proton into some fixed position in space). The difference is quite small though, since the proton is 2000 times as heavy as the electron, and so

$$\mu_H = \frac{m_e m_p}{m_e + m_p} = m_e \frac{1}{1 + \frac{m_e}{m_p}} \approx \frac{2000}{2001} m_e. \qquad (7.198)$$

We introduce some abbreviations in order to simplify (7.197):

$$a = \frac{\hbar^2}{me^2}, \qquad E_R = \frac{\hbar^2}{2ma^2} = \frac{me^4}{2\hbar^2}, \qquad (7.199)$$

$$\rho = r/a, \qquad \kappa_n = \sqrt{-\frac{E_n}{E_R}}. \qquad (7.200)$$

We assume that E_n is negative, since we look for bound states, and since the Coulomb potential is everywhere negative. The quantity a has the dimension of a length. For the hydrogen atom it is called **Bohr radius** and has the size 0.529×10^{-8} cm. E_R has the dimension of an energy. For the hydrogen atom it is called **Rydberg energy**, with the value 13.6 eV. Here eV (that is: electron Volt) is a unit commonly used in atomic and particle physics: 1 eV is the electric energy of an electron in a voltage of 1 V. The quantities ρ and κ are dimensionless.

Inserting (7.200) into (7.197), we obtain

$$\left[\frac{d}{d\rho^2} + \frac{2}{\rho} - \frac{l(l+1)}{\rho^2} - \kappa_n^2\right] U_{nl}(\rho) = 0. \tag{7.201}$$

We already know the behavior for $\rho \to 0$: it must be $U_{nl} \sim \rho^{l+1}$, cf. (7.149). For $\rho \to \infty$ the two middle terms in the brackets can be neglected, and with

$$\left[\frac{d}{d\rho^2} - \kappa_n^2\right] U_{nl}(\rho) \approx 0 \tag{7.202}$$

we obtain the behavior $U_{nl} \sim e^{-\kappa_n \rho}$ for $\rho \to \infty$. This suggests to try the following ansatz:

$$U_{nl}(\rho) = e^{-\kappa_n \rho} \rho^{l+1} f_{nl}(\rho) \tag{7.203}$$

with a function $f_{nl}(\rho)$ for which we assume that it can be expanded in a power series in all of $\mathbb{R}^+$,

$$f_{nl}(\rho) = \sum_{j=0}^{\infty} a_j^{(nl)} \rho^j. \tag{7.204}$$

Plugging this ansatz into (7.201) one obtains (verify this!)

$$f_{nl}'' + 2 f_{nl}' \left(\frac{l+1}{\rho} - \kappa_n\right) + 2 f_{nl} \frac{1 - \kappa_n(l+1)}{\rho} = 0 \tag{7.205}$$

and

$$\sum_{j=0}^{\infty} a_j^{(nl)} \rho^j \left(\frac{j(j-1)}{\rho^2} - \frac{2\kappa_n j}{\rho} + \frac{2(l+1)j}{\rho^2} + \frac{2(1 - \kappa_n(l+1))}{\rho}\right) = 0. \tag{7.206}$$

This equation has to be fulfilled for each power of ρ separately. If we compare the terms for a fixed power ρ^j, we get after a short calculation the following recursive relation between the coefficients $a_j^{(nl)}$:

$$\frac{a_{j+1}^{(nl)}}{a_j^{(nl)}} = 2 \frac{\kappa_n(j+l+1) - 1}{(j+1)(j+2l+2)} \tag{7.207}$$

From this we can conclude that the power series must terminate at some point, i.e. that f_{nl} is a polynomial: Assume that the power series does *not* terminate. Then one has for $j \gg l$

$$\frac{a_{j+1}^{(nl)}}{a_j^{(nl)}} \approx \frac{2\kappa_n}{j}. \tag{7.208}$$

The exponential function $e^{2\kappa_n\rho}$ has a similar behavior:

$$e^{2\kappa_n\rho} = \sum_{j=0}^{\infty} \beta_j \rho^j, \quad \beta_j = \frac{(2\kappa_n^j)}{j!}, \tag{7.209}$$

$$\frac{\beta_{j+1}}{\beta_j} = \frac{2\kappa_n}{j+1} \approx \frac{2\kappa_n}{j} \tag{7.210}$$

for large values of j. For $\rho \to \infty$, where the higher powers of ρ dominate, f_{nl} would therefore look like $e^{2\kappa_n\rho}$. With (7.203) this would lead to $U_{nl} \sim e^{\kappa_n\rho}$ for $\rho \to \infty$, which is of course not normalizable and thus constitutes a contradiction. The power series therefore has to terminate. Due to (7.207), this is exactly the case if

$$\kappa_n = \frac{1}{j_{\max} + l + 1} \tag{7.211}$$

for some integer $j_{\max}$. The power series then terminates at the power $\rho^{j_{\max}}$, since with $\alpha_{j_{\max}+1}^{(nl)}$ all further coefficients vanish. We define

$$n := j_{\max} + l + 1 \tag{7.212}$$

and see that this is an appropriate energy quantum number, due to

$$\kappa_n = \frac{1}{n} \tag{7.213}$$

and so, with (7.200),

$$E_n = -\frac{E_R}{n^2}. \tag{7.214}$$

The termination condition for the power series gave us the energy eigenvalues for the Coulomb potential! For fixed n, l can assume any value from 0 to $n - 1$. In the latter case $j_{\max} = 0$, and f_{nl} is a constant. For each value of l, there are $2l + 1$ possible values for the magnetic quantum number m, namely any integer from $-l$ to $+l$. The total degeneracy of the energy value E_n is therefore

$$g_n = \sum_{l=0}^{n-1} (2l + 1) = n^2. \tag{7.215}$$

The second equation can be easily verified by induction.

Exercise 7.11
Try it!

For the hydrogen atom (and many other systems) the two possible spin orientations have to be taken into account additionally, leading to a further factor of 2 in g_n.

The formula (7.214) confirms and explains a number of observations regarding the hydrogen spectrum. In atomic transitions where the electron falls from a higher energy E_{n_1} to a lower one E_{n_2}, the energy difference is released in the form of a photon. The energy of the photon is

$$\hbar\omega = -E_R\left(\frac{1}{n_1^2} - \frac{1}{n_2^2}\right). \tag{7.216}$$

On the other hand, photons of this energy are absorbed by hydrogen, raising the electron to a higher energy level. Already before the discovery of quantum mechanics it was found that the frequencies of the absorbed light behave like the differences of inverse squares.

Now we want to finish the calculation of the eigenstates, i.e. of the radial functions $R_{nl}(r)$. The recursion formula (7.207) reads, after inserting (7.213),

$$\frac{a_{j+1}^{(nl)}}{a_j^{(nl)}} = -\frac{2}{n}\frac{n - (j+l+1)}{(j+1)(j+2l+2)}. \tag{7.217}$$

If one applies this relation recursively (that's why it is called recursion formula), one obtains

$$
\begin{aligned}
a_j^{(nl)} &= \left(-\frac{2}{n}\right)^j a_0^{(nl)} \frac{n-(l+j)}{j(2l+j+1)} \times \frac{n-(l+j-1)}{(j-1)(2l+j)} \times \cdots \times \frac{n-(l+1)}{1(2l+2)} \\
&= \left(-\frac{2}{n}\right)^j a_0^{(nl)} \frac{(2l+1)!(n-(l+1))}{j!(j+2l+1)!(n-(j+l+1))!}.
\end{aligned}
$$

For f_{nl} this yields

$$f_{nl}(\rho) = a_0^{(nl)} \sum_{j=0}^{n-(l+1)} (-2\kappa\rho)^j \frac{(2l+1)!(n-(l+1))}{j!(j+2l+1)!(n-(j+l+1))!}. \tag{7.218}$$

This can be somewhat abbreviated with the help of the **associated Laguerre polynomials** L_p^k. These are defined by

$$L_p^k(x) = \sum_{j=0}^{p} (-1)^j x^j \frac{((p+k)!)^2}{j!(k+j)!(p-j)!}. \tag{7.219}$$

For $k = 2l + 1$ and $p = n - l - 1$ this implies

$$L_{n-l-1}^{2l+1}(x) = \sum_{j=0}^{n-l-1} (-1)^j x^j \frac{((n+l)!)^2}{j!(j+2l+1)!(n-(j+l+1))!}. \tag{7.220}$$

The denominator is identical to the one in (7.218). The numerators are different, but in both cases j does not occur in them. The ratio of the two numerators,

$$\frac{(2l+1)!(n-(l+1))}{((n+l)!)^2}, \tag{7.221}$$

can therefore be pulled out of the sum as a constant factor. This yields

$$U_{nl}(\rho) \sim e^{-\kappa_n \rho} \rho^{l+1} L_{n-l-1}^{2l+1}(2\kappa_n \rho). \tag{7.222}$$

To determine R_{nl}, one only needs to "backtransform":

$$\kappa_n = \frac{1}{n}, \quad \rho = \frac{r}{a}, \quad R_{nl} = \frac{U_{nl}}{r}, \tag{7.223}$$

$$R_{nl}(r) \sim e^{-\frac{r}{na}} \left(\frac{r}{a}\right)^l L_{n-l-1}^{2l+1}\left(\frac{2r}{na}\right) \tag{7.224}$$

For completeness, one would have to determine the normalization constants. This is indeed possible in terms of a general expression, but quite cumbersome and requires the knowledge of several properties of the associated Laguerre polynomials. We are therefore going to skip this. The result for the first three radial functions is

$$R_{10}(r) = 2a^{-3/2} e^{-r/a} \tag{7.225}$$

$$R_{20}(r) = (2a)^{-3/2} \left(2 - \frac{r}{a}\right) e^{-r/(2a)} \tag{7.226}$$

$$R_{21}(r) = 3^{-1/2}(2a)^{-3/2} \frac{r}{a} e^{-r/(2a)} \tag{7.227}$$

The length a is the **scale** of the radial function, meaning that r occurs only in the combination r/a. Due to the general relation $\int d^3r |\psi|^2 = 1$ any wave function must be of dimension $(\text{length})^{-3/2}$. This explains the factor of $a^{-3/2}$ appearing in each normalization constant. The absolute value of the binding energy $-E_n$ gets smaller for growing n, namely with n^{-2}, so the binding gets looser. The exponential function $e^{-r/(na)}$ shows that the wave function decays slower for higher values of n. In fact one can show that the expectation value of the distance between the two particles is given by

$$\langle r \rangle_{nl} = \frac{a}{2}[3n^2 - l(l+1)]. \tag{7.228}$$

The average distance thus grows even quadratically with n. In the definition of a (7.199), m and e^2 are in the denominator. From this, one can see how a change of particle species affects the size of the two-particle object.

- If we replace the electron with a muon, which is 200 times heavier, to form with the proton a muonic hydrogen atom, then this is 200 times smaller than the normal hydrogen atom.
- For the He^+ ion, a nucleus with charge $2e$ plus one electron, e^2 has to be replaced by $2e^2$. The He^+ ion has therefore half the size of a hydrogen atom.

Exercise 7.12

The values E_n are the binding energies in the naive hydrogen atom, which result from the stationary Schrödinger equation for the wave function of the relative position r_R. For the total energy of the hydrogen atom, the kinetic energy has to be taken into account, which results from the wave function of the center of gravity position r_{CG}. What are the corresponding eigenvalues and eigenfunctions? Make clear to yourself that the spectrum of the hydrogen atom is therefore continuous, in a sense, already from the ground state energy E_0, and that the electron (as well as the proton) in such a pseudo-eigenstate (which is still a bound state!) is uniformly distributed over the entire space (how is r_e related to r_R and r_{CG}?). In reality, r_S is localized by the interaction of the hydrogen atom with its environment.

Nerd's Corner 7.3

The associated Laguerre polynomials L_p^k are related to the "normal" Laguerre polynomials L_p we've mentioned in Sect. 3.2 in a similar way as the associated Legendre functions are related to the "normal" Legendre polynomials: They are solutions of two differential equations, where the L_p^k as solutions of the second equation are connected to the L_p as solutions to the first equation, namely

$$L_p^k(x) = \frac{d^k}{dx^k} L_{p+k}(x). \tag{7.229}$$

The wave functions $\psi_{nlm}(\mathbf{r})$ have to be orthogonal to each other,

$$\int d^3 x \, \psi_{nlm}^*(x) \psi_{n'l'm'}(x) = \delta_{nn'}\delta_{ll'}\delta_{mm'}. \tag{7.230}$$

If (l, m) is different (l', m'), this is given by the orthogonality of the spherical harmonics Y_{lm}. But if $l = l'$ and $m = m'$, the radial functions have to provide the orthogonality,

$$\int dr\, r^2 \, R^*_{nl}(r) R_{n'l}(r) = \delta_{nn'}. \tag{7.231}$$

For the associated Laguerre polynomials this implies

$$\int d\rho\, \rho^{2l+2} e^{-(\kappa_n + \kappa_{n'})\rho} L^{2l+1}_{n-l-1}(2\kappa_n \rho) L^{2l+1}_{n'-l-1}(2\kappa_{n'}\rho) = 0 \tag{7.232}$$

with $n \neq n'$. This orthogonality relation can indeed be proven, see, for example, Boas (2007).

Self-check questions:

1. How do the binding energies E_n of the naive hydrogen atom depend on the parameter n? What does this imply for the absorption and emission spectrum of atomic hydrogen?
2. What is the size of a hydrogen atom (roughly), and what does the word *size* mean here?

References

R. Shankar, Principles of Quantum Mechanics, 2nd edn. (Springer, 2011). (Excellent textbook. Two extensive chapters on the path integral)

M.J. Boas, (2007) Mathematical Methods in the Physical Sciences, 3rd edn. (Wiley, *(A classic on mathematical methods* (Particularly instructive for QM is the treatment of Legendre, Hermite and Laguerre polynomials), 2007)

Chapter 8
Scattering Theory

Abstract The theory of scattering in QM is introduced, with a focus on the meaning of basic notions and general structure.

Scattering theory plays a big role in the investigation of the structure of matter. While the states of the hydrogen atom can be easily calculated and compared with measurements of the absorption spectrum, this is not so easy for more complicated objects like an atomic nucleus. To investigate the structure of such objects, particles are shot on them, which are then scattered. From the statistical distribution of the directions in which these particles are scattered, one can deduce the structure of the object, in particular its interaction potential $V(\mathbf{r})$. The statistical distribution of scattering directions is specified in terms of a certain function, the **differential scattering cross section** $d\sigma/d\Omega$. The task of the theorist is to compute $d\sigma/d\Omega$ for a given model. The task of the experimentalist is to measure $d\sigma/d\Omega$ in an experiment and to compare the result with the functions provided by the theorist, to verify or falsify models about the structure of the object under study.

We will study here only certain kinds of scattering where the following assumptions hold:

- The target (the object under investigation) is at rest and the recoil it experiences by the scattering is negligible. It is either much heavier than the scattered particles or fixed in its place in some other way. As a consequence, no kinetic energy is transferred from the scattered particle to the target.
- The scattering is **elastic**, which means that also no energy is transferred to the *inner* degrees of freedom of the target (for example by raising one of its constituents to a higher energy level).

These assumptions have two advantages:

- The energy of the scattered particle is not changed by the scattering. We can therefore operate with energy eigenstates.
- The potential caused by the target is not changed by the scattering.

On the other hand, some prominent scattering experiments disappear from our view in this way. In particle accelerators, for example, particles are scattered from each other,

© Springer International Publishing Switzerland 2016

J.-M. Schwindt, *Conceptual Basis of Quantum Mechanics*,

Undergraduate Lecture Notes in Physics, DOI 10.1007/978-3-319-24526-3_8

thereby creating cascades of new particles which are then registered by detectors. Here, none of the mentioned assumptions is fulfilled.

At first, we are going to introduce the notion of the scattering cross section, in its classical and its quantum mechanical variant. Then we describe an approximation scheme which allows us to determine approximately the differential scattering cross section $d\sigma/d\Omega$ for a given potential $V(\mathbf{r})$. This is the **Born approximation**. Finally, for spherically symmetric potentials the basic idea of the **partial wave expansion** will be presented, which is another method to compute $d\sigma/d\Omega$.

8.1 Scattering Cross Section

The notion of the scattering cross section originates from classical scattering theory, where the trajectories of particles are deterministic. A ray of particles is shot on a target. This ray has a certain width, a finite cross section. Depending on where in this cross section a particle of this ray is located, it will hit the target in one or another place or even completely miss it. Imagine the target is a sphere. A particle in the middle of the ray hits the sphere in the middle and rebounds backwards. A particle somewhat further outside touches the sphere at the boundary and is only slightly deflected. A particle even further outside misses the sphere completely and keeps flying in its straight direction. A detector at a distance r to the target registers particles deflected to a certain angular range $\Delta\Omega = \Delta\phi\Delta\theta\sin\theta$ (Fig. 8.1).

The particles flying to $\Delta\Omega$ originate from a certain part $\Delta\sigma$ of the cross section of the original ray. One therefore has a map between an *area element* of the incoming ray approaching the target *in parallel* to an *angular element* of the stream of particles leaving the target *in radial direction*. In the limit $\Delta\Omega \to 0$ we obtain the **differential scattering cross section** $d\sigma/d\Omega$, which has the dimension of an area. The **total scattering cross section** is the integral over the differential one

$$\sigma = \int_{-1}^{1} d\cos\theta \int_{0}^{2\pi} d\phi \, \frac{d\sigma}{d\Omega}(\theta, \phi). \tag{8.1}$$

This is the cross section of the part of the ray which is affected by the target at all. In the case of a solid sphere, this is the cross section of the sphere, $\sigma = \pi R^2$, where R is the radius of the sphere.

In QM, the whole thing looks a bit more complicated. Indeed one could describe the particles as wave packets with a certain width and length, and then calculate how these packages are torn apart when they hit the target. But such a calculation would be very complicated. Instead one takes advantage of the fact that the particles don't transfer any energy to the target and analyzes certain energy eigenstates, i.e. stationary states, where the time dependence consists only of phase rotations. This raises the question how the scattering behavior is encoded in these states. The effect of the target is given in terms of a time-independent potential $V(\mathbf{r})$. So, the states

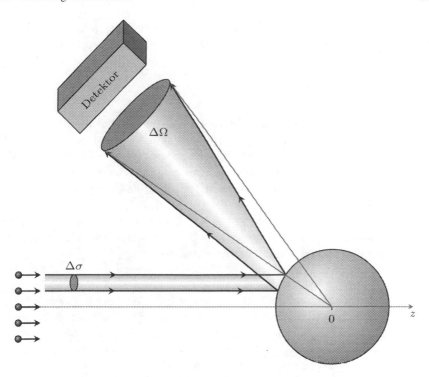

Fig. 8.1 Differential scattering cross section in classical mechanics by means of the example of scattering from a solid sphere

we are looking for are solutions of the stationary Schrödinger equation with this potential. The solutions represent free, not bound states, which are therefore not square-integrable. The particles described by them come from infinity and go back to infinity.

How must such a stationary state look like to represent a scattering? We assume that the potential is restricted to a small region in space, around the origin of the coordinate system we use, or decays sufficiently fast as a function of r that we can regard the particles as free in the limit $r \to \infty$. (This excludes for example the harmonic oscillator which has only bound states, no scattered states.) The wave function then consists of two parts,

$$\psi_k(\mathbf{r}) = \psi_{\text{in}}(\mathbf{r}) + \psi_{\text{sc}}(\mathbf{r}). \tag{8.2}$$

The *incoming* part $\psi_{\text{in}}(\mathbf{r})$ represents the parallel stream of particles before the scattering, and the part of the stream which is unaffected by the target. We assume that this stream moves in z-direction. In the limit $z \to \pm\infty$ one then has (free particle)

$$\psi_{\text{in}}(\mathbf{r}) = e^{ikz}, \tag{8.3}$$

where we again ignore normalization. The *scattered* part $\psi_{sc}(\mathbf{r})$ represents the stream of particles moving radially outwards in all directions due to the scattering,

$$\psi_{sc}(\mathbf{r}) = f(\theta, \phi) \frac{e^{ikr}}{r} \tag{8.4}$$

in the limit $r \to \pm\infty$. The factor r^{-1} takes care of the fact that the density of the scattered particles decreases as r^{-2} (since the size of the spherical shells over which they are distributed increases with r^2). The **scattering amplitude** $f(\theta, \phi)$ represents the dependence of the scattering on the direction and is related to the differential cross section, as we will see. The elasticity of the scattering ensures that the k in ψ_{sc} equals the k in ψ_{in}.

Close to the target, the wave function will look more complicated. But since we assume that the detector which registers the outgoing particles is sufficiently far from the target, we don't need to know more about this part.

There are infinitely many pseudo-eigenstates with energy $E = \hbar^2 k^2/(2m)$. The form of ψ_{in} and ψ_{sc} specifies which one of them is meant. That such a pseudo-state exists and is unique will be shown below. But first we have to clarify how the scattering cross section is defined in QM, and how it can be read off from ψ_{sc}.

The image outlined above is peculiar: The wave function ψ_k describes a *pseudo-state*, and so it cannot be associated with one or a fixed number of particles. It is *stationary*, i.e. there is no time evolution except phase rotations. The incoming and outgoing particles arise only when pseudo-states from a small interval $[k - \epsilon, k + \epsilon]$ are overlapped to form square-integrable wave packets. Such a wave packet ψ_{wp} first runs from $z = -\infty$ towards $r = 0$, where it is torn apart by the potential. A part of it continues moving along the z-axis, towards $z = +\infty$, the remainder moves radially outwards in all directions. If ϵ is chosen small enough, some crucial properties of ψ_k will still hold for the wave packet, in particular the scattering amplitude $f(\theta, \phi)$. For when the uncertainty of momentum $\hbar\epsilon$ is small, the location uncertainty is large, the wave packet thus distributed over a large region. One can then imagine "zooming" into the package, finding that the form of ψ_{wp} deviates only slightly from e^{ikz} for the incoming part, and only slightly from $f(\theta, \phi)\frac{e^{ikr}}{r}$ for the scattered part. Only this connection allows us to associate ψ_k with the scattering of particles.

In order to get the differential cross section, we have to deal with the current densities of ψ_{in} and ψ_{sc}. Remember that the current density $\mathbf{j}$ is defined by

$$\mathbf{j} = \frac{\hbar}{2im}(\psi^* \nabla \psi - \psi \nabla \psi^*). \tag{8.5}$$

In the case of ψ_{in} we are interested in the current in z-direction, in the case of ψ_{sc} the current in r-direction, in both cases far from the origin, where the expressions (8.3) and (8.4) are valid,

$$j_{in} = \mathbf{j}_{in} \cdot \mathbf{e}_z, \quad j_{sc} = \mathbf{j}_{sc} \cdot \mathbf{e}_r. \tag{8.6}$$

Here $\mathbf{e}_z$ and $\mathbf{e}_r$ are the unit vectors in z and r-direction. We investigate the currents $\mathbf{j}_{in}$ associated with ψ_{in} and $\mathbf{j}_{sc}$ associated with ψ_{sc} separately. The current of ψ_k also contains interference terms of the form $\psi_{in}^* \nabla \psi_{sc}$ etc., which are ignored here. This can be justified by the fact that in the end we want to form wave packets again, where incoming and outgoing particle are time-wise and spatially separated and therefore cannot interfere with each other. For large r, j_{in} and j_{sc} are determined as

$$j_{in} = \frac{\hbar k}{m}, \quad j_{sc} = \frac{\hbar k}{mr^2} |f(\theta, \phi)|^2. \tag{8.7}$$

Exercise 8.1

Verify this. For j_{sc} you can use the fact that $\mathbf{e}_r \cdot \nabla = \frac{\partial}{\partial r}$.

We want to justify that

$$\frac{d\sigma}{d\Omega} = \frac{r^2 j_{sc}}{j_{in}} = |f(\theta, \phi)|^2 \tag{8.8}$$

is a useful definition for the differential scattering cross section. Therefore we have to bother ourselves with dimensions, a topic we have ignored so far, which is actually quite disgraceful for physicists. In the definition (8.3), ψ_{in} is dimensionless. For ψ_{sc} to be dimensionless too, $f(\theta, \phi)$ must have the dimension of a length; hence $\frac{d\sigma}{d\Omega}$ has the dimension of an area, which is promising. With the expressions given above $\mathbf{j}$ has the dimension of a velocity: $\hbar k$ is the momentum, and momentum divided by mass is velocity. On the other hand, we in the previous chapter we mentioned that ψ actually is of dimension (length)$^{-3/2}$. To be precise we would have to multiply ψ_k with a constant of this dimension. But for pseudo-states this is not so important, since the normalization condition $\int d^3r |\psi|^2 = 1$, from which the dimension of ψ was derived, is not valid here. When one constructs superpositions,

$$\psi_{wp}(\mathbf{r}) = \int dk\, a(k) \psi_k(\mathbf{r}), \tag{8.9}$$

one can still assign the dimension (length)$^{-3/2}$ to the coefficients $a(k)$ so that the resulting wave packet is correct. Admittedly this is not very nice. It would be better to ensure from the beginning (by multiplication with an appropriate constant) that already ψ_k is of dimension (length)$^{-3/2}$. Then the current density is of dimension (number density times velocity) or equivalently (number per area per time). Then we can understand the quantities in (8.8) in the following way: $d\sigma j_{in}$ is the number of incoming particles passing through the area $d\sigma$ per time unit. Here, the area is always to be understood as orthogonal to the direction of motion, which means in this case: in (xy)-direction. On the other hand, $r^2 d\Omega j_{sc}$ is the number of scattered particles per time unit streaming into the angular direction $d\Omega$, which means, at distance r from the origin: streaming through the area $r^2 d\Omega$. Equation (8.8) therefore has the following interpretation: $d\sigma$ is the area passed by just as many particles as are scattered into the angular direction $d\Omega$. This is the same interpretation as for the classical scattering cross section, except for one difference: In the classical case, where everything is deterministic, this is a mapping from a certain (differential) area $d\sigma$ with fixed boundaries to the angular region $d\Omega$. A particle passing through $d\sigma$ necessarily ends up in $d\Omega$. In QM this is no longer true. Here $d\sigma$ is only the *size* of an area which is passed by the same number of particles as $d\Omega$. But neither one could find specific boundaries for $d\sigma$ ("the area from $x = x_0$ to $x = x_1$ and $y = y_0$ to $y = y_1$"), nor can the particles in $d\sigma$ be uniquely identified with those in $d\Omega$. It is only an equality in numbers.

Self-check questions:

1. How is the differential scattering cross section defined in classical mechanics?
2. What kinds of wave functions are considered in quantum scattering theory?
3. How is the differential scattering cross section defined in QM?

8.2 Born Approximation

After we've defined the differential scattering cross section, we want to see how it can be calculated for a given potential. Don't be too optimistic about this: Exact solutions are not to be expected. We have to resort to approximation schemes.

As always in QM, the starting point of any calculation is the Schrödinger equation, of which one has to choose an appropriate variant. In this case, we take the stationary version,

$$-\frac{\hbar^2}{2m}\Delta\psi_k(\mathbf{r}) + V(\mathbf{r})\psi_k(\mathbf{r}) = E\psi_k(\mathbf{r}), \tag{8.10}$$

insert the energy value

$$E = \frac{\hbar^2 k^2}{2m}, \tag{8.11}$$

bring the energy to the left, the potential to the right hand side and divide by $-\hbar^2/(2m)$:

$$(\Delta + k^2)\psi_k(\mathbf{r}) = \frac{2m}{\hbar^2}V(\mathbf{r})\psi_k(\mathbf{r}) \tag{8.12}$$

Then we define the **Green's function** for the operator $(\Delta + k^2)$ by the relation

$$(\Delta + k^2)G(\mathbf{r}) = \delta^3(\mathbf{r}). \tag{8.13}$$

There are two solutions for this differential equation,

$$G_\pm(\mathbf{r}) = -\frac{1}{4\pi}\frac{e^{\pm ikr}}{r}. \tag{8.14}$$

Now we claim that

$$\psi_k(\mathbf{r}) = \psi_{in}(\mathbf{r}) + \frac{2m}{\hbar^2} \int d^3x' \, G_+(\mathbf{r} - \mathbf{r}') V(\mathbf{r}') \psi_k(\mathbf{r}') \qquad (8.15)$$

solves (8.12). The second term on the right hand side represents ψ_{sc}, the radially outgoing scattered wave. That's why we use G_+, since $G_+(\mathbf{r} - \mathbf{r}')$ for $r \gg r'$ describes radially outgoing waves, in contrast to G_-. It remains to show that (8.15) indeed solves (8.12). We find

$$(\Delta + k^2)\psi_{in} = (\Delta + k^2)e^{ikz} = 0. \qquad (8.16)$$

So, only the second term matters:

$$(\Delta + k^2) \int d^3x' \, G_+(\mathbf{r} - \mathbf{r}') V(\mathbf{r}') \psi_k(\mathbf{r}') \qquad (8.17)$$

$$= \int d^3x' \, (\Delta + k^2) G_+(\mathbf{r} - \mathbf{r}') V(\mathbf{r}') \psi_k(\mathbf{r}') \qquad (8.18)$$

$$= \int d^3x' \, \delta^3(\mathbf{r} - \mathbf{r}') V(\mathbf{r}') \psi_k(\mathbf{r}') \qquad (8.19)$$

$$= V(\mathbf{r}) \psi_k(\mathbf{r}) \qquad (8.20)$$

In the first step we have pushed the operator $(\Delta + k^2)$ into the integral. This is possible because we integrate over $\mathbf{r}'$, but $(\Delta + k^2)$ acts only on the $\mathbf{r}$-dependency of functions. In the second step we have used (8.13), and the fact that the transformation $f(\mathbf{r}) \to f(\mathbf{r} - \mathbf{r}')$ constitutes only a shift by $\mathbf{r}'$, to which derivatives respond with the same shift, similar to

$$\frac{d}{dx} f(x) = g(x) \Rightarrow \frac{d}{dx} f(x - x_0) = g(x - x_0). \qquad (8.21)$$

In the third step we simply used the definition of the delta function. This completes the proof that (8.15) solves (8.12). But of course there is a snag to it: (8.15) is an implicit equation, i.e. ψ_k appears on the left as well as on the right hand side, and the whole thing cannot be simply solved for ψ_k. We can only proceed iteratively, i.e. insert for ψ_k on the right hand side again the same equation, etc. The result is the **Born series**,

$$\psi_k(\mathbf{r}) = \psi_{\text{in}}(\mathbf{r}) + \frac{2m}{\hbar^2} \int d^3r'\, G_+(\mathbf{r} - \mathbf{r}') V(\mathbf{r}') \psi_{\text{in}}(\mathbf{r}') \tag{8.22}$$

$$+ \left(\frac{2m}{\hbar^2}\right)^2 \int d^3r' G_+(\mathbf{r} - \mathbf{r}') V(\mathbf{r}') \int d^3r'' G_+(\mathbf{r}' - \mathbf{r}'') V(\mathbf{r}'') \psi_{\text{in}}(\mathbf{r}'')$$

$$+ \cdots$$

whose convergence behavior we are not going to discuss here (but see related remarks in Chap. 11). The first approximation, terminating the series after the first row, yields the **Born approximation**

$$\psi_k(\mathbf{r}) = \psi_{\text{in}}(\mathbf{r}) + \frac{2m}{\hbar^2} \int d^3r'\, G_+(\mathbf{r} - \mathbf{r}') V(\mathbf{r}') \psi_{\text{in}}(\mathbf{r}') \tag{8.23}$$

$$= \psi_{\text{in}}(\mathbf{r}) - \frac{2m}{\hbar^2} \int d^3r'\, \frac{1}{4\pi} \frac{e^{ik|\mathbf{r}-\mathbf{r}'|}}{|\mathbf{r} - \mathbf{r}'|} V(\mathbf{r}') \psi_{\text{in}}(\mathbf{r}'). \tag{8.24}$$

We are interested in the behavior for large r. Assuming that the values of r', where $V(\mathbf{r}')$ has considerable contributions are much smaller than r (we postulated that V decays sufficiently fast), we use for the expression

$$\frac{e^{ik|\mathbf{r}-\mathbf{r}'|}}{|\mathbf{r} - \mathbf{r}'|} \tag{8.25}$$

an approximation with $r' = |\mathbf{r}'| \ll r = |\mathbf{r}|$. The denominator can be simply replaced by r, but for the numerator we have to be a bit more careful, due to the oscillating exponential function, and take one further term into account; i.e. we neglect terms of order r'^2/r^2, but keep terms of order r'/r:

$$|\mathbf{r} - \mathbf{r}'| = \sqrt{r^2 - 2\mathbf{r} \cdot \mathbf{r}' + r'^2} = r\sqrt{1 - 2\frac{\mathbf{r} \cdot \mathbf{r}'}{r^2} + \frac{r'^2}{r^2}} \tag{8.26}$$

$$\approx r\sqrt{1 - 2\frac{\mathbf{r} \cdot \mathbf{r}'}{r^2}} \approx r\left(1 - \frac{\mathbf{r} \cdot \mathbf{r}'}{r^2}\right) \tag{8.27}$$

$$= r - \mathbf{e}_r \cdot \mathbf{r}' \tag{8.28}$$

Thus (8.25) becomes

$$\frac{e^{ikr}}{r} e^{-ik\mathbf{e}_r \cdot \mathbf{r}'}. \tag{8.29}$$

Inserted into (8.24) this gives

$$\psi_k(\mathbf{r}) = \psi_{\text{in}}(\mathbf{r}) - \frac{m}{2\pi\hbar^2} \frac{e^{ikr}}{r} \int d^3r'\, e^{-ik\mathbf{e}_r \cdot \mathbf{r}'} V(\mathbf{r}') \psi_{\text{in}}(\mathbf{r}'). \tag{8.30}$$

This formula finally justifies our ansatz (8.4) for ψ_{sc}. Comparison with (8.4) yields

$$f(\theta, \phi) = -\frac{m}{2\pi\hbar^2} \int d^3r' \, e^{-ik\mathbf{e}_r \cdot \mathbf{r}'} V(\mathbf{r}')\psi_{in}(\mathbf{r}') \tag{8.31}$$

$$= -\frac{m}{2\pi\hbar^2} \int d^3r' \, e^{-ik\mathbf{e}_r \cdot \mathbf{r}' + ik\mathbf{e}_z \cdot \mathbf{r}} V(\mathbf{r}') \tag{8.32}$$

$$= -\frac{m}{2\pi\hbar^2} \int d^3r' \, e^{-i\mathbf{q} \cdot \mathbf{r}'} V(\mathbf{r}'). \tag{8.33}$$

In the second row we have inserted the known expression for ψ_{in}. In the third row, the vector $\mathbf{q}$ was introduced,

$$\mathbf{q} = k(\mathbf{e}_r - \mathbf{e}_z). \tag{8.34}$$

Here $\hbar\mathbf{q}$ is just the **momentum transfer** that a scattered particle receives during the scattering (at first it has momentum $\hbar k\mathbf{e}_z$, after the scattering $\hbar k\mathbf{e}_r$). Because of

$$\mathbf{q} = k(\sin\theta\cos\phi, \sin\theta\sin\phi, \cos\theta - 1), \tag{8.35}$$

there is a unique relation between $\mathbf{q}$ and (θ, ϕ). We can therefore regard the scattering amplitude f also as a function of $\mathbf{q}$. The final expression for the Born approximation is then:

Born Approximation

$$f(\theta, \phi) = f(\mathbf{q}) = -\frac{m}{2\pi\hbar^2} \int d^3r' \, e^{-i\mathbf{q} \cdot \mathbf{r}'} V(\mathbf{r}') \tag{8.36}$$

So, according to the Born approximation, the scattering amplitude is (up to a constant factor) just the Fourier transformed of the potential!

As an example we want to investigate scattering by a Coulomb potential. Let the scattered particles have charge $Z_1 e$, whereas the target's charge is $Z_2 e$, so that

$$g = Z_1 Z_2 e^2. \tag{8.37}$$

Now we run into a problem: the Coulomb potential is not suitable for the method discussed here, since it does not decay fast enough at infinity and thus the assumptions for ψ_{in} and ψ_{sc} for large r are not fulfilled.

Exercise 8.3

Show that the radial functions $R_{nl}(r)$ of a free state in the Coulomb potential behave for $r \to \infty$ in the following way:

$$R_{nl}(r) \sim \frac{e^{i(kr - \gamma \ln kr)}}{r}, \quad \gamma = \frac{gm}{k\hbar^2} \tag{8.38}$$

Insert the function $U_{nl}(r) = r R_{nl}(r)$ into the corresponding radial equation, neglecting the angular momentum term,

$$\left[-\frac{\hbar^2}{2m} \frac{d^2}{dr^2} + \frac{g}{r} \right] U_{nl}(r) = \frac{\hbar^2 k^2}{2m} U_{nl}(r). \tag{8.39}$$

In your calculation, ignore terms of order r^{-2}.

To "sanitize" scattering by a Coulomb potential, we proceed in the following way: First we replace the Coulomb potential by a **Yukawa potential**,

$$V(r) = g \frac{e^{-\beta r}}{r}, \tag{8.40}$$

which has an additional exponential screening factor, for which we finally take the limit $\beta \to 0$.

According to the Born approximation (8.36),

$$f(\mathbf{q}) = -\frac{mg}{2\pi\hbar^2} \int d^3r \, e^{i\mathbf{q}\cdot\mathbf{r}} \frac{e^{-\beta r}}{r}. \tag{8.41}$$

In order to perform the integration, one can temporarily rotate the z-direction of the coordinate system into the direction of $\mathbf{q}$, $\mathbf{q} = q\mathbf{e}_z$, so that

$$e^{i\mathbf{q}\cdot\mathbf{r}} = e^{iqr\cos\theta}. \tag{8.42}$$

Exercise 8.4

Perform the integration in spherical coordinates. Result:

$$f(\mathbf{q}) = -\frac{2mg}{\hbar^2} \frac{1}{\beta^2 + q^2} \tag{8.43}$$

Now we rotate the z-direction back into its original position, to determine q^2 as a function of θ,

$$q^2 = k^2|\mathbf{e}_r - \mathbf{e}_z|^2 = k^2(\mathbf{e}_r^2 + \mathbf{e}_z^2 - 2\mathbf{e}_r \cdot \mathbf{e}_z) \tag{8.44}$$

$$= 2k^2(1 - \cos\theta) = 4k^2 \sin^2\frac{\theta}{2}. \tag{8.45}$$

Plugging this into (8.43) and taking the limit $\beta \to 0$, this results in the scattering amplitude for the Coulomb potential,

$$f(\theta) = -\frac{mg}{2\hbar^2 k^2 \sin^2\frac{\theta}{2}} = -\frac{g}{4E \sin^2\frac{\theta}{2}}. \tag{8.46}$$

The corresponding differential scattering cross section (with g expanded) is

$$\frac{d\sigma}{d\Omega} = \left(\frac{Z_1 Z_2 e^2}{4E \sin^2\frac{\theta}{2}}\right)^2, \tag{8.47}$$

the **Rutherford scattering formula**.

Self-check questions:

1. How is the Born series derived?
2. What does the Born approximation say about the relation between scattering amplitude and potential?
3. Which special problem arises the case of the Coulomb potential, and how is it solved?

8.3 Partial Wave Expansion

Another method for the treatment of scattering problems is the partial wave expansion. We describe here only the basic idea.

The partial wave expansion works only for spherically symmetric potentials $V(r)$. For the wave function we again assume the form of (8.2)–(8.4). Since the combination of incoming wave ψ_{in} and potential V is symmetric with respect to rotations around the z-axis, the scattering amplitude cannot depend on ϕ, i.e. $f(\theta, \phi) = f(\theta)$.

We use the expansion of $e^{ikr\cos\theta}$ into Legendre polynomials (7.194) to represent ψ_{in}. Furthermore we use that the scattering amplitude (due to the missing ϕ-dependency) can be written as a linear combination of the Y_{l0}, i.e. of the Legendre polynomials,

$$f(\theta) = \sum_{l=0}^{\infty} a_l P_l(\cos\theta). \tag{8.48}$$

This yields for large r

$$\psi_k(\mathbf{r}) = \sum_{l=0}^{\infty} \left[(2l + 1)i^l j_l(kr) + a_l \frac{e^{ikr}}{r} \right] P_l(\cos\theta). \tag{8.49}$$

One can easily show that the definition (7.187) implies

$$j_l(kr) \approx \frac{\sin(kr - l\pi/2)}{kr} \tag{8.50}$$

for $r \to \infty$: For large r the dominating term is the one where the derivative always acts on the sine function; for with each action of a derivative on the r in the denominator, another factor of r enters the denominator. The derivative of $\sin(\rho)$ is $\cos\rho = \sin(\rho - \pi/2)$. The second derivative gives $-\sin\rho = \sin(\rho - 2\pi/2)$ etc., which by induction results in (8.50).

Instead of splitting ψ_k into ψ_{in} and ψ_{sc}, one can also directly expand in Legendre polynomials,

$$\psi_k(\mathbf{r}) = \sum_{l=0}^{\infty} b_l g_l(kr) P_l(\cos\theta). \tag{8.51}$$

Here, g_l are so far undetermined radial functions. One can show that for $r \to \infty$, where the potential vanishes, g_l also has to behave like

$$g_l(kr) \approx \frac{\sin(kr - \lambda_l)}{kr} \tag{8.52}$$

with a so far undetermined phase λ_l. To simplify comparison with j_l later, this is rewritten as $\lambda_l = l\pi/2 - \delta_l$,

$$g_l(kr) \approx \frac{\sin(kr - l\pi/2 + \delta_l)}{kr} \tag{8.53}$$

for $r \to \infty$. Here δ_l is the so-called **scattering phase shift**.

Comparison between (8.49) and (8.51), taking into account the behavior of j_l and g_l, yields after a short calculation the relation

$$f(\theta) = k^{-1} \sum_{l=0}^{\infty} (2l + 1)e^{i\delta_l} \sin\delta_l P_l(\cos\theta), \tag{8.54}$$

which determines the connection between the scattering phase shift and the scattering amplitude. What is gained by this? Not much in the first place. The unknown function $f(\theta)$ is reduced to the infinitely many unknown numbers δ_l. An advantage comes up if ψ_k can be easily brought into the form (8.51). This is for example the case for scattering by a solid sphere ($V(r) = \infty$ for $r < r_0$) or for scattering by a spherical potential well. Then the δ_l can be determined, and from them the scattering amplitude.

In general one can at least say that the δ_l will decrease rapidly for growing l. Indeed, a higher angular momentum for a fixed momentum $\hbar k$ implies a higher distance to the target, hence a lower $\hbar k$ implies a higher distance to the target, hence a lower influence of V, hence a smaller deviation between g_l and the radial function j_l of a free particle, and hence a smaller δ_l.

For unknown $g_l(kr)$, there is an approximation scheme for δ_l: By a tricky comparison between the free solutions ($V = 0$) and ψ_k one can apply an iterative method, similar to the Born series, and gets to first approximation

$$\delta_l \approx \frac{2mk}{\hbar^2} \int_0^\infty dr \, V(r) r^2 j_l^2(kr). \tag{8.55}$$

Here we only want to use (8.54) to derive an interesting result. The total scattering cross section is (with $\int d\Omega = \int_{-1}^1 d\cos\theta \int_0^{2\pi} d\phi$):

$$\sigma = \int d\Omega |f(\theta)|^2$$

$$= k^{-2} \int d\Omega \left| \sum_l (2l+1) e^{i\delta_l} \sin \delta_l P_l(\cos\theta) \right|^2$$

$$= k^{-2} \int d\Omega \left| \sum_l \sqrt{4\pi(2l+1)} e^{i\delta_l} \sin \delta_l Y_{l0}(\theta) \right|^2$$

$$= \frac{4\pi}{k^2} \int d\Omega \sum_{l,l'} \left[\sqrt{2l+1} e^{i\delta_l} \sin \delta_l Y_{l0}(\theta) \right]^* \left[\sqrt{2l'+1} e^{i\delta_{l'}} \sin \delta_{l'} Y_{l'0}(\theta) \right]$$

$$= \frac{4\pi}{k^2} \sum_{l,l'} \sqrt{(2l+1)(2l'+1)} e^{i(\delta_{l'}-\delta_l)} \sin \delta_l \sin \delta_{l'} \int d\Omega Y_{l0}^*(\theta) Y_{l'0}(\theta)$$

$$= \frac{4\pi}{k^2} \sum_{l,l'} \sqrt{(2l+1)(2l'+1)} e^{i(\delta_{l'}-\delta_l)} \sin \delta_l \sin \delta_{l'} \delta_{ll'}$$

$$= \frac{4\pi}{k^2} \sum_l (2l+1) \sin^2 \delta_l$$

$$= \frac{4\pi}{k} \operatorname{Im} f(\theta = 0)$$

In the third row we have used (7.123), in the third last one the orthonormality of Y_{lm}. In the last row once again (8.54) was used, with $\operatorname{Im} e^{i\delta_l} = \sin \delta_l$ and the relation $P_l(1) = 1$ which holds for all Legendre polynomials.

The result is the **Optical Theorem**: The total scattering cross section equals, up to a constant factor, the imaginary part of the scattering amplitude in forward direction.

Self-check questions:

1. Can you explain the idea of the partial wave expansion?
2. What does the Optical Theorem say?

In Part II we have investigated a very specific kind of quantum systems:

- Each system contained only one single quantum object ("particle"); for spherically symmetric potentials it could be a two-body problem, but this could be reduced to a one-body problem.
- The state of the quantum object could be expressed exclusively in terms of a wave function.
- The forces acting on the object could be represented by a time-independent potential $V(\mathbf{r})$.

These restrictions had the consequence that the Schrödinger equation could be formulated in a very specific way, namely

$$\left(-\frac{\hbar^2}{2m}\Delta + V(\mathbf{r})\right)\psi(\mathbf{r}) = E\psi(\mathbf{r}) \tag{8.56}$$

for the stationary and

$$i\hbar\frac{\partial}{\partial t}\psi(\mathbf{r}, t) = \left(-\frac{\hbar^2}{2m}\Delta + V(\mathbf{r})\right)\psi(\mathbf{r}, t) \tag{8.57}$$

for the time-dependent equation. But we have to emphasize that this is a special case, which is only rarely justified in such a pure form!

In this 3rd part we are going to reveal more general situations where the Schrödinger equation no longer has the form assumed above. The most general form of the Schrödinger equation, as it was written down in the postulates of QM,

$$H|\psi\rangle = E|\psi\rangle \tag{8.58}$$

as an eigenvalue equation and

$$i\hbar\frac{\partial}{\partial t}|\psi(t)\rangle = H|\psi(t)\rangle, \tag{8.59}$$

remain unchanged, of course. But $|\psi\rangle$ is not necessarily only a wave function, and H does not necessarily have the form $\left(-\frac{\hbar^2}{2m}\Delta + V(\mathbf{r})\right)$.

As a first example, we are going to meet **spin** again, this time focusing more on its physical context (so far we have introduced it as a purely mathematical example). In order to describe the spin of a quantum object, the wave function must be extended by a spin state.

In a second example we discuss **electromagnetism**, which can no longer (or only in the electrostatic case) be represented by a scalar potential alone, but uses an additional vector potential **A**. The Hamiltonian operator is thereby modified.

Later we investigate **N-particle systems** and learn that one has to distinguish between two fundamentally different types of particles: **fermions** with half-integer and **bosons** with integer spin.

Furthermore we will consider the approximation schemes of **perturbation theory**, for both the stationary and the time-dependent Schrödinger equation. The **path integral** is introduced as an alternative approach to QM. And finally we have a look into the relativistic QM of the electron, which leads us to the **Dirac equation**, a relativistic generalization of the Schrödinger equation.

Part III
Advanced Topics

Chapter 9
Spin

Abstract It is shown how to generalize the concept of angular momentum and how to combine several observables of that type. Some mathematical background is given regarding Lie groups and Lie algebras.

9.1 Spin 1/2 and Spin 1

In Sect. 7.1 on angular momentum we succeeded in deriving the possible eigenvalues of $\mathbf{L}^2$ and L_z only from the commutator relations

$$[L_i, L_j] = i\hbar \sum_{k=1}^{3} \epsilon_{ijk} L_k, \quad [\mathbf{L}^2, L_i] = 0. \tag{9.1}$$

When we were looking for eigenfunctions is position space, we only had to make one restriction. The algebraic relations allowed for the quantum number l integer as well as half-integer values. But during the investigation of possible eigenfunctions it became clear that only integer values could be realized. The reason was that the eigenfunctions $e^{im\phi}$ of L_z required a periodicity of 2π, so that only integer m-values and hence only integer l-values were possible. But the question arises whether the half-integer values may still be realized if we drop the assumption that a state $|\psi\rangle$ can be expressed by a wave function alone.

It turns out that this is connected to a different question: For the motion of a planet around the sun, there are two kinds of angular momentum: the orbital angular momentum $\mathbf{l} = \mathbf{r} \times \mathbf{p}$ of the motion, and the angular momentum $\mathbf{s}$ resulting from the planet's own rotation. The atom was for a long time considered as a kind of quantized mini solar system. For the orbital angular momentum $\mathbf{l}$ we found an equivalence in the quantum numbers l, m, which do not exactly provide an angular momentum vector, but at least eigenvalues for its norm and its z-component; that's all we could bargain out of it. One may ask if the self-rotation $\mathbf{s}$ also has a QM analogy, e.g. for an electron. But please note that these can be only analogies. An electron is not a small spherical body. Neither does it orbit around the nucleus in a literal sense, nor does it rotate around its own axis.

© Springer International Publishing Switzerland 2016   233
J.-M. Schwindt, *Conceptual Basis of Quantum Mechanics*,
Undergraduate Lecture Notes in Physics, DOI 10.1007/978-3-319-24526-3_9

Already in the 1920s there were two indications that such an equivalent of the self-rotation exists. Both were based on the fact that a rotating charge causes a magnetic moment μ which interacts with an external magnetic field. The first indicator was the famous **Stern-Gerlach experiment** where rays of atoms were deflected by a z-dependent external magnetic field $\mathbf{B} = B(z)\mathbf{e}_z$. The deflection is proportional to the z-component of the magnetic moment, μ_z, and the ray is split by its diverse values of μ_z. (In Chap. 10 on electromagnetism we will discuss the details of this interaction.) The experiment shows that even for $l = 0$ there is still a magnetic moment, with two possible z-components, indicating a self-rotation, a **spin** of the electron, with properties expected for the quantum number $l = 1/2$: two possible m-values, $m = \pm 1/2$.

The other indication resulted from the **anomalous Zeeman effect**: In a constant external magnetic field $\mathbf{B} = B\mathbf{e}_z$, the energy levels of the hydrogen atom are further split. Different m-values generate different μ_z-components and thus to different interaction energies with $\mathbf{B}$. But the split does not fit the expectations from the possible orbital angular momenta. Again everything looks like there is a kind of self-rotation of the electron, with two possible values for the component in a given direction.

For generalized angular momenta, the following notation became a widely used convention: One uses as before the letter L for operators representing the orbital angular momentum, and S for the operators of spin. In addition, the letter J (J as in Joker) can represent any kind of angular momentum: spin, orbital angular momentum, or a combination of both, as long as the characteristic commutator relations

$$[J_i, J_j] = i\hbar \sum_{k=1}^{3} \epsilon_{ijk} J_k, \quad [\mathbf{J}^2, J_i] = 0 \qquad (9.2)$$

hold. Depending on which letter is used for the operator, the quantum number l is replaced by j or s, whereas the name of the quantum number for $L_z/S_z/J_z$ always remains m.

In the following, we will several times need the relations (7.61), (7.62) for the raising and lowering operators. Therefore, we repeat them here—with J instead of L:

$$J_+|j, m, \alpha\rangle = \hbar\sqrt{(j - m)(j + m + 1)}|j, m + 1, \alpha\rangle \qquad (9.3)$$

$$J_-|j, m, \alpha\rangle = \hbar\sqrt{(j + m)(j - m + 1)}|j, m - 1, \alpha\rangle \qquad (9.4)$$

Let's figure out which properties a spin with $s = 1/2$ must have (so, now we use s instead of l or j). There are two possible values of m, namely $\pm 1/2$. Let's forget about the wave function for a moment and note that a Hilbert space with enough room for a spin value of $s = 1/2$ needs to be at least two-dimensional. Let's assume for a moment that the Hilbert space is characterized by spin alone, i.e. that it is in fact *only* two-dimensional, $\mathcal{H} = \mathbb{C}^2$. Then we can choose the two eigenstates for $m = \pm 1/2$ as basis states and call them $|z+\rangle$ and $|z-\rangle$. The operator S_z has the eigenvalues $\pm\hbar/2$, so S_z has in this basis the form

$$S_z = \frac{\hbar}{2} \begin{pmatrix} 1 & 0 \\ 0 & -1 \end{pmatrix}. \tag{9.5}$$

For the raising and lowering operators one obtains from (9.3) to (9.4):

$$S_-|z+\rangle = \hbar|z-\rangle, \quad S_-|z-\rangle = 0, \quad S_+|z+\rangle = 0, \quad S_+|z-\rangle = \hbar|z+\rangle \tag{9.6}$$

and so

$$S_+ = \hbar \begin{pmatrix} 0 & 1 \\ 0 & 0 \end{pmatrix}, \quad S_- = \hbar \begin{pmatrix} 0 & 0 \\ 1 & 0 \end{pmatrix}. \tag{9.7}$$

It follows

$$S_x = \frac{1}{2}(S_+ + S_-) = \frac{\hbar}{2} \begin{pmatrix} 0 & 1 \\ 1 & 0 \end{pmatrix}, \quad S_y = \frac{1}{2i}(S_+ - S_-) = \frac{\hbar}{2} \begin{pmatrix} 0 & -i \\ i & 0 \end{pmatrix}. \tag{9.8}$$

We have derived the Pauli matrices!

The operator $S^2 = S_x^2 + S_y^2 + S_z^2$ must for both basis states have the eigenvalue $\hbar^2 s(s+1) = \hbar^2 \frac{3}{4}$, hence

$$S^2 = \frac{3\hbar^2}{4} \mathbf{1}. \tag{9.9}$$

So, we have reproduced another result.

To account for the experimental evidence, the $m = \pm 1/2$ states have to be somehow incorporated into the total Hilbert space of the electron. They cannot be in the part though which is described by the wave function, since that part knows only integer m-values. Instead one has to combine the two-dimensional spin vector space $\mathcal{H}_\chi = \mathbb{C}^2$ with the space $\mathcal{H}_\psi$ of wave functions via a tensor product to a larger Hilbert space of the electron:

$$\mathcal{H} = \mathcal{H}_\psi \otimes \mathcal{H}_\chi \tag{9.10}$$

If $|x\rangle$ is chosen as a pseudo-basis for the space of wave functions $\mathcal{H}_\psi$, then a basis of $\mathcal{H}$ is given by

$$|x\pm\rangle = |x\rangle \otimes |z\perp\rangle. \tag{9.11}$$

The state $|\Psi\rangle$ of the electron is then given by *two* wave functions $(\psi_+(\mathbf{r}), \psi_-(\mathbf{r}))$ (or, in other words, by a wave function with two components). The normalization condition $\langle \Psi | \Psi \rangle = 1$ reads:

$$\int d^3r \, (\psi_+^*(\mathbf{r}), \psi_-^*(\mathbf{r})) \cdot \begin{pmatrix} \psi_+(\mathbf{r}) \\ \psi_-(\mathbf{r}) \end{pmatrix} = \int d^3r \left(|\psi_+(\mathbf{r})|^2 + |\psi_-(\mathbf{r})|^2 \right) = 1 \tag{9.12}$$

In the simplest case one considers only states which can be written as a tensor product of an element of $\mathcal{H}_\psi$ and one of $\mathcal{H}_\chi$,

$$|\Psi\rangle = |\psi\rangle \otimes |\chi\rangle, \quad |\chi\rangle = \begin{pmatrix} \alpha \\ \beta \end{pmatrix} \tag{9.13}$$

$$\psi_+(\mathbf{r}) = \alpha\psi(\mathbf{r}), \quad \psi_-(\mathbf{r}) = \beta\psi(\mathbf{r}). \tag{9.14}$$

All other, "entangled" states can be constructed from them as linear combinations. The word "entangled" is in quotation marks, because it is normally used for a combination of several particles, not for the combination of factors of a single particle. With (9.13) the normalization condition can be formulated separately for $|\psi\rangle$ and $|\chi\rangle$,

$$\int d^3r \, |\psi(\mathbf{r})|^2 = 1, \quad |\alpha|^2 + |\beta|^2 = 1. \tag{9.15}$$

The phenomenon of spin occurs not only for electrons, but for almost all elementary particles. The only exception is the only recently discovered Higgs boson. It is the first and only know elementary particle without spin. For all other particles the following statement holds: the elementary constituents of matter (electron, muon, tauon, quarks, neutrinos) all have $s = 1/2$, they are "spin-$\frac{1}{2}$ particles". The quanta carrying the elementary interactions (photon for electromagnetism, Z- and W-particle for the weak, gluon for the strong nuclear force) have $s = 1$, they are " spin-1 particles". For each particle the spin quantum number d is fixed, it cannot be changed by anything, in contrast to l.

Exercise 9.1
For a spin-1 particle, S_z has three possible eigenvalues $\hbar, 0, -\hbar$. The corresponding spin Hilbert space is three-dimensional. As a basis one can again choose the eigenstates of S_z, $|z+\rangle$, $|z0\rangle$, $|z-\rangle$. Apply the same procedure which was used above for spin $\frac{1}{2}$ to determine the spin operators $S_x, S_y, S_z, \mathbf{S}^2$ in this space. Result:

$$S_x = \frac{\hbar}{\sqrt{2}} \begin{pmatrix} 0 & 1 & 0 \\ 1 & 0 & 1 \\ 0 & 1 & 0 \end{pmatrix}, \quad S_y = \frac{\hbar}{\sqrt{2}} \begin{pmatrix} 0 & -i & 0 \\ i & 0 & -i \\ 0 & i & 0 \end{pmatrix}, \tag{9.16}$$

$$S_z = \hbar \begin{pmatrix} 1 & 0 & 0 \\ 0 & 0 & 0 \\ 0 & 0 & -1 \end{pmatrix}, \quad \mathbf{S}^2 = 2\hbar^2 \mathbf{1} \tag{9.17}$$

9.2 Addition of Angular Momenta

An often occurring problem in quantum physics is the following: Given are two different sets of angular momentum operators J_{1i} and J_{2i} with $i \in \{x, y, z\}$, acting on different Hilbert spaces $\mathcal{H}_1$ and $\mathcal{H}_2$, so that the entire Hilbert space $\mathcal{H}$ of the system under investigation is the tensor product of the two,

$$\mathcal{H} = \mathcal{H}_1 \otimes \mathcal{H}_2. \tag{9.18}$$

As operators in the total Hilbert space, J_{1i} and J_{2i} act in the form $J_{1i}^{(1)} = J_{1i} \otimes \mathbf{1}$ and $J_{2i}^{(2)} = \mathbf{1} \otimes J_{2i}$, respectively. One abbreviates here and writes simply J_{1i} and J_{2i}, but one should always keep in mind that the operator acts only on one factor of the Hilbert space, leaving the other one untouched. Since they act in different spaces, all J_{1i} commute with all J_{2i}. In $\mathcal{H}$, one can therefore simultaneously diagonalize the operators $\mathbf{J}_1^2, J_{1z}, \mathbf{J}_2^2, J_{2z}$ and obtains the eigenstates

$$|j_1, m_1; j_2, m_2\rangle := |j_1 m_1\rangle \otimes |j_2 m_2\rangle. \tag{9.19}$$

For simplicity we assume that the quantum numbers of angular momentum uniquely characterize a state. In the case of orbital angular momentum this means that we take only the space of square-integrable functions of (θ, ϕ) as our Hilbert space, not the space of all square-integrable functions in three dimensions; i.e. we ignore the radial functions for the moment. The square-integrable functions of (θ, ϕ) can be uniquely expanded in spherical harmonics. Now we consider the operators

$$J_i := J_{1i} + J_{2i} := J_{1i} \otimes \mathbf{1} + \mathbf{1} \otimes J_{2i}, \tag{9.20}$$
$$\mathbf{J}^2 := J_x^2 + J_y^2 + J_z^2. \tag{9.21}$$

Exercise 9.2

(a) Show that the J_i are again angular momentum operators, i.e.

$$[J_i, J_j] = i\hbar \sum_k \epsilon_{ijk} J_k, \quad [\mathbf{J}^2, J_i] = 0. \tag{9.22}$$

(b) Show that

$$[\mathbf{J}^2, \mathbf{J}_1^2] = [\mathbf{J}^2, \mathbf{J}_2^2] = [J_i, \mathbf{J}_1^2] = [J_i, \mathbf{J}_2^2] = 0, \tag{9.23}$$

but

$$[\mathbf{J}^2, J_{1z}] \neq 0, \quad [\mathbf{J}^2, J_{2z}] \neq 0. \tag{9.24}$$

Expand all tensor products and use

$$(A \otimes B)(C \otimes D) = (AC) \otimes (BD). \qquad (9.25)$$

The exercise shows that $\mathbf{J}^2$ can be simultaneously diagonalized with J_z, $\mathbf{J}_1^2$ and $\mathbf{J}_2^2$, but not with J_{1z} and J_{2z}. This results therefore in a basis of the Hilbert space $\mathcal{H}$ different from the one is (9.19). The notation for this new basis is

$$|(j_1, j_2)j, m\rangle. \qquad (9.26)$$

It remains to show that the four operators $\mathbf{J}^2, J_z, \mathbf{J}_1^2, \mathbf{J}_2^2$ constitute a *complete* system of commuting observables for $\mathcal{H}$, i.e. that no further operator is necessary to specify the basis states. Also the question arises which values j can take on for given $(j_1 j_2)$. We will come back to that in a moment.

The elements of one basis can be written as linear combinations of the other. The quantum numbers j_1, j_2 appear in both bases. The linear combinations therefore have the following form:

$$|(j_1, j_2)j, m\rangle = \sum_{m_1=-j_1}^{j_1} \sum_{m_2=-j_2}^{j_2} \alpha(j_1, j_2, m_1, m_2, j, m)|j_1, m_1; j_2, m_2\rangle \qquad (9.27)$$

The problem announced in the beginning of this section is: Find the **Clebsch-Gordan coefficients**

$$\alpha(j_1, j_2, m_1, m_2, j, m) = \langle j_1, m_1; j_2, m_2|(j_1, j_2)j, m\rangle. \qquad (9.28)$$

The solution turns out to be quite cumbersome and cannot be written in a closed general form. However there is a generally valid algorithm that can be used to determine the coefficients successively. Before we study this algorithm, we want to give a few examples for physical situations where this problem is relevant:

- When two particles surround a common center, e.g. the two electrons of a Helium atom, it makes sense to ask for the total orbital angular momentum $\mathbf{L} = \mathbf{L}_1 + \mathbf{L}_2$, and how this is composed of the orbital angular momenta of the individual electrons. $\mathcal{H}_1$ and $\mathcal{H}_2$ are then function spaces spanned by the spherical harmonics. The elements of $\mathcal{H}_1$ are functions of the angular coordinates (θ_1, ϕ_1) of the first electron, the elements of $\mathcal{H}_2$ functions of the angular coordinates of the second electron (θ_2, ϕ_2).
- A quark and an antiquark are combined into a meson. The spin of the meson results from the spins of the individual quarks, $\mathbf{S} = \mathbf{S}_1 + \mathbf{S}_2$. $\mathcal{H}_1$ and $\mathcal{H}_2$ are the two-dimensional spin spaces of the first and the second quark, respectively.
- In the Hamiltonian operator of the real (no longer naive!) hydrogen atom, there is a correction term due to the magnetic interaction between the electron's spin and

the magnetic field created by the electron's orbit around the nucleus,

$$H_{mag} \sim \mathbf{L} \cdot \mathbf{S} \tag{9.29}$$

where $\mathbf{L}$ and $\mathbf{S}$ are the orbital angular momentum and the spin of the electron. We will come back to that in the chapter on perturbation theory. For the treatment of this system it is useful to consider eigenstates of $\mathbf{J}^2$ and J_z, where $\mathbf{J} = \mathbf{L} + \mathbf{S}$. The Hilbert spaces $\mathcal{H}_1$ and $\mathcal{H}_2$ are the space of square-integrable functions of (θ, ϕ) and the two-dimensional spin space of the electron, respectively.

If more than two generalized angular momenta are to be combined, the procedure has to be applied successively,

$$\mathbf{J}_A = \mathbf{J}_1 + \mathbf{J}_2, \quad \mathbf{J}_B = \mathbf{J}_A + \mathbf{J}_3, \quad \text{etc.} \tag{9.30}$$

This is the case for hadrons, for example, particles combined out of three quarks (like proton and neutron), or when for two particles spins as well as angular momenta are to be combined.

Now we want to describe the algorithm to determine the Clebsch-Gordan coefficients. The subspace of $\mathcal{H}$ containing only states with given j_1, j_2 is closed under the operations of J_i, $\mathbf{J}^2$, J_{1i}, $\mathbf{J}_1^2$, J_{2i}, $\mathbf{J}_2^2$ since all these operators commute with $\mathbf{J}_1^2$ and $\mathbf{J}_2^2$. This follows from (9.23) and the known properties of angular momentum. (Please clarify that for yourself.) Each subspace of this kind,

$$\mathcal{H}(j_1 j_2) = \mathcal{H}_1(j_1) \otimes \mathcal{H}_2(j_2) \tag{9.31}$$

can thus be considered separately for our Clebsch-Gordan problem. We therefore assume in the following j_1 and j_2 as given and fixed.

At first we want to figure out which values j and m can have. For m it is relatively simple: due to $J_z = J_{1z} + J_{2z}$ and $[J_{1z}, J_{2z}] = 0$ (why is this latter condition necessary?) the eigenvalues are simply added, $m = m_1 + m_2$, implying

$$\alpha(j_1, j_2, m_1, m_2, j, m) = 0 \quad \text{for} \quad m \neq m_1 + m_2. \tag{9.32}$$

This also satisfies our "classical" expectation that the z-components of angular momentum are simply added. The largest possible value of m is thus $j_1 + j_2$, the smallest $-j_1 - j_2$. What about j? Without loss of generality we assume $j_1 \geq j_2$. Classically, the norm of the sum of two vectors is maximally the sum, minimally the difference of the individual norms:

$$|\mathbf{u}| + |\mathbf{v}| \geq |\mathbf{u} + \mathbf{v}| \geq |\mathbf{u}| - |\mathbf{v}|, \tag{9.33}$$

if $|\mathbf{u}| \geq |\mathbf{v}|$. In QM, the closest you can get to such a statement is;

$$j_1 + j_2 \geq j \geq j_1 - j_2 \tag{9.34}$$

We want to show that this is indeed satisfied, by counting the number of independent states for given m. The value $m = j_1 + j_2$ can be formed in exactly one way: from $m_1 = j_1$ and $m_2 = j_2$. Therefore, the maximal value of j must be exactly $j_1 + j_2$. If it were larger, then larger values of m would occur; if it were smaller, $m = j_1 + j_2$ could not occur at all. Hence

$$|(j_1 j_2) j_1 + j_2, j_1 + j_2\rangle = |j_1, j_1; j_2, j_2\rangle \tag{9.35}$$

We have thus found our first Clebsch-Gordan coefficient,

$$\alpha(j_1, j_2, m_1 = j_1, m_2 = j_2, j = j_1 + j_2, m = j_1 + j_2) = 1. \tag{9.36}$$

With the state $|(j_1 j_2) j_1 + j_2, j_1 + j_2\rangle$, due to the action of the lowering operator J_- the entire range of states $|(j_1 j_2) j_1 + j_2, m\rangle$ with $j_1 + j_2 \geq m \geq -j_1 - j_2$ must occur in $\mathcal{H}(j_1 j_2)$, and that exactly once. We emphasize "exactly once", because the question is still open whether the operators $\mathbf{J}^2, J_z, \mathbf{J}_1^2, \mathbf{J}_2^2$ constitute a *complete* system of commuting observables for $\mathcal{H}$. Since we see here and in the following with given values of the four quantum numbers occur exactly once in $\mathcal{H}$, we conclude that no further quantum number is needed to specify a state in $\mathcal{H}$, i.e. that the system of these four operators is indeed complete in $\mathcal{H}$.

The value $m = j_1 + j_2 - 1$ can be obtained in two ways: with $(m_1 = j_1, m_2 = j_2 - 1)$ and $(m_1 = j_1 - 1, m_2 = j_2)$. Therefore, the value $j = j_1 + j_2 - 1$ must occur, such that in the $|(j_1 j_2) jm\rangle$-basis a second state with $m = j_1 + j_2 - 1$ exists, but no second one for $m = j_1 + j_2$. The lowering operator then generates the entire range of states $|(j_1 j_2) j_1 + j_2 - 1, m\rangle$ with $j_1 + j_2 - 1 \geq m \geq -(j_1 + j_2 - 1)$, which therefore exist in $\mathcal{H}(j_1 j_2)$, exactly once.

It goes on just like that the value $m = j_1 + j_2 - 2$ can be obtained in three ways: with $(m_1 = j_1, m_2 = j_2 - 2)$ and $(m_1 = j_1 - 1, m_2 = j_2 - 1)$ and $(m_1 = j_1 - 2, m_2 = j_2)$. Hence the value $j = j_1 + j_2 - 2$ must occur, such that in the $|(j_1 j_2) jm\rangle$-basis a third state occurs having $m = j_1 + j_2 - 2$, but not a third one having $m = j_1 + j_2 - 1$. The lowering operator then generates the entire range of states $|(j_1 j_2) j_1 + j_2 - 2, m\rangle$ with $j_1 + j_2 - 2 \geq m \geq -(j_1 + j_2 - 2)$, which therefore exist in $\mathcal{H}(j_1 j_2)$, exactly once. And so forth.

Things only change when we reach $m = j_1 - j_2 - 1$. Now on the one hand a new possibility to obtain this value appears at the "bottom end" of m_1, namely $(m_1 = j_1 - 2j_2 - 1, m_2 = j_2)$, but on the other hand a possibility on the "upper end" disappears, since $(m_1 = j_1, m_2 = -j_2 - 1)$ is no longer possible: the value of m_2 is lower than allowed. The number of states with $m = j_1 - j_2 - 1$ is therefore the same as the number of states with $m = j_1 - j_2$. This means that while $j = j_1 - j_2$ was still required, $j = j_1 - j_2 - 1$ is not, simply because no further states are needed. It goes on like that, the smaller values of j are also not required. The inequality (9.34) has been verified.

Exercise 9.3

The basis $\{|j_1, m_1; j_2, m_2\rangle\}$ of $\mathcal{H}(j_1 j_2)$ consists of $(2j_1 + 1)(2j_2 + 1)$ states. Verify that this also holds for the basis $\{|(j_1, j_2)jm\rangle\}$, i.e. show

$$\sum_{j=j_1-j_2}^{j_1+j_2} (2j + 1) = (2j_1 + 1)(2j_2 + 1). \tag{9.37}$$

Now we get to the actual algorithm. Starting from (9.35), we apply the operator $J_- = J_{1-} + J_{2-}$ on both sides:

$$J_-|(j_1 j_2)j_1 + j_2, j_1 + j_2\rangle = (j_{1-}|j_1, j_1\rangle) \otimes |j_2, j_2\rangle + |j_1, j_1\rangle \otimes (j_{2-}|j_2, j_2\rangle). \tag{9.38}$$

With (9.4) follows

$$\hbar\sqrt{2(j_1 + j_2)} \, |(j_1 j_2)j_1 + j_2, j_1 + j_2 - 1\rangle \tag{9.39}$$
$$= \hbar\sqrt{2j_1} \, |j_1, j_1 - 1; j_2, j_2\rangle + \hbar\sqrt{2j_2} \, |j_1, j_1; j_2, j_2 - 1\rangle$$

or

$$|(j_1 j_2)j_1 + j_2, j_1 + j_2 - 1\rangle \tag{9.40}$$
$$= \sqrt{\frac{j_1}{j_1 + j_2}} \, |j_1, j_1 - 1; j_2, j_2\rangle + \sqrt{\frac{j_2}{j_1 + j_2}} \, |j_1, j_1; j_2, j_2 - 1\rangle.$$

This gives us the coefficients for $j = j_1 + j_2$, $m = j_1 + j_2 - 1$ and all values of m_1 and m_2, where as expected only $(m_1 = j_1, m_2 = j_2 - 1)$ and $(m_1 = j_1 - 1, m_2 = j_2)$ contribute. By further repeated application of J_-, we then obtain the coefficients for $j = j_1 + j_2$ and all values of m, m_1 and m_2, i.e. for the entire range of states $|(j_1 j_2)j_1 + j_2, m\rangle$ with $j_1 + j_2 \geq m \geq -j_1 - j_2$.

Next we consider $|(j_1 j_2)j_1 + j_2 - 1, j_1 + j_2 - 1\rangle$. Just as $|(j_1 j_2)j_1 + j_2, j_1 + j_2 - 1\rangle$ this can have only two contributions, $(m_1 = j_1, m_2 = j_2 - 1)$ and $(m_1 = j_1 - 1, m_2 = j_2)$:

$$|(j_1 j_2)j_1 + j_2, j_1 + j_2 - 1\rangle = \beta_1 \, |j_1, j_1 - 1; j_2, j_2\rangle + \beta_2 \, |j_1, j_1; j_2, j_2 - 1\rangle$$
$$|(j_1 j_2)j_1 + j_2 - 1, j_1 + j_2 - 1\rangle = \gamma_1 \, |j_1, j_1 - 1; j_2, j_2\rangle + \gamma_2 \, |j_1, j_1; j_2, j_2 - 1\rangle$$

Since the basis states have to be orthogonal,

$$\langle(j_1 j_2)j_1 + j_2, j_1 + j_2 - 1|(j_1 j_2)j_1 + j_2 - 1, j_1 + j_2 - 1\rangle = 0, \tag{9.41}$$

it follows

$$\beta_1^* \gamma_1 + \beta_2^* \gamma_2 = 0. \tag{9.42}$$

Due to the normalization condition

$$|\gamma_1|^2 + |\gamma_2|^2 = 1 \tag{9.43}$$

γ_1 and γ_2 can be derived, up to a phase, from (9.40):

$$\gamma_1 = \sqrt{\frac{j_2}{j_1 + j_2}}, \quad \gamma_2 = -\sqrt{\frac{j_1}{j_1 + j_2}} \tag{9.44}$$

So, we have determined $|(j_1 j_2) j_1 + j_2 - 1, j_1 + j_2 - 1\rangle$. Now we can apply J_- again, to derive $|(j_1 j_2) j_1 + j_2 - 1, m\rangle$ for all possible values of m.

The next state to tackle is $|(j_1 j_2) j_1 + j_2 - 2, j_1 + j_2 - 2\rangle$. This state has now three contributions, $(m_1 = j_1, \ m_2 = j_2 - 2)$ and $(m_1 = j_1 - 1, \ m_2 = j_2 - 1)$ and $(m_1 = j_1 - 2, \ m_2 = j_2)$, i.e. three unknown coefficients. But there are also three conditions to determine them: $|(j_1 j_2) j_1 + j_2 - 2, j_1 + j_2 - 2\rangle$ must be orthogonal to $|(j_1 j_2) j_1 + j_2, j_1 + j_2 - 2\rangle$ and also to $|(j_1 j_2) j_1 + j_2 - 1, j_1 + j_2 - 2\rangle$; furthermore, a normalization condition has to be obeyed. Then again J_- takes us down the range of m-values, then we proceed with the next j-value etc (Fig. 9.1).

In this way all coefficients for a given (j_1, j_2) are determined. As a remark, we could have also started with

$$|(j_1 j_2) j_1 + j_2, -j_1 - j_2\rangle = |j_1, -j_1; j_2, -j_2\rangle \tag{9.45}$$

and proceed from there via J_+.

We are going to try out the algorithm only for the simplest nontrivial example, the combination of two spins with $s_1 = s_2 = 1/2$. We expect three states having $j = 1$ and one having $j = 0$. Equation (9.35) yields

$$|(\tfrac{1}{2}, \tfrac{1}{2}) 1, 1\rangle = |z+; z+\rangle, \tag{9.46}$$

Equation (9.40) yields

$$|(\tfrac{1}{2}, \tfrac{1}{2}) 1, 0\rangle = \frac{1}{\sqrt{2}} (|z+; z-\rangle + |z-; z+\rangle) \tag{9.47}$$

and (9.45) yields

$$|(\tfrac{1}{2}, \tfrac{1}{2}) 1, -1\rangle = |z-; z-\rangle. \tag{9.48}$$

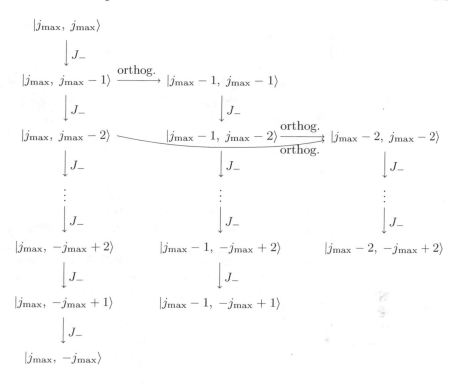

Fig. 9.1 Graphical representation of the algorithm to determine the Clebsch-Gordan coefficients. The states refer to the $|(j_1 j_2)jm\rangle$-basis, where we have dropped the indicators $(j_1 j_2)$ and set $j_{max} = j_1 + j_2$ for better readability

Finally, $|(\frac{1}{2}, \frac{1}{2})0, 0\rangle$ has to be orthogonal to $|(\frac{1}{2}, \frac{1}{2})1, 0\rangle$ and so

$$|(\frac{1}{2}, \frac{1}{2})0, 0\rangle = \frac{1}{\sqrt{2}}(|z+; z-\rangle - |z-; z+\rangle). \qquad (9.49)$$

We conclude that the **triplet** with $j = 1$ is symmetric under the exchange of the two spins, whereas the **singlet** is antisymmetric. We already found this constellation in Sect. 2.10 on tensor products.

Exercise 9.4

Apply the algorithm for the determination of Clebsch-Gordan coefficients to other values of (j_1, j_2), e.g. $(1, \frac{1}{2})$, $(1, 1)$, $(\frac{3}{2}, \frac{1}{2})$ etc. You can find the solutions on the internet:

http://en.wikipedia.org/wiki/Table_of_Clebsch-Gordan_coefficients

Self-check questions:

1. Which two orthonormal bases of which Hilbert spaces are connected by the Clebsch-Gordan coefficients?
2. Which values can the quantum number j take on, given j_1 and j_2?
3. Given two spin-$\frac{1}{2}$ systems, how are the states with $j = 1$ and $j = 0$ combined from the individual spins?

9.3 SO(3) and SU(2)

One of the most beautiful topics a theoretical physicist or mathematician may deal with is that of **Lie groups** and **Lie algebras**. The connection between spin and angular momentum is deeply rooted in this topic. Also in the Standard Model of Particle Physics and in the attempts to construct a "Grand Unified Theory" out of it, Lie groups and Lie algebras play a central role. We want to develop these two notions here briefly and explain what they mean for spin and angular momentum. By the way, Lie was a Norwegian.

From your mathematical studies, you certainly remember the notion of a **group**. This is a certain kind of mathematical structure: a set G, on which an operation

$$G \times G \to G, \quad (g_1, g_2) \to g_3 = g_1 g_2, \tag{9.50}$$

is defined which maps two elements g_1, g_2 to an element g_3. The operation can be written in terms of a multiplication, $g_3 = g_1 g_2$ ("multiplicative group"), or in terms of an addition, $g_3 = g_1 + g_2$ ("additive group"). For abstract groups, where the elements are not numbers, the notation as a multiplication is most commonly used. The operation has to fulfill the following axioms:

- **Associativity**, $g_1(g_2 g_3) = (g_1 g_2) g_3$.
- There is a **neutral element** e, such that for all $g \in G$ one has $ge = eg = g$. For multiplicative groups one denotes e as 1 (for obvious reasons), for additive groups as 0.
- For each $g \in G$ there is an **inverse** g^{-1} (for additive groups: $-g$), such that $gg^{-1} = g^{-1}g = e$ (for additive groups: $g + (-g) = (-g) + g = 0$).

For example, $\mathbb{Z}, \mathbb{Q}, \mathbb{R}, \mathbb{C}$ are groups with respect to addition, $\mathbb{Q}\backslash\{0\}, \mathbb{R}\backslash\{0\}, \mathbb{C}\backslash\{0\}$ are groups with respect to multiplication. (The zero had to be removed, since it has no inverse.) If the operation is also commutative, $g_1 g_2 = g_2 g_1$ (or $g_1 + g_2 = g_2 + g_1$), the group is calles **abelian**. For all the groups of numbers mentioned above this is the case.

A group is called **Lie group**, if it is also a manifold. We want to avoid introducing the notion of a manifold (for that, you better attend a course on differential geometry or general relativity) we simplify this somewhat and simply say: it is continuous (in contrast to the finite groups or the discrete $\mathbb{Z}$) and has "sufficiently nice" properties regarding continuity and differentiability, details of which we are not concerned with.

What is crucial is the notion of a representation of a Lie group and the notion of the Lie algebra associated with a Lie group. We first write down the definitions of these in one piece, and will then explain them by means of examples.

A **representation** of a group G is a combination (G, V) of G with a some vector space V, on which the elements of G act as linear operators. More precisely: A representation $T(G)$ is a map from G into the space of automorphisms $\mathrm{Hom}(V, V)$ of a vector space V:

$$T : G \rightarrow \mathrm{Hom}(V, V), \tag{9.51}$$

with the requirement that

$$T(g_1 g_2) = T(g_1) T(g_2). \tag{9.52}$$

For finite-dimensional vector spaces, $T(g)$ can be thought of as matrices. The left hand side of (9.52) involves a group multiplication, the right hand side a matrix multiplication. The trivial representation is given by $T(g) = \mathbf{1}$ for all $g \in G$.

A **Lie algebra** L is a vector space whose elements are linear operators and which is closed under the commutation of these operators, i.e.

$$A, B \in L \Rightarrow [A, B] \in L. \tag{9.53}$$

(One can formulate this in a more abstract way, but we don't want to.) Any $A \in L$ can be written as a linear combination if n basis operators E_i (we consider only finite-dimensional Lie algebras),

$$A = \sum_{i=1}^{n} \alpha_i E_i. \tag{9.54}$$

If L is a vector space over $\mathbb{R}$ (or $\mathbb{C}$), with the coefficients α_i thus real (or complex), L is called a **real (or complex) Lie algebra**. In physics, we are mostly interested in real Lie algebras. We did not specify on which vector space V the operators act; for L is supposed to be determined by the specification of the commutator relations of its elements, $[A, B]$ for $A, B \in L$. If all commutators vanish, L is called **abelian**. On which vector space V the elements of L act, is again the matter of a **representation**. This is defined for L in the same way as for G, as a map which associates with each "abstract" operator A a linear map $T(A) \in \mathrm{Hom}(V, V)$. Warning: If L is a real Lie algebra, this does not imply that V is a real vector space or that the components of the matrix $T(A)$ are real!

One can show that for each Lie group G there is a Lie algebra L_G with the following property: Every element $g \in G$ which is connected to the unit element e by a continuous path (i.e. which "belongs to the connected component of unity"), can be written in the form $g = \exp(A)$ for some $A \in L_G$. In turn, also $\exp(A) \in G$ for any $A \in L_G$. The basis operators E_i of L_G are then also called the **generators** of G.

This all sounds quite complicated, but it will appear much less wild when you see it in action for a few examples.

1st example: $G = \mathbb{R} \backslash \{0\}$

The nonzero real numbers form a Lie group regarding multiplication. Within this group, the set of positive real numbers $\mathbb{R}^+$ is the connected component of unity. For each positive real number is connected to the number 1 by a continuous path; the negative numbers however are not, because there is an—infinitely small—gap between the positive and the negative numbers, due to the missing zero. Any positive number g can be written in the form $g = \exp \lambda$ with $\lambda \in \mathbb{R}$, but the negative numbers cannot. $\mathbb{R}$ is a one-dimensional real vector space, the Lie algebra L_G of the Lie group G. The number 1, as the basis vector of the one-dimensional vector space $L_G = \mathbb{R}$, is the generator of G. The multiplication of real numbers commutes, hence G and thus also L_G is abelian.

If we had chosen the complex numbers $G = \mathbb{C} \backslash \{0\}$ instead, the negative numbers would have been also contained in the connected component of unity; for then one can draw a path in the complex plane avoiding the zero. Indeed, for complex numbers also the negative numbers can be written in exponential form, $z = \exp(a + i\pi) = -\exp(a)$.

Back to the real G. A representation can be chosen as any vector space $V = \mathbb{R}^n$ or $V = \mathbb{C}^n$, defining $T(g) = g\mathbf{1}$ for $g \in G$, i.e. g acts as g times the unit matrix on V; vectors in V stretched by $T(g)$ by a factor of g. The same holds for the representations of the Lie algebra, $T(\lambda) = \lambda \mathbf{1}$ for $\lambda \in L_G$.

Already $\mathbb{R}^+$ by itself is a Lie group, since multiplication and division are closed in it. The Lie algebra of $\mathbb{R}^+$ is the same as the one for $\mathbb{R} \backslash \{0\}$, namely $\mathbb{R}$ (for the group, only the other connected component is missing, the one containing the negative numbers).

2nd example: $U(1), SO(2), O(2)$

Real Lie groups (that is, Lie groups associated with real Lie algebras) are in general more complicated than complex ones. This is because one gets a different Lie group if some generators E_n are replaced by iE_n (for complex Lie algebras, iE_n is automatically contained if E_n is, there is no difference). For example, the set of multiples of i,

$$L = u(1) = \{\lambda i | \lambda \in \mathbb{R}\}, \tag{9.55}$$

is a one-dimensional *real* vector space (the vector space is real, although it contains imaginary numbers!), since it consists of real multiples of the basis vector i. As a vector space, L is isomorphic to $\mathbb{R}$; as a Lie algebra too, since all commutators vanish in any case. This is in general no longer true for higher-dimensional Lie algebras; there it makes a difference for the commutators if an operator is multiplied by i.

The Lie group of $u(1)$ is $G = U(1)$, the set of complex numbers of norm 1,

$$U(1) = \{\exp(i\lambda) | \lambda \in \mathbb{R}\}. \tag{9.56}$$

$U(1)$ is *not* isomorphic to $\mathbb{R}^+$ or $\mathbb{R}\backslash\{0\}$, since in contrast to those $U(1)$ is periodic. All $\lambda + 2\pi n$ generate the same group element. The simplest nontrivial representation of $U(1)$ acts on the vector space $\mathbb{C}^1$, simply by multiplication, $T(g) = g$. This means, $\exp(i\lambda)$ acts on $\mathbb{C}$ as a multiplication operator, multiplying any complex number by $\exp(i\lambda)$, thus causing a phase shift by the angle λ.

Another interesting representation is given by the rotation matrices in two dimensions, $V = \mathbb{R}^2$ and

$$T(\exp(i\lambda)) = \begin{pmatrix} \cos\lambda & \sin\lambda \\ -\sin\lambda & \cos\lambda \end{pmatrix}. \tag{9.57}$$

The corresponding representation of the Lie algebra is

$$T(i\lambda) = \lambda \begin{pmatrix} 0 & 1 \\ -1 & 0 \end{pmatrix}. \tag{9.58}$$

Exercise 9.5
Verify that $T(g_1 g_2) = T(g_1)T(g_2)$ for $g_1, g_2 \in U(1)$ and $T(\exp(i\lambda)) = \exp(T(i\lambda))$ for $i\lambda \in u(1)$.

If $U(1)$ appears in this form, the group is also called $SO(2)$. But both are one and the same, $U(1) \cong SO(2)$. If in addition to the rotations also reflections are included, one gets the Lie group $O(2)$. This is the group of all linear maps in two dimensions which leave the length of all vectors unchanged. $SO(2)$ is the subgroup of those elements of $O(2)$ having determinant 1 (reflections have determinant -1). $SO(2)$ is related to $O(2)$ like $\mathbb{R}^+$ to $\mathbb{R}\backslash\{0\}$. $O(2)$ contains two connected components: the one of the rotation matrices, which have determinant 1, and the one of reflection matrices, which have determinant -1.

An even more interesting representation of $U(1)$ or $SO(2)$ arises when the group acts on an infinite-dimensional space of functions, for example on the space $\mathrm{PR}_\phi(\mathbb{R}^2)$ of all functions $\mathbb{R}^2 \to \mathbb{C}$ that can, in polar coordinates, at each point (r, ϕ) expanded into a Taylor series w.r.t. ϕ,

$$f(r, \phi + \lambda) = \sum_{n=0}^{\infty} \frac{\lambda^n}{n!} \frac{\partial^n}{\partial\phi^n} f(r, \phi). \tag{9.59}$$

Then we define

$$T(i\lambda) = \lambda \frac{\partial}{\partial\phi}, \tag{9.60}$$

implying

$$T(\exp(i\lambda)) = \exp(T(i\lambda)) = \sum_{n=0}^{\infty} \frac{\lambda^n}{n!} \frac{\partial^n}{\partial\phi^n} \tag{9.61}$$

which is just the operator that rotates a function f on $\mathbb{R}^2$ by the angle λ,

$$T(\exp(i\lambda))f = \tilde{f}, \quad \tilde{f}(r, \phi) = f(r, \phi + \lambda). \tag{9.62}$$

The generator $E_1 = i \in u(1)$ of the Lie group appears here in the form of $T(E_1) = \frac{\partial}{\partial\phi}$. This is—up to a factor $-i\hbar$—the angular momentum operator in two dimensions. One therefore says that the angular momentum operator generates rotations.

3rd example: translations

The translations (shifts of the coordinate origin) form an *additive* group, where the neutral element is the shift by the null vector. In contrast to rotations, translations of $\mathbb{R}^2$ *cannot* be regarded as the representation of a Lie group, since these by definition act via matrix *multiplication*. This changes when we operate on function space again.

The simplest two-dimensional real Lie algebra is the abelian one, spanned by two commuting basis operators $E_1, E_2, [E_1, E_2] = 0$. We construct a representation on the infinite-dimensional vector space $V = \mathrm{PR}_{x,y}(\mathbb{R}^2)$ of function $\mathbb{R}^2 \to \mathbb{C}$ which can be expanded into a Taylor series at any point (x, y), this time regarding cartesian coordinates x and y. The representation of the Lia algebra is now defined via

$$T(E_1) = \frac{\partial}{\partial x}, \quad T(E_2) = \frac{\partial}{\partial y}. \tag{9.63}$$

This is justified, since the two partial derivatives commute,

$$\frac{\partial}{\partial x}\frac{\partial}{\partial y} = \frac{\partial}{\partial y}\frac{\partial}{\partial x}. \tag{9.64}$$

The associated Lie group is obtained by exponentiating the Lie algebra representation (i.e. we specify the Lie group via a representation):

$$T(\exp(\alpha E_1 + \beta E_2)) = \exp(T(\alpha E_1 + \beta E_2)) \tag{9.65}$$

$$= \exp\left(\alpha\frac{\partial}{\partial x} + \beta\frac{\partial}{\partial y}\right) \tag{9.66}$$

$$= \exp\left(\alpha\frac{\partial}{\partial x}\right)\exp\left(\beta\frac{\partial}{\partial y}\right) \tag{9.67}$$

$$= \sum_{k=0}^{\infty}\frac{\lambda^k}{k!}\frac{\partial^k}{\partial x^k}f(x, y)\sum_{l=0}^{\infty}\frac{\lambda^l}{l!}\frac{\partial^l}{\partial\phi^l}f(x, y) \tag{9.68}$$

Similar to the Taylor expansion w.r.t. ϕ, this operator causes a translation (shift) of all functions on $\mathbb{R}^2$ by (α, β),

$$T(\exp(\alpha E_1 + \beta E_2))f = \tilde{f}, \quad \tilde{f}(x, y) = f(x + \alpha, y + \beta). \tag{9.69}$$

The generators E_1, E_2 of the Lie group appear here in terms of partial derivatives w.r.t. x and y. These are—up to a factor $-i\hbar$—the momentum operators P_x, P_y in two dimensions. One therefore says that the momentum operators generate translations. Of course, the same holds in three or more dimensions.

4th example: angular momentum and spin

So far we were dealing with abelian Lie groups and algebras. Our first non-abelian example is already the one which is responsible for our main concern: the connection between rotations, angular momentum and spin.

We define the Lie algebra $so(3)$ as the three-dimensional Lie algebra whose three basis operators obey the following commutation relations:

$$[E_i, E_j] = -\epsilon_{ijk} E_k, \tag{9.70}$$

or

$$[E_1, E_2] = -E_3, \quad [E_2, E_3] = -E_1, \quad [E_3, E_1] = -E_2. \tag{9.71}$$

You can easily verify that a possible representation on $V = \mathbb{R}^3$ is given by

$$T_3(E_1) = \begin{pmatrix} 0 & 0 & 0 \\ 0 & 0 & 1 \\ 0 & -1 & 0 \end{pmatrix}, \quad T_3(E_2) = \begin{pmatrix} 0 & 0 & -1 \\ 0 & 0 & 0 \\ 1 & 0 & 0 \end{pmatrix}, \quad T_3(E_3) = \begin{pmatrix} 0 & 1 & 0 \\ -1 & 0 & 0 \\ 0 & 0 & 0 \end{pmatrix}. \tag{9.72}$$

(We denote it T_3, in order to distinguish this representation from others to come.) Analogous to the connection between (9.58) and (9.57) in our $SO(2)$ example, one has

$$\exp(\lambda T_3(E_1)) = R_x(\lambda), \quad \exp(\lambda T_3(E_2)) = R_y(\lambda), \quad \exp(\lambda T_3(E_3)) = R_z(\lambda), \tag{9.73}$$

where R_x, R_y, R_z are the rotation matrices from (2.140). A Lie group associated with the Lie algebra $so(3)$ is thus the group $SO(3)$ of rotations in three dimensions. Any rotation matrix R can be written in the form

$$R = \exp[\alpha_1 T_3(E_1) + \alpha_2 T_3(E_2) + \alpha_3 T_3(E_3)] \tag{9.74}$$

Here, the α_i are components of the vector in $\mathbb{R}^3$ whose direction is the axis of rotation and whose norm is the angle of rotation. Note that due to the missing commutativity the exponential functions in general cannot be separated,

$$\exp[\alpha_1 T_3(E_1) + \alpha_2 T_3(E_2) + \alpha_3 T_3(E_3)]$$
$$\neq \exp[\alpha_1 T_3(E_1)] \exp[\alpha_2 T_3(E_2)] \exp[\alpha_3 T_3(E_3)]. \tag{9.75}$$

Similar to $SO(2)$ we can again find a representation in an infinite-dimensional function space (this time for functions $\mathbb{R}^3 \rightarrow \mathbb{C}$):

$$T_\infty(E_1) = y\frac{\partial}{\partial z} - z\frac{\partial}{\partial y} \tag{9.76}$$

$$T_\infty(E_2) = z\frac{\partial}{\partial x} - x\frac{\partial}{\partial z} \tag{9.77}$$

$$T_\infty(E_3) = x\frac{\partial}{\partial y} - y\frac{\partial}{\partial x} \tag{9.78}$$

You can easily verify that the commutation relations (9.71) hold, and recognize the connection to the components L_k of angular momentum:

$$L_k = -i\hbar T_\infty(E_k) \tag{9.79}$$

Just as in two dimensions, one has $T_\infty(E_3) = \frac{\partial}{\partial\phi}$, and $\exp(\alpha T_\infty(E_3))$ generates rotations of functions about the z-axis and by the angle α. Since the situation is rotation symmetric—the choice of the z-direction was arbitrary—we conclude that the representation of a generic group element

$$T_\infty(g) = \exp[\alpha_1 T_\infty(E_1) + \alpha_2 T_\infty(E_2) + \alpha_3 T_\infty(E_3)] \tag{9.80}$$

causes a rotation of the functions about the axis $\boldsymbol{\alpha}/|\boldsymbol{\alpha}|$ and by the angle $|\boldsymbol{\alpha}|$. The statement that angular momentum generates rotations therefore still holds in three dimensions.

As a remark, note that $SO(3)$ can also be extended by adding reflections with determinant -1, resulting in a Lie group $O(3)$ which again has two connected components, again defined by the sign of the determinant.

Another representation of the Lie algebra $so(3)$ is given by $V = \mathbb{C}^2$ and

$$T_2(E_1) = \frac{1}{2}\begin{pmatrix} 0 & i \\ i & 0 \end{pmatrix}, \quad T_2(E_2) = \frac{1}{2}\begin{pmatrix} 0 & 1 \\ -1 & 0 \end{pmatrix}, \quad T_2(E_3) = \frac{1}{2}\begin{pmatrix} i & 0 \\ 0 & -i \end{pmatrix}. \tag{9.81}$$

(Verify the commutation relations!) One immediately recognizes: Up to a factor $i/2$, these are the Pauli matrices,

$$T_2(E_k) = \frac{i}{2}\sigma_k. \tag{9.82}$$

Exponentiation of the generators in this representation yields (verify!):

$$\exp[\alpha T_2(E_1)] = \begin{pmatrix} \cos\frac{\alpha}{2} & i\sin\frac{\alpha}{2} \\ i\sin\frac{\alpha}{2} & \cos\frac{\alpha}{2} \end{pmatrix} \tag{9.83}$$

$$\exp[\alpha T_2(E_2)] = \begin{pmatrix} \cos\frac{\alpha}{2} & \sin\frac{\alpha}{2} \\ -\sin\frac{\alpha}{2} & \cos\frac{\alpha}{2} \end{pmatrix} \tag{9.84}$$

$$\exp[\alpha T_2(E_3)] = \begin{pmatrix} e^{i\frac{\alpha}{2}} & 0 \\ 0 & e^{-i\frac{\alpha}{2}} \end{pmatrix} \tag{9.85}$$

It is crucial that α on the right hand side always occurs with a factor $1/2$. As a consequence, the such generated Lie group is periodic with a period of 4π, in contrast to $SO(3)$ which is periodic with a period of 2π. The Lie algebra is the same in both cases, but the Lie group is different! It is called $SU(2)$. It is the group of unitary transformations, i.e. of those linear maps which leave the norm of each complex vector invariant.

Already in Sect. 2.6, we investigated such unitary maps. If you compare, you will find that $\exp[2\alpha T_2(E_i)]$ corresponds to $U_i^\dagger(\alpha) = \exp(i\alpha\sigma_i)$. (For a passive transformation, the components of a spin state transform via $U_i^\dagger$, not via U_i!) We found that such a unitary "rotation" is associated with a rotation of the spin vector by the *doubled* angle. Now we see the deeper reason: The same Lie algebra element $\sum \alpha_i E_i$—turned into an element of the Lie group by exponentiation—causes a rotation in both representations (or its unitary equivalent) but in the case of $SU(2)$ is only half the size. In other words: the angle of the corresponding rotation in the real dimensions is twice as large.

It is experimentally confirmed that spin generates a magnetic moment μ, a vector in our three-dimensional space. If the coordinate system in this three-dimensional space is rotated via the matrix $\exp(\sum \alpha_i T_3(E_i))$, then one has to apply a corresponding transformation $\exp(\sum \alpha_i T_2(E_i))$ to the spin state, so that the relation between spin and magnetic moment is unchanged. Hence the vector β occurring in (2.139) is indeed a vector in the three-dimensional state we live in (and not just in an abstract space of "internal" degrees of freedom, like the "color" of the quarks). Due to the periodicity of 4π one could say that an electron needs to do a 360 *twice* for its spin vector to be the same again. But this does not hold for the spin *state*. Here the difference between vector and state becomes important. A rotation by 2π turns the spin vector $|\chi\rangle$ into $-|\chi\rangle$. But these two vectors represent the same spin state. Everything is consistent: After doing a 360 with the electron, it looks just as before.

Symmetries and conserved quantities

We have seen that certain operators generate certain transformations in function space. For example, L_z generates rotations about the z-axis,

$$e^{\frac{i}{\hbar}\alpha L_z}f(r, \theta\phi) = f(r, \theta, \phi + \alpha), \tag{9.86}$$

and P_x generates translations in x-direction,

$$e^{\frac{i}{\hbar}\alpha P_x}f(x, y, z) = f(x + \alpha, y, z). \tag{9.87}$$

This is also true for wave functions, of course, and yields a relation between symmetries and conserved quantities, which you may be familiar with from classical physics, under the name of Noether's theorem. In QM, a quantity is conserved if the associated hermitian operator A commutes with the Hamiltonian operator, $[A, H] = 0$. For two states $|\psi_1\rangle$, $|\psi_2\rangle$ this implies:

$$\langle \psi_1 e^{-\frac{i}{\hbar}\alpha A}|H|e^{\frac{i}{\hbar}\alpha A}\psi_2\rangle = \langle \psi_1|H|e^{-\frac{i}{\hbar}\alpha A}e^{\frac{i}{\hbar}\alpha A}\psi_2\rangle = \langle \psi_1|H|\psi_2\rangle \qquad (9.88)$$

This means: If all states are transformed via $\exp(i\alpha A/\hbar)$,

$$|\psi\rangle \rightarrow |\tilde{\psi}\rangle = e^{\frac{i}{\hbar}\alpha A}|\psi\rangle, \qquad (9.89)$$

all matrix elements of H remain unchanged,

$$\langle \tilde{\psi}_1|H|\tilde{\psi}_2\rangle = \langle \psi_1|H|\psi_2\rangle. \qquad (9.90)$$

Such an invariance of H under certain transformations is called a **symmetry** of the system. We have therefore shown that a conserved quantity always generates a symmetry. The inverse statement is also true, for if (9.88) holds for all $|\psi_1\rangle$, $|\psi_2\rangle$, it necessarily follows that $[A, H] = 0$.

So, if H commutes with L_z, then the system has to be invariant under the transformations generated by L_z, i.e. under rotations about the z-axis. If H commutes with all momenta P_i (which is only the case for a constant potential), then the system is invariant under translations (this is just what the constancy of the potential means).

Chapter 10
Electromagnetic Interaction

Abstract It is shown how electromagnetism enters QM. This provides some insights into a deeper meaning of gauge invariance. The Aharanov-Bohm effect demonstrates how the vector potential is much more real in QM than in classical electromagnetism.

The interaction of a particle having charge q with a time-independent electric field $\mathbf{E}$ can still be covered by a scalar potential

$$V(\mathbf{r}) = q\phi(\mathbf{r}), \quad \mathbf{E}(\mathbf{r}) = -\nabla\phi(\mathbf{r}). \tag{10.1}$$

But as soon as time dependency or magnetic fields come into play, a vector potential $\mathbf{A}(\mathbf{r})$ is required, for which we haven't found a place yet in the Schrödinger equation.

In the first section, we will derive the classical Hamiltonian function of a charged particle in a space- and time-dependent electromagnetic field, and from that the Hamiltonian operator. The vector potential appears there as a modification of the momentum operator,

$$\mathbf{P} \rightarrow \mathbf{P} - \frac{q}{c}\mathbf{A}. \tag{10.2}$$

As a next step, we will show what the gauge symmetry of electromagnetism means for QM. This opens up an intriguing new view of electromagnetism. It looks like nature necessarily had to create an electromagnetic field in order to provide the Schrödinger equation with a certain symmetry. The invariance under *local* phase shifts of the wave function.

Thirdly, we will investigate the magnetic moment created by the angular momentum and spin of a charged particle. This magnetic moment plays a big role for the interaction between atoms, within atoms, and between atoms and external magnetic fields.

Finally, we will study some effects and experiments which confirm these considerations:

- the Zeeman effect, which generates a split of the energy levels of the hydrogen atom when exposed to an external magnetic fields;

© Springer International Publishing Switzerland 2016   253
J.-M. Schwindt, *Conceptual Basis of Quantum Mechanics*,
Undergraduate Lecture Notes in Physics, DOI 10.1007/978-3-319-24526-3_10

- the Stern-Gerlach experiment, which demonstrates the existence of the electron's spin by splitting a ray of atoms in a spatially varying magnetic field;
- the Aharanov-Bohm effect, which describes a shift of the interference pattern in the double slit experiment, if an inductor is placed behind the wall with the double slit.

Particularly fascinating is the Aharanov-Bohm effect, since the particles are influenced by the magnetic field inside the inductor, although they don't even get in contact with it on their way from the slit to the screen.

Note that the electromagnetic field occurs here as a *classical* field. The theory which incorporates the quantum nature of both the charged particles and the electromagnetic field is called quantum electrodynamics (**QED**), which is however beyond the scope of this book.

10.1 Hamiltonian Operator

We repeat the classical physics of a particle with q in an electromagnetic field. Due to the electric force, the particle experiences an acceleration parallel to the **E**-field, and due to the Lorentz force an acceleration orthogonal to the **B**-field and to its own direction of motion,

$$m\ddot{\mathbf{r}} = q \left(\mathbf{E}(\mathbf{r}, t) + \frac{\dot{\mathbf{r}}}{c} \times \mathbf{B}(\mathbf{r}, t) \right). \tag{10.3}$$

The fields are expressed via a scalar potential ϕ and a vector potential **A**,

$$\mathbf{B} = \nabla \times \mathbf{A}, \quad \mathbf{E} = -\nabla\phi - \frac{1}{c} \frac{\partial \mathbf{A}}{\partial t}. \tag{10.4}$$

Maybe you have already seen the Hamiltonian function h of the particle in a course on electrodynamics. It reads

$$h = \frac{1}{2m} \left(\mathbf{p} - \frac{q}{c} \mathbf{A} \right)^2 + q\phi. \tag{10.5}$$

Exercise 10.1
Show that Hamilton's equations associated with the Hamiltonian function (10.5) lead to the equation of motion (10.3). Proceed component-wise, i.e. calculate $\ddot{x}_i$. After insertion of (10.4) into (10.3) and expansion of the double vector product $\dot{\mathbf{r}} \times (\nabla \times \mathbf{A})$ with the rule

$$\mathbf{a} \times (\mathbf{b} \times \mathbf{c}) = \mathbf{b}(\mathbf{a} \cdot \mathbf{c}) - \mathbf{c}(\mathbf{a} \cdot \mathbf{b}), \tag{10.6}$$

(10.3) has the form

$$m\ddot{x}_i = q\left(-\frac{1}{c}\frac{\partial}{\partial t}A_i - \frac{\partial}{\partial x_i}\phi + \frac{1}{c}\sum_{j=1}^{3}(\dot{x}_j\frac{\partial}{\partial x_i}A_j - \dot{x}_j\frac{\partial}{\partial x_j}A_i)\right). \quad (10.7)$$

The same should be obtained from

$$\dot{p}_i = -\frac{\partial}{\partial x_i}h, \quad \dot{x}_i = \frac{\partial}{\partial p_i}h \quad (10.8)$$

At one point, you have to be careful: When you differentiate the last equation with respect to time, in order to get $\ddot{x}_i$, then this is a *total* time derivative, i.e. a time derivative along the direction of motion of the particle, and this means, according to the chain rule,

$$\dot{A}_i = \frac{\partial}{\partial t}A_i + \sum_{j=1}^{3}\dot{x}_j\frac{\partial}{\partial x_j}A_i. \quad (10.9)$$

The associated Hamiltonian operator is obtained by replacing $\mathbf{p} \to -i\hbar\nabla$:

$$H = \frac{1}{2m}\left(-i\hbar\nabla - \frac{q}{c}\mathbf{A}\right)^2 + q\phi \quad (10.10)$$

Exercise 10.2
Show that the continuity equation for the probability density is modified in the presence of electromagnetic fields;

$$\frac{\partial}{\partial t}(\psi^*\psi) + \nabla\cdot\mathbf{j} = 0 \quad (10.11)$$

now holds for

$$\mathbf{j} = \frac{\hbar}{2im}\left(\psi^*\nabla\psi - \psi\nabla\psi^*\right) - \frac{q}{mc}\mathbf{A}\psi^*\psi. \quad (10.12)$$

10.2 Gauge Invariance

The Maxwell equations are invariant under the **gauge transformation**

$$\phi \rightarrow \phi' = \phi - \frac{1}{c}\frac{\partial \chi}{\partial t}, \quad \mathbf{A} \rightarrow \mathbf{A}' + \nabla \chi \tag{10.13}$$

with an arbitrary scalar field $\chi(\mathbf{r}, t)$. (At this point we assume that you already know this transformation from a course on electrodynamics. If not, you can easily verify, that $\mathbf{E}$ and $\mathbf{B}$ are not modified by (10.13).) Now, what happens to the Schrödinger equation

$$i\hbar\frac{\partial\psi(\mathbf{r}, t)}{\partial t} = \left[\frac{1}{2m}\left(-i\hbar\nabla - \frac{q}{c}\mathbf{A}(\mathbf{r}, t)\right)^2 + q\phi(\mathbf{r}, t)\right]\psi(\mathbf{r}, t) \tag{10.14}$$

under such gauge transformations?

We want to take a different starting point for our discussion, a starting point which is seemingly unrelated to gauge transformations. We consider the Schrödinger equation without an electromagnetic field,

$$i\hbar\frac{\partial\psi(\mathbf{r}, t)}{\partial t} = -\frac{\hbar^2}{2m}\Delta\psi(\mathbf{r}, t). \tag{10.15}$$

We know that this equation is invariant under *global* phase shifts. That means, if $\psi(\mathbf{r}, t)$ is a solution, then due to the linearity of the Schrödinger equation

$$\psi'(\mathbf{r}, t) = e^{i\lambda}\psi(\mathbf{r}, t) \tag{10.16}$$

is also a solution. But what if we make λ space- and time-dependent, i.e. introduce a *locally varyiing* phase shift,

$$\psi'(\mathbf{r}, t) = e^{i\lambda(\mathbf{r},t)}\psi(\mathbf{r}, t)? \tag{10.17}$$

This is no longer a solution of the Schrödinger equation, for now we get an additional term with a time derivative of λ on the left hand side, and several terms with spatial derivatives of λ on the right hand side. These terms don't cancel each other, as long as there are no restrictions for the function $\lambda(\mathbf{r}, t)$.

But what if we absolutely want the ψ' in (10.17) to solve the Schrödinger equation, i.e. require (10.17) to be a symmetry transformation for arbitrary functions $\lambda(\mathbf{r}, t)$? This can only be achieved by adding additional objects to the Schrödinger equation and transforming them simultaneously with ψ, and that in such a way that the resulting additional term exactly cancel the additional terms from the transformation of ψ.

It turns out that electromagnetism is the solution to exactly this problem. One introduces a scalar function $\phi(\mathbf{r}, t)$ and a vectorial function $\mathbf{A}(\mathbf{r}, t)$ into the Schrödinger equation as in (10.14). Then one requires that a transformation (10.17) of the wave function is always accompanied by a transformation (10.13), defining

$$\chi(\mathbf{r}, t) := \frac{\hbar c}{q}\lambda(\mathbf{r}, t). \tag{10.18}$$

Under these combined transformations, the Schrödinger equation is invariant.

Exercise 10.3
Insert the transformed quantities into the Schrödinger equation (10.14) and show that the extra term from the transformation of ϕ cancels the change in $\frac{\partial}{\partial t}\psi$, whereas the extra term from the transformation of $\mathbf{A}$ cancels the change in $\nabla\psi$.

It remains to provide classical field equations which describe the behavior of the new functions ϕ and $\mathbf{A}$; and which are also required to be invariant under the transformation. It turns out that the Maxwell equations are the simplest possibility. Now we get the impression **that the entire electromagnetism is nothing but a consequence of a symmetry requirement for the Schrödinger equation**, namely the invariance under local phase transformations (10.17).

Nerd's Corner 10.1
This point of view has proven extraordinarily fruitful in particle physics. It turns out that the weak and strong nuclear forces can be similarly derived from certain symmetry requirements for wave functions (or quantum fields). As a consequence, the notion of symmetry has been established as the most fundamental of all principles in physics. Since then (i.e. since the 1970s) hordes of theoretical physicists have been occupied looking for the symmetry which unifies all three interactions relevant for particle physics (electromagnetism, weak and strong nuclear force) into one Grand Unified Theory. In fact, these symmetries are expressed in terms of Lie groups.

As you know from the section on Lie groups, (10.17) represents the action of an element of $U(1)$ on ψ, separately at each point. This means that ψ is now not addressed by $U(1)$ in the infinite-dimensional representation (which would lead to a rotation in space); instead, at each point in space, ψ is taken as a number, as a vector in the one-dimensional vector space $\mathbb{C}^1$, and each point in space is associated with an individual element $e^{i\lambda} \in U(1)$, acting on

ψ in the one-dimensional representation. In the theories of particle physics (**Yang-Mills theories**), $U(1)$ is replaced by other Lie groups. Since these no longer have nontrivial one-dimensional representations, ψ at each point in space needs to be an element of a higher-dimensional vector space, i.e. the state $|\psi\rangle$ belongs to a Hilbert space which is a tensor product of the space of wave functions and this "internal" space, similar to our treatment of spin.

The symmetry group of the strong nuclear force, for instance, is $SU(3)$. The wave functions of the quarks belong to a three-dimensional representation of $SU(3)$ and have therefore three components, which in the flowery language of particle physics are associated with three colors.

Exercise 10.4

One can add to $\mathbf{A}$ and ϕ constants $\mathbf{C}$ and η, respectively, without changing $\mathbf{E}$ or $\mathbf{B}$,

$$\mathbf{A}'(\mathbf{r}, t) = \mathbf{A}(\mathbf{r}, t) + \mathbf{C}, \qquad \phi'(\mathbf{r}, t) = \phi(\mathbf{r}, t) + \eta. \tag{10.19}$$

Hence, this defines a certain gauge transformation. Show that this gauge transformation transforms the wave function in the following way:

$$\psi'(\mathbf{r}, t) = e^{i(\alpha\eta t + \beta\mathbf{C}\cdot\mathbf{r} + \gamma)}\psi(\mathbf{r}, t) \tag{10.20}$$

where α and β are constants to be determined, and an arbitrary constant γ. The first term describes an artificial shift of energy by a constant value, due to the electric potential, whereas the second one describes an artificial shift of momentum by a constant vector, due to the vector potential. The third term is a global phase shift. Verify that the difference between ψ and ψ' has no physical relevance. In particular, the momentum operator $\mathbf{P} = -i\hbar\nabla$ is now associated with the canonical momentum $\mathbf{p}$, which is *not* $\mathbf{p} = m\dot{\mathbf{r}}$, but, according to Hamilton's equation,

$$m\dot{\mathbf{r}} = \mathbf{p} - \frac{q}{c}\mathbf{A}. \tag{10.21}$$

Show that the expectation value of velocity $\langle\dot{\mathbf{r}}\rangle$ is unchanged.

Self-check questions:

1. How is the Schrödinger equation modified by the presence of a vector potential $\mathbf{A}$?
2. What are gauge transformations of ϕ and $\mathbf{A}$?
3. How does ψ need to transform for the Schrödinger equation to be invariant?

10.3 Magnetic Moment

In order to obtain the magnetic moment of a charged quantum object, we do our calculations in **Coulomb gauge**, $\nabla \cdot \mathbf{A} = 0$. (At this point we assume that you are already familiar with this gauge from a course on electrodynamics. If this is not the case: For a given $\mathbf{A}$, one can always find a gauge transformation which annihilates the divergence of $\mathbf{A}$, and the result are ϕ and $\mathbf{A}$ in Coulomb gauge.) In Coulomb gauge, the Hamiltonian operator (10.10) can be rewritten as

$$H = -\frac{\hbar^2}{2m}\Delta + \frac{iq\hbar}{mc}\mathbf{A} \cdot \nabla + \frac{q^2}{2mc^2}\mathbf{A}^2 + q\phi. \tag{10.22}$$

For when multiplying out the brackets in (10.10), an expression proportional to $(\nabla \cdot \mathbf{A} + \mathbf{A} \cdot \nabla)$ arises, which has to be understood in terms of its effect on a wave function ψ:

$$(\nabla \cdot \mathbf{A} + \mathbf{A} \cdot \nabla)\psi := \sum_{i=1}^{3}\left(\frac{\partial}{\partial x_i}(A_i\psi) + A_i\frac{\partial}{\partial x_i}\psi\right) \tag{10.23}$$

$$= \sum_{i=1}^{3}\left[\left(\frac{\partial}{\partial x_i}A_i\right)\psi + A_i\frac{\partial}{\partial x_i}\psi + A_i\frac{\partial}{\partial x_i}\psi\right] \tag{10.24}$$

$$= \sum_{i=1}^{3}0 + 2A_i\frac{\partial}{\partial x_i}\psi \tag{10.25}$$

$$= 2(\mathbf{A} \cdot \nabla)\psi \tag{10.26}$$

In the third row, the definition of Coulomb gauge was used. Now let $\mathbf{B}$ be a space- and time-independent magnetic field. Then the vector potential in Coulomb gauge is, up to physically irrelevant additive constants (see Exercise 10.4) which can be freely chosen,

$$\mathbf{A}(\mathbf{r}) = -\frac{1}{2}\mathbf{r} \times \mathbf{B}, \qquad \phi = 0 \tag{10.27}$$

Exercise 10.5
Verify this by showing:

$$\nabla \times (\mathbf{r} \times \mathbf{B}) = -2\mathbf{B}, \qquad \nabla \cdot (\mathbf{r} \times \mathbf{B}) = 0. \tag{10.28}$$

For realistic magnetic fields as they can be produced in laboratories, $\mathbf{A}$ is small compared to the momentum expectation value of a particle,

$$\frac{q}{c}|\mathbf{A}| \ll |\langle \mathbf{P} \rangle|, \tag{10.29}$$

and so the expression $\mathbf{A}^2$ in (10.22) can be neglected in general. Plugging (10.27) into (10.22), we obtain

$$H = -\frac{\hbar^2}{2m}\Delta + H_{\text{mag}} \tag{10.30}$$

where

$$H_{\text{mag}} = \frac{iq\hbar}{mc}\mathbf{A} \cdot \nabla = -\frac{iq\hbar}{2mc}(\mathbf{r} \times \mathbf{B}) \cdot \nabla \tag{10.31}$$

$$= \frac{iq\hbar}{2mc}\mathbf{B}(\mathbf{r} \times \nabla) = -\frac{q}{2mc}\mathbf{B} \cdot \mathbf{L}. \tag{10.32}$$

We have used the vector identity

$$(\mathbf{u} \times \mathbf{v}) \cdot \mathbf{w} = -\mathbf{v} \cdot (\mathbf{u} \times \mathbf{w}) \tag{10.33}$$

and identified in $-i\hbar\mathbf{r} \times \nabla$ the angular momentum operator $\mathbf{L}$. In classical electrodynamics, the interaction energy of a charge and current distribution is to first approximation given by

$$W_{\text{mag}} = -\boldsymbol{\mu} \cdot \mathbf{B}, \tag{10.34}$$

where $\boldsymbol{\mu}$ is the **magnetic moment**. Comparison of (10.32) and (10.34) shows that the operator $\mathbf{M}$ of the magnetic moment can be defined as

$$\mathbf{M} = \frac{q}{2mc}\mathbf{L}, \tag{10.35}$$

such that

$$H_{\text{mag}} = -\mathbf{B} \cdot \mathbf{M}. \tag{10.36}$$

So, now we know the relation between angular momentum and magnetic moment. If the magnetic field is in z-direction, only the z-component is relevant, and the corresponding eigenvalues are

$$\mu_z = \frac{q}{2mc}\hbar m. \tag{10.37}$$

Careful! Here m occurs as a mass in the denominator and as a magnetic quantum number in the numerator. (Finally we know why it is called magnetic.) The best would be (if possible) to provide masses with an index for the particle species, e.g. m_e for an electron, to avoid confusion.

The magnetic moment generated by spin cannot be derived so easily, unfortunately. We can assume that the associated operator $\mathbf{M}_s$ has the form

$$\mathbf{M}_s = \frac{qg}{2mc}\mathbf{S}, \tag{10.38}$$

where g is a dimensionless constant called the g-factor (how imaginative!) which expresses how much the **gyromagnetic ratio**

$$\gamma := \frac{qg}{2mc} \tag{10.39}$$

deviates from the corresponding factor in (10.35). Only in the relativistic Dirac theory, g can be determined for an electron, with the result $g_e = 2$, see Exercise 14.2. In QED, small corrections to this value arise, which were verified by experiments to an extreme precision.

For the proton one finds experimentally $g_p \approx 5.6$. For the neutron, one would not expect a magnetic moment at all, since it is electrically neutral. In experiments, however, one finds a non-vanishing μ also for the neutron, of a similar size as for the electron. This was one of the first indications that the neutron cannot be elementary, but has a substructure of "smaller" charged particles, the quarks. The same holds for the proton. If it were elementary, Dirac theory would imply that $g_p = 2$.

Self-check questions:

1. What is the relation between the orbital angular momentum and the magnetic moment of a particle?
2. Was is the g-factor? What is its value for an electron?

10.4 Effects

In the last three sections we have developed the theory of a charged quantum object in an electromagnetic field. Now we want to see some applications of this theory.

10.4.1 Normal Zeeman Effect

The Zeeman effect is concerned with a hydrogen atom exposed to a constant magnetic field $\mathbf{B} = B\mathbf{e}_z$. How are the energy eigenvalues affected by that? The Hamiltonian operator is

$$H = H_0 + H_{\text{mag}}, \quad H_0 = -\frac{\hbar^2}{2\mu_H}\Delta - \frac{e^2}{r}, \quad H_{\text{mag}} = \frac{eB}{2m_ec}L_z. \tag{10.40}$$

We already know the eigenstate $|nlm\rangle$ of H_0. Since H_{mag} contains only the operator L_z, the $|nlm\rangle$ are also eigenstates of H_{mag}. The eigenvalues can be read off directly:

$$H_{\mathrm{mag}}|nlm\rangle = \frac{eB}{2m_ec}\hbar m|nlm\rangle \tag{10.41}$$

Altogether we have

$$(H_0 + H_{\mathrm{mag}})|nlm\rangle = (E_n^{(0)} + E_m^{(\mathrm{mag})})|nlm\rangle, \tag{10.42}$$

$$E_n^{(0)} = \frac{m_e e^4}{2\hbar^2 n^2}, \quad E_m^{(\mathrm{mag})} = \frac{eB}{2m_ec}\hbar m. \tag{10.43}$$

The n^2-fold degenerate energy level E_n is therefore split by the magnetic field into $2n-1$ levels. For the highest l-value for given n is $l = n-1$, and this corresponds to $2l+1 = 2n-1$ possible values m (Fig. 10.1). The energies for fixed n, m are still degenerate, since an m-value in general occurs for several l-values, namely for all l with $m \le l < n$.

This calculation does not yet take spin into account. In the experimental results, the splitting goes further, in a way that was not understood before the dynamics of spin was known. That's why the part of the effect presented here is called **normal Zeeman-Effekt** whereas the part caused by the spin is called **anomalous Zeeman effect**.

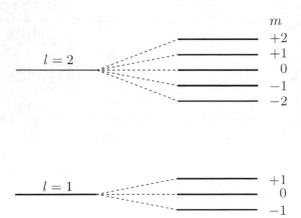

Fig. 10.1 Normal Zeeman effect: splitting of energy levels for the hydrogen atom

10.4.2 Stern-Gerlach Experiment

The Stern-Gerlach experiment also uses a magnetic field in z-direction, but its strength depends on z, $\mathbf{B} = B(z)\mathbf{e}_z$. A ray of atoms is sent in x-direction through this magnetic field. A screen registers the point of impact of each atom after it leaves the field (Fig. 10.2). For simplicity we assume that the ray consists of hydrogen atoms in the ground state $(n, l, m) = (0, 0, 0)$. (in the original experiment, silver atoms were used). The magnetic moment of the atoms is thus only caused by the spin (the proton is shielded by the electron shell),

$$\mathbf{M}_s = -\frac{eg_e}{2m_e c}\mathbf{S}, \quad g_e \approx 2. \tag{10.44}$$

The interaction energy is given by the following operator:

$$H_{\text{mag}} = -M_z B(z), \quad M_z := (\mathbf{M}_s)_z = -\frac{eg_e}{2m_e c}S_z \tag{10.45}$$

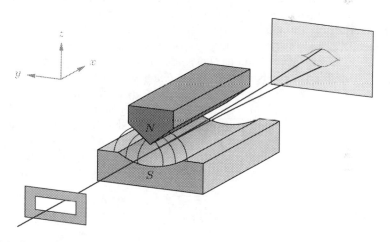

Fig. 10.2 Stern-Gerlach experiment: split of a ray of atoms in an inhomogeneous magnetic field

For an eigenstate of spin, H_{mag} describes a z-dependent potential,

$$V(z) = \pm \frac{e g_e}{2 m_e c} \frac{\hbar}{2} B(z). \qquad (10.46)$$

This causes a force $F_z = -\partial V(z)/\partial z$. Since the expectation values of the atom positions $\langle \mathbf{r} \rangle$ behave just as in classical mechanics, the atoms are accelerated upwards or downwards, depending on the direction of the spin. The ray is thus split into two parts. The experiment represents a *measurement* of the spin by the magnetic field. Each of the electrons is "forced" to choose one of the two eigenstates of S_z.

Due to $l = 0$, a split caused by orbital angular momentum was impossible, so that the existence of a spin with two possible values in a given direction could be inferred from the the measurement result.

10.4.3 Aharanov-Bohm Effect

We consider a double slit experiment with electrons, where in addition a coil is placed behind the wall with the slits. The coil contains a magnetic field as soon as a voltage is applied. The magnetic flux is orthogonal to the plane in which the electrons move. We are going to show that the interference pattern on the screen is shifted when the magnetic field is turned on. And this also holds for points on the screen where none of the two possible paths of the electron got in touch with the magnetic field, i.e. both paths miss the coil (cf. Fig. 10.3). The reason for this is that electrons, in contrast to classical objects, directly interact with the vector potential $\mathbf{A}$, see (10.14). The vector potential does not vanish outside the coil, other than the magnetic field $\mathbf{B}$. In classical electrodynamics, $\mathbf{A}$ is only an auxiliary quantity. In QM, $\mathbf{A}$ becomes "real" in a sense.

At first we want to remind ourselves of Stokes' theorem and its application to the vector potential. In general,

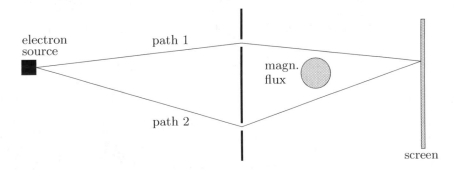

Fig. 10.3 Experimental setup for the Aharanov-Bohm effect

$$\oint_{\partial S} d\mathbf{r} \cdot \mathbf{u}(\mathbf{r}) = \int_S d\mathbf{S} \cdot (\nabla \times \mathbf{u})(\mathbf{r}). \tag{10.47}$$

Here S is a surface, $d\mathbf{S}$ the corresponding area element, ∂S the boundary of the surface and $d\mathbf{r}$ the line element of the boundary. With the definition of the **magnetic flux**,

$$\Phi_m = \int_S d\mathbf{S} \cdot \mathbf{B} \tag{10.48}$$

this implies for $\mathbf{A}$

$$\oint_{\partial S} d\mathbf{r} \cdot \mathbf{A} = \Phi_m. \tag{10.49}$$

In our setup, $\mathbf{B}$ vanishes outside the coil. By a theorem from vector analysis, in this region exists a function $\chi(\mathbf{r})$ such that $\mathbf{A}(\mathbf{r})$ in this region can be written as

$$\mathbf{A} = \nabla \chi. \tag{10.50}$$

Now, can't we just perform a gauge transformation with the function $-\chi$ to make $\mathbf{A}$ vanish? If the magnetic field is turned off, yes. When it is turned on, no. For $\mathbf{A}$ is supposed to be continuous and even differentiable at the boundary of the coil. If $\mathbf{A}$ inside the coil can't be chosen such that it vanishes at the boundary, it can't be annihilated from outside. Equation (10.49) holds for any gauge, and therefore $\mathbf{A}$ cannot vanish at the boundary of the coil. We therefore set $\mathbf{A} = 0$ while the magnetic field is turned off, and $\mathbf{A} = \nabla \chi$ while the magnetic field is turned on.

Now χ can be expressed through $\mathbf{A}$ by taking the line integral of (10.50), starting from an arbitrary point $\mathbf{r}_0$,

$$\chi(\mathbf{r}) = \int_{\mathbf{r}_0}^{\mathbf{r}} d\mathbf{r}' \cdot \mathbf{A}(\mathbf{r}'). \tag{10.51}$$

One can easily show, using the analogy with gauge transformations, that if $\psi(\mathbf{r}, t)$ solves the Schrödinger equation for $\mathbf{A} = 0$ (magnetic field turned off), then

$$\psi'(\mathbf{r}, t) = \exp\left(-i\frac{e}{\hbar c}\chi(\mathbf{r})\right)\psi(\mathbf{r}, t) \tag{10.52}$$

solves the Schrödinger equation for $\mathbf{A} = \nabla \chi$. We choose for $\mathbf{r}_0$ the position of the electron source.

At first the magnetic field is turned off. The double slit causes behind the wall a split of the wave function in two parts: ψ_1 is the part that results from the wave passing through the first slit, ψ_2 the one from the second slit. The observed interference pattern is a consequence of the probabilities to find an electron at a certain position on the screen, which is given by the interference of the two parts,

$$\psi(\mathbf{r}) = \psi_1(\mathbf{r}, t) + \psi_2(\mathbf{r}, t). \tag{10.53}$$

When the magnetic field is turned on, ψ changes in the following way, according to (10.52), where $\int_1$ and $\int_2$ are the line integrals along path 1 and 2, respectively:

$$
\begin{aligned}
\psi' &= \psi_1' + \psi_2' \\
&= \exp\left(-i\frac{e}{\hbar c}\int_1 d\mathbf{r}' \cdot \mathbf{A}\right)\psi_1 + \exp\left(-i\frac{e}{\hbar c}\int_2 d\mathbf{r}' \cdot \mathbf{A}\right)\psi_2 \\
&= \exp\left(-i\frac{e}{\hbar c}\int_1 d\mathbf{r}' \cdot \mathbf{A}\right)\left\{\psi_1 + \exp\left[i\frac{e}{\hbar c}\left(\int_1 d\mathbf{r}' \cdot \mathbf{A} - \int_2 d\mathbf{r}' \cdot \mathbf{A}\right)\right]\psi_2\right\} \\
&= \exp\left(-i\frac{e}{\hbar c}\int_1 d\mathbf{r}' \cdot \mathbf{A}\right)\left[\psi_1 + \exp\left(i\frac{e}{\hbar c}\oint d\mathbf{r}' \cdot \mathbf{A}\right)\psi_2\right] \\
&= \exp\left(-i\frac{e}{\hbar c}\int_1 d\mathbf{r}' \cdot \mathbf{A}\right)\left[\psi_1 + \exp\left(i\frac{e}{\hbar c}\Phi_m\right)\psi_2\right]
\end{aligned}
$$

The relative phase between ψ_1' and ψ_2' has been shifted by $\frac{e}{\hbar c}\Phi_m$, leading to a shift in the interference pattern.

Self-check questions:

1. What is the normal Zeeman effect?
2. How does the Stern-Gerlach experiment work? What does it prove?
3. What is the Aharanov-Bohm effect?

Chapter 11
Perturbation Theory

Abstract We make use of divergent power series that pretend to be convergent. They help us to solve some QM problems that cannot be solved exactly. In the end, this even leads us to the Golden Rule.

There are only a few problems in quantum mechanics which can be solved exactly by a closed expression. In most cases, approximation schemes are required. The most widely used one among them is perturbation theory. One starts from a simple Hamiltonian operator H_0 whose exact eigenvalues and eigenstates are known. A second, "smaller" Hamiltonian operator H_i is then added, describing an interaction that "perturbs" H_0. We write H_i in the form $H_i = \lambda H_1$, with a dimensionless parameter λ, in which we are going to expand (as a power series) the perturbations of states and eigenvalues. These power series in general do not converge (they often have a convergence radius of 0), but still provide excellent approximations after a few terms, i.e. they are **asymptotic series**.

One distinguishes between **stationary** and **time-dependent perturbation theory**. In the first case, H_1 is constant in time. Here, the matter of interest are the changes of eigenvalues and eigenstates compared to those of H_0. We have already met a simple example of this kind with the normal Zeeman effect. In the second case, H_1 is time-dependent, and the matter of interest are now transition rates between eigenstates of H_0.

There are other approximation schemes, which we don't investigate here though. The most prominent are the WKB approximation and the variational method. We refer you to the literature.

11.1 Stationary Perturbation Theory

11.1.1 Expansion in the Perturbation Parameter

Given is an "unperturbed" Hamiltonian operator H_0 whose eigenstates $|j^{(0)}\rangle$ and eigenvalues $E_j^{(0)}$ are known. We take into account that the energy levels can be

© Springer International Publishing Switzerland 2016

J.-M. Schwindt, *Conceptual Basis of Quantum Mechanics*,

Undergraduate Lecture Notes in Physics, DOI 10.1007/978-3-319-24526-3_11

degenerate, i.e. it could be $E_{j_1}^{(0)} = E_{j_2}^{(0)}$. We use j to enumerate the states, not the eigenvalues. Then we add a perturbation to H_0,

$$H = H_0 + H_i = H_0 + \lambda H_1. \tag{11.1}$$

Here H_i describes an interaction. An example is H_{mag} in the normal Zeeman effect, (10.40). It is assumed that H_i is "small" compared to H_0, in a yet to be specified sense. In particular, we extract a dimensionless parameter λ from H_i which we use to produce a power series for the states and eigenvalues, $H_{\text{ww}} = \lambda H_1$. In the case of the normal Zeeman effect we might set $H_1 = \frac{eB_0}{2m_e c} L_z$ with an arbitrary magnetic field strength B_0, and $\lambda = B/B_0$, where B is the actual magnetic field strength. The parameter λ has no physical significance, it doesn't even have to be "small". It is used only as an auxiliary variable for our power series expansion.

The eigenvalue equation for the total Hamiltonian operator reads

$$(H_0 + \lambda H_1)|j\rangle = E_n|j\rangle. \tag{11.2}$$

We imagine that at first $\lambda = 0$ and is then continuously increased or lowered (λ can be negative) by a finite value, thereby shifting eigenvalues from $E_j^{(0)}$ to E_j and eigenvectors from $|j^{(0)}\rangle$ to $|j\rangle$. E_j and $|j\rangle$ thus become functions of λ, and we attempt a power series expansion

$$E_j = E_j^{(0)} + \lambda E_j^{(1)} + \lambda^2 E_j^{(2)} + \cdots \tag{11.3}$$

$$|j\rangle = N(\lambda)\left(|j^{(0)}\rangle + \lambda|j^{(1)}\rangle + \lambda^2|j^{(2)}\rangle + \cdots\right). \tag{11.4}$$

Here $N(\lambda)$ is a normalization constant, which we extracted in a way such that inside the brackets the coefficient of $|j^{(0)}\rangle$ always remains 1. Each $|j^{(i)}\rangle$ can again be expressed in the basis $|j^{(0)}\rangle$, i.e. for $i > 0$ one has

$$|j^{(i)}\rangle = \sum_{k \neq j} \alpha_k^{(i)}|k^{(0)}\rangle. \tag{11.5}$$

Here we have also assumed that $|j^{(i)}\rangle$ has no contribution from $|j^{(0)}\rangle$. For the $|j^{(i)}\rangle$ are supposed to be *corrections* to $|j^{(0)}\rangle$, which are therefore orthogonal to $|j^{(0)}\rangle$. A contribution $\sim|j^{(0)}\rangle$ can always be absorbed by the normalization constant $N(\lambda)$.

We plug this ansatz into the eigenvalue equation (11.2) and compare terms of the same order in λ. The expression of zeroth order is the equation of the unperturbed Hamiltonian operator,

$$H_0|j^{(0)}\rangle = E_j^{(0)}|j^{(0)}\rangle. \tag{11.6}$$

More interesting is the first order. Here the left hand side gives a contribution from $H_0|j^{(1)}\rangle$ and one from $H_1|j^{(0)}\rangle$, the right hand side one from $E_j^{(0)}|j^{(1)}\rangle$ and one from $E_j^{(1)}|j^{(0)}\rangle$, altogether

$$H_0 \sum_{k\neq j} \alpha_k^{(1)}|k^{(0)}\rangle + H_1|j^{(0)}\rangle = E_j^{(0)} \sum_{k\neq j} \alpha_k^{(1)}|k^{(0)}\rangle + E_j^{(1)}|j^{(0)}\rangle. \qquad (11.7)$$

In second order, we get contributions from $H_0|j^{(2)}\rangle$ and $H_1|j^{(1)}\rangle$ on the left, contributions from $E_j^{(0)}|j^{(2)}\rangle$, $E_j^{(1)}|j^{(1)}\rangle$ and $E_j^{(2)}|j^{(0)}\rangle$ on the right hand side, altogether

$$H_0 \sum_{k\neq j} \alpha_k^{(2)}|k^{(0)}\rangle + H_1|j^{(1)}\rangle$$
$$= E_j^{(0)} \sum_{k\neq j} \alpha_k^{(2)}|k^{(0)}\rangle + E_j^{(1)} \sum_{k\neq j} \alpha_k^{(1)}|k^{(0)}\rangle + E_j^{(2)}|j^{(0)}\rangle. \qquad (11.8)$$

And so forth. In order to extract $E_j^{(i)}$ and $\alpha_k^{(i)}$ iteratively from these equations (these are the unknowns), we use the orthogonality of the basis states. At first we multiply (11.7) from the left by $\langle j^{(0)}|$ and receive

$$E_j^{(1)} = \langle j^{(0)}|H_1|j^{(0)}\rangle. \qquad (11.9)$$

In many cases this is already everything one wants to know. But there is a little problem we will discuss in a moment. To determine $\alpha_k^{(1)}$ for $k \neq j$ we multiply (11.7) from the left by $\langle k^{(0)}|$ and receive

$$\alpha_k^{(1)} E_k^{(0)} + \langle k^{(0)}|H_1|j^{(0)}\rangle = \alpha_k^{(1)} E_j^{(0)} \qquad (11.10)$$

or

$$\alpha_k^{(1)}(E_j^{(0)} - E_k^{(0)}) = \langle k^{(0)}|H_1|j^{(0)}\rangle. \qquad (11.11)$$

And here is the problem! For degenerate energy levels, $E_j^{(0)} = E_k^{(0)}$, $\langle k^{(0)}|H_1|j^{(0)}\rangle$ has to vanish, otherwise the equation becomes inconsistent. We therefore have to choose the basis $|j^{(0)}\rangle$ from the beginning such that H_1 is diagonal within the eigenspace of H_0 for the eigenvalue $E_j^{(0)}$. This can be always achieved, since inside an eigenspace $H_0 = E_j^{(0)}\mathbf{1}$ is a multiple of the unit matrix and therefore commutes with H_1 (only inside the eigenspace, not as a whole!). For $E_j^{(0)} = E_k^{(0)}$ then both sides of (11.11) vanish, and for $E_j^{(0)} \neq E_k^{(0)}$ one obtains

$$\alpha_k^{(1)} = \frac{\langle k^{(0)}|H_1|j^{(0)}\rangle}{E_j^{(0)} - E_k^{(0)}}. \qquad (11.12)$$

In second order we multiply (11.8) with $\langle j^{(0)}|$ and obtain

$$E_j^{(2)} = \sum_{k \neq m} \alpha_k^{(1)} \langle j^{(0)}|H_1|k^{(0)}\rangle. \tag{11.13}$$

Inserting (11.12) results in

$$E_j^{(2)} = \sum_{k|E_j^{(0)} \neq E_k^{(0)}} = \frac{|\langle k^{(0)}|H_1|j^{(0)}\rangle|^2}{E_j^{(0)} - E_k^{(0)}}. \tag{11.14}$$

Next we can multiply (11.8) from the left with $\langle k^{(0)}|$, $k \neq j$, to determine $\alpha_k^{(2)}$ and then proceed with the third order. This can be continued to eternity.

Now, is this a good approximation? When can it be used? The question seems to be whether the power series in λ converge and what their radius of convergence is. The bad news is: In most cases there is either nothing known about convergence, or it is even known that the series does not converge, i.e. has a vanishing radius of convergence. Perturbation theory often has to deal with **asymptotic series**, which after a number of terms gets surprisingly close to the exact result, seemingly convergent, but afterwards diverges hopelessly. One can here only take a pragmatic approach: One computes the first few terms, checks that it looks convergent so far, and compares the result with experiments. If this works, one declares $H_i = \lambda H_1$ as "sufficiently small" compared to H_0. The good news is: It works surprisingly often and surprisingly well.

The approach depends strongly on the behavior of the matrix elements $\langle k^{(0)}|H_1|j^{(0)}\rangle$. The more of them vanish for a given j, the simpler the calculation becomes. It is therefore helpful to choose the basis $|j^{(0)}\rangle$ such that H_1 looks "as diagonal as possible". It helps if an operator A is known which commutes with both H_0 and H_1. Then $|j^{(0)}\rangle$ can be chosen as eigenstates of A (simultaneous diagonalizability with H_0) and gets $\langle k^{(0)}|H_1|j^{(0)}\rangle = 0$ whenever $|j^{(0)}\rangle$ and $|k^{(0)}\rangle$ belong to different eigenvalues of A. For then $H_1|j_0\rangle$ belongs to the same A-eigenspace as $|j_0\rangle$, cf. (7.33).

11.1.2 Stark Effect

The Stark effect is the electric equivalent of the Zeeman effect: A hydrogen atom is exposed to a constant electric field $\mathbf{E} = E\mathbf{e}_z$, and we want to determine the effect on the energy levels. The perturbation H_{el} is given by the electric potential $\phi = Ez$ to which the electron of the atom is exposed (the proton is shielded by the electron and does not notice the electric field),

$$H_{el} = eEZ. \tag{11.15}$$

We set $H_1 = eE_0Z$ with an arbitrary field strength E_0 and thus $\lambda = E/E_0$. (It doesn't matter if we use in H a lower-case z or an upper-case Z. Z is the position operator in z-direction, acting on a wave function by multiplication with z.) In contrast to H_{mag} from the Zeeman effect, H_{el} commutes neither with H_0, the Hamiltonian operator of the naive hydrogen atom, nor with $\mathbf{L}^2$.

Exercies 11.1
Show that

$$[H_0, H_1] = -\frac{ieE_0\hbar}{m}P_z, \quad [\mathbf{L}^2, H_1] = 2ieE_0\hbar(XL_y - L_xY). \quad (11.16)$$

If we use the basis $|nlm\rangle$, all kinds of non-diagonal entries $\langle n'l'm|H_1|nlm\rangle \neq 0$ are to be expected. We left m in the bra part unprimed, since H_1 at least commutes with L_z. So, matrix elements between different m-values vanish.

We restrict ourselves to the calculation of the energy corrections to first order, and even that only for $n = 1$ and 2. For that, we take advantage of a nice property of the wave functions $\psi_{nlm}(\mathbf{r})$: they all have a **defined parity**, i.e. they are even or odd functions,

$$\psi_{nlm}(-\mathbf{r}) = \pm\psi_{nlm}(\mathbf{r}). \quad (11.17)$$

This follows from the properties of the spherical harmonics. The norm squared of a function with defined parity is always an even function,

$$\psi_{nlm}^*(-\mathbf{r})\psi_{nlm}(-\mathbf{r}) = \psi_{nlm}^*(\mathbf{r})\psi_{nlm}(\mathbf{r}). \quad (11.18)$$

The function $f(\mathbf{r}) = z$ is odd, the product of an even and an odd function is odd, and the integral over an odd function vanishes. It follows

$$\langle nlm|H_1|nlm\rangle = eE_0 \int d^3r\ \psi_{nlm}^*(\mathbf{r})\psi_{nlm}(\mathbf{r})z = 0. \quad (11.19)$$

In the $|nlm\rangle$-basis therefore all diagonal entries of H_1 vanish! Checking the expression (11.9) for the first energy correction, we could come to the false conclusion that all corrections vanish to first order. For $n = 1$ this is indeed true, but for $n = 2$ our little "problem" becomes relevant. The second energy level is fourfold degenerate with the four $|nlm\rangle$-states $|2, 0, 0\rangle$, $|2, 1, 1\rangle$, $|2, 1, 0\rangle$ and $|2, 1, -1\rangle$. Our method works only in a basis of this four-dimensional subspace where H_1 is diagonal. We have to find this basis. For that, we first write H_1 as a 4×4-matrix in the basis $\{|2, 0, 0\rangle, |2, 1, 1\rangle, |2, 1, 0\rangle, |2, 1, -1\rangle\}$. This is relatively easy, since we already know that all diagonal entries and all entries with different m-values vanish. It remains only

$$H_1^{(n,l,m)}\big|_{n=2} = eE_0 \begin{pmatrix} 0 & 0 & \zeta & 0 \\ 0 & 0 & 0 & 0 \\ \zeta & 0 & 0 & 0 \\ 0 & 0 & 0 & 0 \end{pmatrix} \qquad (11.20)$$

with

$$\zeta = \langle 2, 1, 0|Z|2, 0, 0\rangle = \int d^3r\, \psi_{2,1,0}^*(\mathbf{r})\psi_{2,0,0}(\mathbf{r})z. \qquad (11.21)$$

Actually, in the third row of the matrix there should be a ζ^*, but we know from our investigation of the naive hydrogen atom that $\psi_{n,l,0}$ is real: the radial function is a real function of r, and Y_{lm} is a real polynomial of $\cos\theta$ times $e^{im\phi}$. But for $m = 0$, $e^{im\phi} = 1$. The integral on the right hand side of (11.21) is therefore also real, hence also ζ. Now we can easily diagonalize $H_1|_{n=2}$, which results in a basis $|j^{(0)}\rangle$, $j = 1, 2, 3, 4$:

$$|1^{(0)}\rangle = \frac{1}{\sqrt{2}}(|2, 0, 0\rangle + |2, 1, 0\rangle) \qquad (11.22)$$

$$|2^{(0)}\rangle = \frac{1}{\sqrt{2}}(|2, 0, 0\rangle - |2, 1, 0\rangle) \qquad (11.23)$$

$$|3^{(0)}\rangle = |2, 1, 1\rangle \qquad (11.24)$$

$$|4^{(0)}\rangle = |2, 1, -1\rangle \qquad (11.25)$$

In this basis, $H_1|_{n=2}$ is given by the matrix

$$H_1^{(j)}\big|_{n=2} = eE_0 \begin{pmatrix} \zeta & 0 & 0 & 0 \\ 0 & -\zeta & 0 & 0 \\ 0 & 0 & 0 & 0 \\ 0 & 0 & 0 & 0 \end{pmatrix}. \qquad (11.26)$$

So, among the four states there is one whose eigenvalue is shifted by $\lambda e E_0\zeta = eE\zeta$, and one whose eigenvalue is shifted by $-\lambda e E_0\zeta = -eE\zeta$. The two other states keep their original eigenvalue, to first order.

Exercies 11.2
Determine ζ by integration in spherical coordinates. Use

$$\psi_{2,0,0}(\mathbf{r}) = (32\pi a^3)^{-1/2}\left(2 - \frac{r}{a}\right)e^{-r/(2a)} \qquad (11.27)$$

$$\psi_{2,1,0}(\mathbf{r}) = (32\pi a^3)^{-1/2}\frac{r}{a}e^{-r/(2a)}\cos\theta. \qquad (11.28)$$

Result:

$$\zeta = -3a \qquad (11.29)$$

In summary, of the four unperturbed stated with the energy $E_2^{(0)}$, two keep this energy to first order, one state gets the energy $E_2^{(0)} - 3eEa$ and the last one gets the energy $E_2^{(0)} + 3eEa$. These shifts are of course independent of the choice of E_0.

11.1.3 Fine and Hyperfine Structure of Hydrogen

Already without an external **E**- or **B**-field, the degeneracy of the energy eigenstates in the hydrogen atom is partially removed. In our discussion of the naive hydrogen atom we have neglected some interactions inside the atom, in particular the effects of spin. There are further corrections, for example due to the fact that the momentum expectation value of the electron is so large that Special Relativity needs to be taken into account.

There are some effects which lead to energy shifts of the order of magnitude 10^{-4} times the "naive" energy, and others with the order of magnitude 10^{-7}. The former are known as the **fine structure** of the hydrogen atom, the latter as **hyperfine structure**. The fine structure involves the following effects:

- **Relativistic correction**: The relativistic energy of the electron consists of the mass energy and the kinetic energy,

$$H_{m,kin} = \sqrt{P^2 c^2 + m^2 c^4} \approx mc^2 + \frac{P^2}{2m} + \frac{P^4}{8m^3 c^2}. \tag{11.30}$$

In the last step we expanded the square root to second order in P^2. The first term is an irrelevant constant, the second is the known non-relativistic kinetic term, the third one is the relativistic correction,

$$H_{\text{rel}} = \frac{P^4}{8m^3 c^2}. \tag{11.31}$$

- **Spin-orbit coupling**: The interaction between the magnetic moment generated by the orbital angular momentum of the electron and the magnetic moment generated by the spin of the electron leads to a correction term

$$H_{\text{LS}} = \frac{e^2}{m^2 c^2 r^3} \mathbf{L} \cdot \mathbf{S}. \tag{11.32}$$

- A further correction is due to the so-called **Darwin term** H_{Darwin}, whose origin is a bit more difficult to explain; we refer you to the literature.

For all these terms, the shifts of the energy levels can be calculated via first-order perturbation theory.

The hyperfine structure finally results from the interaction between the magnetic moment of the electron (of both spin and orbital angular momentum) and the magnetic moment generated by the spin of the proton. Of the same order of magnitude is also the so-called Lamb shift, an effect which can only be explained and computed in QED.

The anomalous Zeeman effect which takes the electron's spin into account can be also treated with stationary perturbation theory. If you are interested into these things, you should read the corresponding chapter in (Cohen-Tannoudji et al. 1992) where the topic of stationary perturbation theory is expanded on 200 pp.

Self-check questions:

1. When does the Hamiltonian operator of a perturbation H_i count as "small" compared to the unperturbed Hamiltonian operator H_0?
2. What are the corrections to the energy eigenvalues in first order?
3. What needs to be considered regarding the choice of basis for the approach to be well-defined?

11.2 Time-Dependent Perturbation Theory

11.2.1 Expansion in the Perturbation Parameter

Again we have an unperturbed Hamiltonian operator H_0 whose eigenstates $|j\rangle$ and eigenvalues E_j are known, and a perturbation which now depends on time though,

$$H(t) = H_0 + \lambda H_1(t). \tag{11.33}$$

For any time t we could iteratively determine the current eigenvalues and eigenstates of H, as in the stationary perturbation theory. But this is useless, because the values and states immediately change again. One therefore eschews this approach, stays with the eigenstates of H_0 and calculates the **transition probabilities** between these states, which are induced by H_1. Since we don't determine any corrections to $|j\rangle$ and E_j, we dropped the superscript (0).

Our task is defined as follows: At time $t = 0$, the system is in a state $|i\rangle$ (an eigenstate of H_0). What is the probability to find it at time t in the state $|f\rangle$ (another eigenstate of H_0)? Here i and f stand for initial and final, respectively. We make the following ansatz:

$$|\psi(t)\rangle = \sum_j \alpha_j(t) e^{-i\omega_j t}|j\rangle, \quad \alpha_j(0) = \delta_{ij}, \quad \omega_j = \frac{E_j}{\hbar} \tag{11.34}$$

To understand the form of this ansatz, note that the factor $e^{-i\omega_j t}$ arises from the unperturbed time evolution induced by H_0. If $\lambda = 0$, all α_j are constant in time. The time evolution of $\alpha_j(t)$ arises only due to $\lambda H_1(t)$. Plugging the ansatz (11.34) into the time-dependent Schrödinger equation

$$i\hbar \frac{d}{dt}|\psi(t)\rangle = (H_0 + \lambda H_1(t))|\psi(t)\rangle, \tag{11.35}$$

the time derivative of $e^{-i\omega_j t}$ on the left hand side immediately cancels the H_0 term on the right hand side, and it remains

$$i\hbar \sum_j \dot{\alpha}_j(t)e^{-i\omega_j t}|j\rangle = \sum_j \alpha_j(t)e^{-i\omega_j t}\lambda H_1(t)|j\rangle. \tag{11.36}$$

Again we use the orthogonality of the states, multiply from the left with $\langle f|e^{i\omega_f t}$ and obtain

$$i\hbar\dot{\alpha}_f(t) = \sum_j \langle f|\lambda H_1(t)|j\rangle \alpha_j(t)e^{i(\omega_f - \omega_j)t}. \tag{11.37}$$

We can now proceed similar as for the Born series: To zeroth order, $\alpha_j^{(0)}(t) = \delta_{ij}$. Plugged into the right hand side of (11.37), this yields an expression for the first approximation $\alpha_f^{(1)}(t)$,

$$i\hbar\dot{\alpha}_f^{(1)}(t) = \langle f|\lambda H_1(t)|i\rangle e^{i(\omega_f - \omega_i)t}. \tag{11.38}$$

This can be integrated to

$$\alpha_f^{(1)}(t) = \delta_{fi} - \frac{i\lambda}{\hbar} \int_0^t dt_1 \, \langle f|H_1(t_1)|i\rangle e^{i(\omega_f - \omega_i)t_1}. \tag{11.39}$$

Of course, this expression is not only valid for f, but for any j, and thus we can insert it on the right hand side of (11.37), in order to get an equation for the second approximation $\alpha_f^{(2)}(t)$. This can be integrated again, and one obtains (please verify!)

$$\alpha_f^{(2)}(t) = \delta_{fi} - \frac{i\lambda}{\hbar} \int_0^t dt_1 \, \langle f|H_1(t_1)|i\rangle \, e^{i(\omega_f - \omega_i)t_1} \tag{11.40}$$

$$- \frac{\lambda^2}{\hbar^2} \int_0^t dt_1 \int_0^{t_1} dt_2 \sum_j \langle f|H_1(t_1)|j\rangle\langle j|H_1(t_2)|i\rangle \, e^{i(\omega_f - \omega_j)t_1} e^{i(\omega_j - \omega_i)t_2}.$$

Again, this can be continued to eternity. Again, we don't know much about the convergence properties of the series in λ that results in this way. The probability to find the system at time t in the state $|f\rangle$, after it has been at time 0 in the state $|i\rangle$, is according to our ansatz (11.34) given by

$$W_{fi}(t) = |\alpha_f(t)|^2. \tag{11.41}$$

At the latest when we get $W_{fi} > 1$, we know that we have left the region of validity of this approach. In many cases only the first approximation $\alpha_f^{(1)}(t)$ is required. With a physicist's optimism one assumes that the approximation is good as long as $\alpha_f^{(1)}(t) \ll 1$ (for $f \neq i$). Or that the approximation is the better the smaller t is, for then the system hasn't had much time to deviate from the original state. The results are in most cases very good. The physicist's optimism is usually rewarded. Eventually, the method of **Feynman graphs** in quantum field theory is based on the same approach and provides results to a precision of many decimals.

11.2.2 Dirac Picture

Maybe you have already noticed that (11.40) has a certain similarity with the first two rows of the expression (2.175) for the propagator $U(t, t_0)$. This is not accidental and can be made more concrete if one moves to the so-called **interaction picture**, also called **Dirac picture**.

The Dirac picture is an intermediate thing between the Schrödinger picture and the Heisenberg picture. In the Heisenberg picture, the complete time dependency was shifted from the states to the operators. In the Dirac picture, this is done only for the time evolution induced by H_0. The propagator $U(t, t_0)$ is thus split into two factors,

$$U(t, t_0) = U_0(t, t_0)U_1(t, t_0). \tag{11.42}$$

Remember that in the Schrödinger picture we had

$$|\psi_S(t)\rangle = U(t, t_0)|\psi_S(t_0)\rangle \tag{11.43}$$

and therefore

$$i\hbar \frac{d}{dt}U(t, t_0) = (H_{0S} + \lambda H_{1S})U(t, t_0). \tag{11.44}$$

Now that we switch forth and back between the pictures, we again use the subscript S for operators and states in the Schrödinger picture. $U_0(t, t_0)$ is defined such that

$$i\hbar \frac{d}{dt}U_0(t, t_0) = H_0 U_0(t, t_0), \tag{11.45}$$

it is thus the propagator of the unperturbed system. It follows

$$i\hbar U_0(t, t_0)\frac{d}{dt}U_1(t, t_0) = i\hbar\left[\frac{d}{dt}U(t, t_0) - \left(\frac{d}{dt}U_0(t, t_0)\right)U_1(t, t_0)\right] \tag{11.46}$$
$$= (H_{0S} + \lambda H_{1S}(t))U(t, t_0) - H_{0S}U_0(t, t_0)U_1(t, t_0)$$
$$= \lambda H_{1S}(t)U_0(t, t_0)U_1(t, t_0).$$

The Dirac picture is now defined via

$$|\psi_D(t)\rangle = U_0^\dagger(t, t_0)|\psi_S(t)\rangle, \tag{11.47}$$

$$A_D(t) = U_0^\dagger(t, t_0)A_S(t)U_0(t, t_0), \tag{11.48}$$

where A is an arbitrary operator. The definition of $|\psi_D(t)\rangle$ yields

$$|\psi_D(t)\rangle = U_0^\dagger(t, t_0)U(t, t_0)|\psi_S(t_0)\rangle \tag{11.49}$$

$$= U_0^\dagger(t, t_0)U_0(t, t_0)U_1(t, t_0)|\psi_S(t_0)\rangle \tag{11.50}$$

$$= U_1(t, t_0)|\psi_S(t_0)\rangle \tag{11.51}$$

$$= U_1(t, t_0)|\psi_D(t_0)\rangle. \tag{11.52}$$

The Dirac state $|\psi_D(t)\rangle$ therefore evolves according to the propagator U_1. From (11.48) follows

$$H_{1S}U_0(t, t_0) = U_0(t, t_0)H_{1D}(t) \tag{11.53}$$

and so (11.46) becomes

$$i\hbar U_0(t, t_0)\frac{d}{dt}U_1(t, t_0) = \lambda U_0(t, t_0)H_{1D}(t)U_1(t, t_0) \tag{11.54}$$

$$\Rightarrow \quad i\hbar\frac{d}{dt}U_1(t, t_0) = \lambda H_{1D}(t)U_1(t, t_0). \tag{11.55}$$

Combined with (11.52), we obtain the equivalent of the Schrödinger equation in the Dirac picture (no, this is not the Dirac equation!),

$$i\hbar|\psi_D(t)\rangle = \lambda H_{1D}(t)|\psi_D(t)\rangle. \tag{11.56}$$

The part of the time dependency of the states induced by H_0 is thus no longer present in this picture, it was shifted into the operators.

Equation (11.55) has the same form as (2.165) and can therefore be solved in the same way. One obtains a series similar to (2.175):

$$U_1(t, t_0) = 1 - \frac{i\lambda}{\hbar}\int_{t_0}^{t} dt_1\, H_{1D}(t_1) \tag{11.57}$$

$$- \frac{\lambda^2}{\hbar^2}\int_{t_0}^{t} dt_1 \int_{t_0}^{t_1} dt_2\, H_{1D}(t_1)H_{1D}(t_2)$$

$$+ \frac{i\lambda^3}{\hbar^3}\int_{t_0}^{t} dt_1 \int_{t_0}^{t_1} dt_2 \int_{t_0}^{t_2} dt_3\, H_{1D}(t_1)H_{1D}(t_2)H_{1D}(t_3)$$

$$+ \cdots$$

Now we return to the Schrödinger picture and translate (11.57) into a series for $U(t, t_0)$. We therefore have to

- multiply the entire equation from the left with $U_0(t, t_0)$;
- replace each $H_{1D}(t')$ with $U_0^\dagger(t', t_0)H_{1S}(t')U_0(t', t_0)$.
- This yields products of the form $U_0(t'', t_0)U_0^\dagger(t', t_0)$. These can be replaced with the help of

$$U_0(t'', t_0)U_0^\dagger(t', t_0) = U_0(t'', t_0)U_0^{-1}(t', t_0) \tag{11.58}$$

$$= U_0(t'', t')U_0(t', t_0)U_0^{-1}(t', t_0) \tag{11.59}$$

$$= U_0(t'', t'). \tag{11.60}$$

After these steps, we obtain, to second order in λ,

$$U(t, t_0) = U_0(t, t_0) - \frac{i\lambda}{\hbar} \int_{t_0}^t dt_1 \, U_0(t, t_1)H_{1S}(t_1)U_0(t_1, t_0) \tag{11.61}$$

$$- \frac{\lambda^2}{\hbar^2} \int_{t_0}^t dt_1 \int_{t_0}^{t_1} dt_2 \, U_0(t, t_1)H_{1S}(t_1)U_0(t_1, t_2)H_{1S}(t_2)U_0(t_2, t_0)$$

$$+ \cdots$$

This equation can be interpreted in the following way: In the term of zeroth order, the system propagates without perturbation from t_0 to t. In the first order term, the system propagates without perturbation from t_0 to t_1. At the time t_1, the perturbation Hamiltonian λH_1 strikes, i.e. it transforms the state $|\psi(t_1)\rangle$ into another state $|\psi'(t_1)\rangle$. This new state then propagates again unperturbed from t_1 to t. The sudden strike of λH_1 can happen at any time between t_0 and t, therefore the integral. In the second order term, λH_1 strikes twice, and apart from these two moments, the system propagates without perturbation. In the next order, λH_1 appears three times etc.

How do we retrieve from (11.61) for the propagator our (11.40) for the second order transition amplitude $\alpha_f^{(2)}$? From (11.34) follows

$$\langle f|\psi(t)\rangle = \alpha_f(t)e^{-i\omega_f t}. \tag{11.62}$$

This implies, with $|\psi(0)\rangle = |i\rangle$,

$$\alpha_f(t) = \langle f|\psi(t)\rangle \, e^{i\omega_f t} \tag{11.63}$$

$$= \langle f|U(t, 0)|i\rangle \, e^{i\omega_f t}. \tag{11.64}$$

Here we insert our expression for U. We only have to use

$$U_0(t', t'')|j\rangle = e^{-i\omega_j(t'-t'')}|j\rangle. \tag{11.65}$$

The term of zeroth order in λ is

$$\langle f|U_0(t,0)|i\rangle\, e^{i\omega_f t} \tag{11.66}$$

$$= \langle f|i\rangle\, e^{-i\omega_f t} e^{i\omega_f t} = \delta_{fi}. \tag{11.67}$$

This is identical to the zeroth order term for α_f. For the first order term we insert two additional unit operators in the form $\mathbf{1} = \sum_j |j\rangle\langle j|$ between the operators:

$$\frac{-i\lambda}{\hbar}\int_0^t dt_1\, \langle f|U_0(t,t_1)H_1(t_1)U_0(t_1,0)|i\rangle e^{i\omega_f t}$$

$$= \frac{-i\lambda}{\hbar}\sum_{j,k}\int_0^t dt_1\, \langle f|U_0(t,t_1)|j\rangle\langle j|H_1(t_1)|k\rangle\langle k|U_0(t_1,0)|i\rangle e^{i\omega_f t} \tag{11.68}$$

$$= \frac{-i\lambda}{\hbar}\sum_{j,k}\int_0^t dt_1\, \delta_{fj}e^{-i\omega_j(t-t_1)}\langle j|H_1(t_1)|k\rangle\delta_{ki}e^{-i\omega_i t_1} e^{i\omega_f t} \tag{11.69}$$

$$= \frac{-i\lambda}{\hbar}\int_0^t dt_1\, \langle f|H_1(t_1)|i\rangle e^{-i\omega_f(t-t_1)}e^{-i\omega_i t_1} e^{i\omega_f t} \tag{11.70}$$

$$= \frac{-i\lambda}{\hbar}\int_0^t dt_1\, \langle f|H_1(t_1)|i\rangle e^{i(\omega_f-\omega_i)t} \tag{11.71}$$

This is identical to the first order term of α_f.

Exercies 11.3
The second order term works in the same way, you only have to insert four unit operators this time. Perform the calculation and reproduce the second order term of α_f. This is a very good exercise, so don't miss it!

11.2.3 Periodic Perturbation and Fermi's Golden Rule

An important application of time-dependent perturbation theory is the calculation of atomic transitions. An atom is in the energy eigenstate $|i\rangle$. What is the probability to find it in the energy eigenstate $|f\rangle$ at time t when the atom is exposed to a time-dependent external field? The most important case is excitation by electromagnetic radiation; the transition from $|i\rangle$ to $|f\rangle$ is then induced by the absorption of a photon. The exact meaning of *absorption* can be only understood in QED. Here in QM, we can model the situation by describing the radiation as a classical electromagnetic wave of frequency ω. The electric and magnetic fields of the wave oscillate with this frequency, and the interaction between atom and wave can be written in the form

$$\lambda H_1(t) = \lambda \bar{H}_1 e^{-i\omega t} \tag{11.72}$$

One speaks of a **periodic perturbation**. The energy of the corresponding photons is $\hbar\omega$.

To be precise, (11.72) does not describe a proper Hamiltonian operator, since it is not hermitian. To render it hermitian, one has to add the hermitian conjugate $\lambda \bar{H}_1 e^{+i\omega t}$. This term represents the spontaneous *emission* of a photon with energy $\hbar\omega$. In the following, we address to the absorption case alone, but have to understand that it is always accompanied by the emission term. The calculation for the emission term is equivalent, the only difference being that a photon is emitted instead of absorbed.

In first order perturbation theory we obtain, according to (11.39), the following transition amplitude between two different states $f \neq i$:

$$\alpha_f^{(1)}(t) = -\frac{i\lambda}{\hbar} \int_0^t dt_1 \langle f|\bar{H}_1|i\rangle e^{i(\omega_f - \omega_i - \omega)t_1} \tag{11.73}$$

$$= -\frac{i\lambda}{\hbar} \langle f|\bar{H}_1|i\rangle \frac{e^{i(\omega_f - \omega_i - \omega)t} - 1}{i(\omega_f - \omega_i - \omega)}. \tag{11.74}$$

With the abbreviation

$$\delta_\omega = \omega_f - \omega_i - \omega \tag{11.75}$$

this can be further simplified to

$$\alpha_f^{(1)}(t) = -\frac{i\lambda}{\hbar} \langle f|\bar{H}_1|i\rangle \frac{e^{i\delta_\omega t} - 1}{i\delta_\omega} \tag{11.76}$$

$$= -\frac{i\lambda}{\hbar} \langle f|\bar{H}_1|i\rangle \frac{2e^{i\delta_\omega \frac{t}{2}} \sin(\delta_\omega \frac{t}{2})}{\delta_\omega}. \tag{11.77}$$

The corresponding **transition probability** is

$$W_{fi}^{(1)}(t) = |\alpha_f^{(1)}(t)|^2 \tag{11.78}$$

$$= \frac{\lambda^2}{\hbar^2} |\langle f|\bar{H}_1|i\rangle|^2 \frac{4\sin^2(\delta_\omega \frac{t}{2})}{(\delta_\omega)^2}. \tag{11.79}$$

One therefore obtains to first order an oscillating behavior in time. For small t, $W_{fi}^{(1)}(t) \sim t^2$, and for large t the transition probability, averaged over time, is the larger the smaller δ_ω is. If δ_ω is too small though, at the latest when

$$\frac{4\lambda^2}{\hbar^2} |\langle f|\bar{H}_1|i\rangle|^2 (\delta_\omega)^{-2} > 1, \tag{11.80}$$

we know that the first order approximation became useless.

In practice, ω is never exact. Instead we have to assume that the radiation has a certain bandwidth, i.e. the photons have slightly different frequencies. Each frequency contributes a first order transition probability $W_{fi}^{(1)}(t, \omega)$ of the form (11.79), which is now also a function of ω. The total transition probability then results as the integral

$$W_{fi}^{(1)}(t) = \int d\omega \, \rho(\omega) W_{fi}^{(1)}(t, \omega). \tag{11.81}$$

Also ω_i and ω_f are not exact if the atoms are in motion and therefore have different kinetic energies which have to be added to the internal energy eigenvalues. We are going to ignore this effect here though.

If we look at $W_{fi}^{(1)}(t, \omega)$ for fixed t as a function of ω or δ_ω (cf. Fig. 11.1), we find that the function is strongly peaked at $\delta_\omega = 0$ and that the peak has an approximate width of $2\pi/t$. This can be understood as a variant of the **Energy-Time Uncertainty Relation**: δ_ω expresses a violation of energy conservation: a photon of energy $\hbar\omega$ is absorbed to raise the energy of an atom from $\hbar\omega_i$ to $\hbar\omega_f$. The difference (atomic energy after the process minus photon energy minus atomic energy before the process) is just $\Delta E = \hbar\delta_\omega$. The image shows: if the process has only a time Δt available, a certain energy change with

$$\Delta E \approx \frac{2\pi\hbar}{\Delta t} \tag{11.82}$$

is allowed.

Fig. 11.1 Dependency of transition probabilities on δ_ω

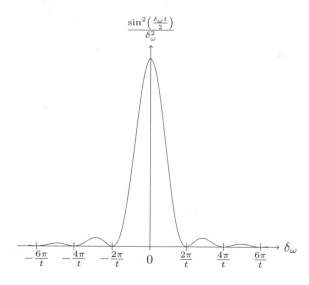

One can show that the family of functions g_ϵ

$$g_\epsilon(x) = \frac{\epsilon}{\pi} \frac{\sin^2(x/\epsilon)}{x^2} \tag{11.83}$$

for $\epsilon \to 0$ converges to the delta distribution,

$$\lim_{\epsilon \to 0} g_\epsilon(x) = \delta(x). \tag{11.84}$$

One therefore needs to prove that $\lim_{\epsilon \to 0} g_\epsilon(x) = 0$ for $x \neq 0$ and that $\int dx\, g_\epsilon(x) = 1$, which we leave to the diligent reader. From (11.84) follows

$$\lim_{t \to \infty} \frac{4 \sin^2(\delta_\omega t/2)}{(\delta_\omega)^2} = 2\pi t \delta(\delta_\omega). \tag{11.85}$$

As expected, this means that for large times the energy conservation law is valid, and the transition probability evolves proportional to t,

$$W_{fi}^{(1)}(t) \to \int d\omega\, \rho(\omega) \frac{2\pi t \lambda^2}{\hbar^2} |\langle f|\bar{H}_1|i\rangle|^2 \delta(\delta_\omega) \tag{11.86}$$

$$= \rho(\omega_f - \omega_i) \frac{2\pi t \lambda^2}{\hbar^2} |\langle f|\bar{H}_1|i\rangle|^2. \tag{11.87}$$

It makes no sense to speak of a limit $t \to \infty$ here, as the right hand side becomes infinite then. The expression is valid for times large enough that frequency differences $\delta_\omega = 2\pi/t$ can no longer be registered by detectors (no deviation from the delta distribution can be found), but small enough that the right hand side remains $\ll 1$. Whether such a time interval exists at all depends on whether $\lambda^2 |\langle f|\bar{H}_1|i\rangle|^2$ is small enough.

The **transition rate** is in general defined as

$$P_{fi}(t) = \frac{dW_{fi}(t)}{dt}. \tag{11.88}$$

In our case P_{fi} is time-independent (at least for times in the interval mentioned above). The result is:

Fermi's Golden Rule

$$P_{fi} = \rho(\omega_f - \omega_i) \frac{2\pi \lambda^2}{\hbar^2} |\langle f|\bar{H}_1|i\rangle|^2 \tag{11.89}$$

It plays a prominent role not only for atomic transitions, but also for scattering experiments in particle physics.

Self-check questions:

1. What quantities are determined in time-dependent perturbation theory?
2. How is the Dirac picture defined?
3. What does Fermi's Golden Rule say, and under what circumstances can it be used?

Reference

C. Cohen-Tannoudji, B. Diu, F. Laloe, *Quantum Mechanics*, 2 vols (Wiley, 1992). (Good textbook with detailed explanations. Particularly extensive when it comes to stationary perturbation theory and the real hydrogen atom)

Chapter 12
N-Particle Systems

Abstract The weird behavior of indistinguishable particles is discussed. They found a way to completely hide their individual identities. Then we show how particles can simultaneously exist and not exist in Fock space. Finally, the density operator is used to make QM even more probabilistic.

In this chapter we investigate quantum systems involving several particles (quantum objects). A particularly peculiar behavior occurs for indistinguishable particles: their state is either completely symmetric or completely antisymmetric w.r.t. exchange of two particles. In the first case they are called bosons, in the second case fermions. This implies that it is impossible to "mark" such a particle in some way in order to identify it later. For fermions furthermore follows the Pauli principle which says that two fermions of the same kind cannot be in the same state. Without this property the entire chemistry would be inconceivable, it is absolutely fundamental for our existence.

Then we briefly discuss the Fock space, a Hilbert space where states with different particle numbers overlap, and where creation and annihilation operators raise or lower the number of particles. This space is fundamental for the further development of QM to quantum field theory.

Finally we introduce the density operator which enables us to compute expectation values in situations with two nested probabilities: the probability to find a certain quantum state in a statistical ensemble of particles, and the probability to find a certain measurement value within this quantum state.

12.1 Bosons and Fermions

12.1.1 Distinguishable and Indistinguishable Particles

We already know from Sects. 2.10 and 3.8, that for systems with several quantum objects tensor products come into play. Consider a situation with two particles, with associated Hilbert spaces $\mathcal{H}^{(1)}$ and $\mathcal{H}^{(2)}$. Then the state of the entire system live in

© Springer International Publishing Switzerland 2016

J.-M. Schwindt, *Conceptual Basis of Quantum Mechanics*,
Undergraduate Lecture Notes in Physics, DOI 10.1007/978-3-319-24526-3_12

the tensor product $\mathcal{H}^{(1)} \otimes \mathcal{H}^{(2)}$. Assume we have an electron and a photon. They can be easily distinguished, since one of them is charged, the other one not, one of them has spin $1/2$, the other one spin 1, one of them has a mass, the other one is massless. Due to the different spins, already the Hilbert spaces are different, we can therefore easily say which one of the spaces $\mathcal{H}^{(1)}$, $\mathcal{H}^{(2)}$ belongs to which particle. If we have an electron and a muon instead, the only difference is in the mass. Charge and spin are identical. But due to the different masses, you can still determine which of the two particles you have in your hand. The Hilbert spaces are identical copies of each other though, and only by convention you can decide which one belongs to which particle.

What happens when we have two electrons? Now there are no differences at all. Imagine we caught an electron in a box. Later we place another one in the same box. Even later we fish one of them out. Then we have no possibility to determine whether this is the first or the second one. It is impossible to mark an electron in order to identify it later. As we will see, QM ensures this in the most radical way. According to some interpretation of quantum field theory, one can say that within each interaction, the old electron is destroyed and a new one created. Then the electron we fished out is none of the two we put in the box.

One may try to define an electron's identity by its state or its role. For example, regarding the atom of an alkali metal, one speaks of the valence electron, the only electron in the outermost shell, as if this defined an identity (although we don't know whether it is really "the same" electron as the one which occupied this shell a microsecond before). This is as if we assigned an individual identity to the role of the American president, ignoring that the role was occupied by many men, speaking of the president as an individual which is now more than 200 years old. For the Dalai Lama it actually works more or less like that. For an electron, it makes sense to some extent because it is the *only* way to give it an individual identity. To avoid a further drift off into philosophy, we now turn to the hard (i.e. mathematical) facts.

12.1.2 Two Particles

Given are two particles of the same sort, i.e. two indistinguishable particles. The Hilbert space of such a particle is $\mathcal{H}_1$, the two-particle state is thus an element of the Hilbert space

$$\mathcal{H}_2 = \mathcal{H}_1^{(1)} \otimes \mathcal{H}_1^{(2)}. \tag{12.1}$$

Here we use the following convention: the number in the lower index of a Hilbert space denotes the number of particles described by it. On the left hand side we therefore have a two-particle Hilbert space, on the right hand side two one-particle Hilbert spaces.

The superscript (1), (2) only serves to enumerate the copies of $\mathcal{H}_1$. For the product states we abbreviate:

$$|\psi_1; \psi_2\rangle := |\psi_1\rangle \otimes |\psi_2\rangle \tag{12.2}$$

To demonstrate how QM enforces the absolute indistinguishability of particles, we have to make use of the **transposition operator** T_{12} and its eigenspaces. This operator exchanges the states of the two particles,

$$T_{12}|\psi_1; \psi_2\rangle = |\psi_2; \psi_1\rangle. \tag{12.3}$$

Obviously $T_{12}^2 = 1$. Therefore, T_{12} can only have two eigenvalues, 1 and -1. Eigenstates of the eigenvalue 1 are called **symmetric**, those of the eigenvalue -1 **antisymmetric**. The symmetric and the antisymmetric states constitute subspaces of $\mathcal{H}_2$. For one easily shows that the linear combination of (anti)symmetric states is again (anti)symmetric. The symmetric subspace is denoted $\mathcal{H}_2^{(+)}$, the antisymmetric $\mathcal{H}_2^{(i)}$. A state of the form

$$|\psi_1; \psi_2\rangle^{(+)} := N^+ \left(|\psi_1; \psi_2\rangle + |\psi_2; \psi_1\rangle\right) \tag{12.4}$$

is always symmetric (N^+ is a normalization constant), a state of the form

$$|\psi_1; \psi_2\rangle^{(-)} := N^- \left(|\psi_1; \psi_2\rangle - |\psi_2; \psi_1\rangle\right) \tag{12.5}$$

always antisymmetric. Due to

$$|\psi_1; \psi_2\rangle = \frac{1}{2}\left(\frac{1}{N^+}|\psi_1; \psi_2\rangle^{(+)} + \frac{1}{N^-}|\psi_1; \psi_2\rangle^{(-)}\right), \tag{12.6}$$

any two-particle state can be written as a linear combination of a symmetric and an antisymmetric state. (If $|\psi_1\rangle = |\psi_2\rangle$, the antisymmetric part vanishes.) So, $\mathcal{H}_2^{(+)}$ and $\mathcal{H}_2^{(-)}$ span the entire Hilbert space $\mathcal{H}_2$. But this holds only for two particles, not for any higher number! Furthermore, symmetric and antisymmetric states are orthogonal to each other,

$$|\Psi^+\rangle \in \mathcal{H}_2^{(+)}, \quad |\Psi^-\rangle \in \mathcal{H}_2^{(-)}$$

$$\Rightarrow \quad \langle\Psi^+|\Psi^-\rangle = \langle\Psi^+|T_{12}T_{12}|\Psi^-\rangle = \langle\Psi^+|(+1)(-1)|\Psi^-\rangle = -\langle\Psi^+|\Psi^-\rangle$$

$$\Rightarrow \quad \langle\Psi^+|\Psi^-\rangle = 0.$$

For $\mathcal{H}_1 = \mathbb{C}^2$, we met the two spaces twice before $\mathcal{H}_2^{(\pm)}$, in Sects. 2.10 and 9.2. $\mathcal{H}_2^{(+)}$ is the three-dimensional space spanned by the **triplet**

$$\{|z+; z+\rangle, \frac{1}{\sqrt{2}}(|z+; z-\rangle + |z-; z+\rangle), |z-; z-\rangle\}; \tag{12.7}$$

$\mathcal{H}_2^{(-)}$ is one-dimensional and consists of the **singlet**

$$\frac{1}{\sqrt{2}}(|z+; z-\rangle + |z-; z+\rangle). \tag{12.8}$$

In general: if $\{|j\rangle\}$ is an orthonormal basis of $\mathcal{H}_1$, then the states

$$|j_1; j_2\rangle \quad \text{for} \quad j_1 = j_2 , \tag{12.9}$$

$$\frac{1}{\sqrt{2}}(|j_1; j_2\rangle + |j_2; j_1\rangle) \quad \text{for} \quad j_1 \neq j_2 \tag{12.10}$$

form an orthonormal basis of $\mathcal{H}_2^{(+)}$, the states

$$\frac{1}{\sqrt{2}}(|j_1; j_2\rangle - |j_2; j_1\rangle) \quad \text{for} \quad j_1 \neq j_2 \tag{12.11}$$

an orthonormal basis of $\mathcal{H}_2^{(-)}$.

Exercise 12.1
What is the dimension of $\mathcal{H}_2^{(\pm)}$ for $\mathcal{H}_1 = \mathbb{C}^n$?

Now back to physics. QM (or nature) ensures that indistinguishable particles are really *absolutely* indistinguishable by requesting that two-particle states of such particles live only in the eigenspaces of the transposition operator T_{12}. **Bosons** are particles whose two-particle states live in $\mathcal{H}_2^{(+)}$, **fermions** those whose two-particle states live in $\mathcal{H}_2^{(-)}$. Experiments show that particles of integer spin are bosons, those of half-integer spin fermions. The **spin-statistics theorem** of quantum field theory provides a theoretical foundation to these observations.

Due to their antisymmetry, two fermions can never have the same state. This is the **Pauli principle**, also called **exclusion principle**.

Let's consider the case where the one-particle Hilbert space $\mathcal{H}_1$ is already a tensor product of two Hilbert spaces, namely the space $\mathcal{H}_{1\psi}$ of wave functions and the space $\mathcal{H}_{1\chi}$ of spin states,

$$\mathcal{H}_1 = \mathcal{H}_{1\psi} \otimes \mathcal{H}_{1\chi}. \tag{12.12}$$

Then

$$\mathcal{H}_2 = \left(\mathcal{H}_{1\psi}^{(1)} \otimes \mathcal{H}_{1\chi}^{(1)}\right) \otimes \left(\mathcal{H}_{1\psi}^{(2)} \otimes \mathcal{H}_{1\chi}^{(2)}\right) \tag{12.13}$$

$$= \left(\mathcal{H}_{1\psi}^{(1)} \otimes \mathcal{H}_{1\psi}^{(2)}\right) \otimes \left(\mathcal{H}_{1\chi}^{(1)} \otimes \mathcal{H}_{1\chi}^{(2)}\right) \tag{12.14}$$

$$= \mathcal{H}_{2\psi} \otimes \mathcal{H}_{2\chi} \tag{12.15}$$

using the definition

$$\mathcal{H}_{2\psi} = \mathcal{H}_{1\psi}^{(1)} \otimes \mathcal{H}_{1\psi}^{(2)}, \quad \mathcal{H}_{2\chi} = \mathcal{H}_{1\chi}^{(1)} \otimes \mathcal{H}_{1\chi}^{(2)}. \tag{12.16}$$

The transposition operator acts then separately on both spaces in (12.15),

$$T_{12} = T_{12}^{(\psi)} \otimes T_{12}^{(\chi)}, \tag{12.17}$$

and the eigenvalues are multiplied. If τ_ψ is an eigenvalue of $T_{12}^{(\psi)}$ and τ_χ one of $T_{12}^{(\chi)}$, then the corresponding eigenvalue τ of T_{12} is

$$\tau = \tau_\psi \tau_\chi. \tag{12.18}$$

It follows

$$\mathcal{H}_2^{(+)} = \left(\mathcal{H}_{2\psi}^{(+)} \otimes \mathcal{H}_{2\chi}^{(+)}\right) \oplus \left(\mathcal{H}_{2\psi}^{(-)} \otimes \mathcal{H}_{2\chi}^{(-)}\right) \tag{12.19}$$

$$\mathcal{H}_2^{(-)} = \left(\mathcal{H}_{2\psi}^{(+)} \otimes \mathcal{H}_{2\chi}^{(-)}\right) \oplus \left(\mathcal{H}_{2\psi}^{(-)} \otimes \mathcal{H}_{2\chi}^{(+)}\right). \tag{12.20}$$

For bosons this implies that wave functions and spin states are either both symmetric or both antisymmetric; for fermions one of them is symmetric, the other antisymmetric.

A standard example is given by the **helium atom** with its two electrons. Either their wave functions are symmetric and their spins antisymmetric, implying that the spin combination of the two electrons is in the singlet state. A helium atom in such a state is called **parahelium**. Or the wave functions are antisymmetric and the spins symmetric, implying that the spin combination belongs to the space spanned by the triplet. A helium atom in such a state is called **orthohelium**. Since the wave functions are differently combined for ortho- and parahelium, they have different energy eigenstates. The ground state can be only occupied by parahelium. Only there, both electrons can occupy the state $|nlm\rangle = |100\rangle$. For orthohelium, this is forbidden by the exclusion principle.

12.1.3 N Particles

If there are N particles instead of just two, the product Hilbert space is initially given by

$$\mathcal{H}_N = \mathcal{H}_1^{(1)} \otimes \mathcal{H}_1^{(2)} \otimes \cdots \otimes \mathcal{H}_1^{(N)}. \tag{12.21}$$

We introduce the following notation for the N-particle states to clarify the positions of the states:

$$|1 : \psi_1; 2 : \psi_2; \ldots; N : \psi_n\rangle := |\psi_1\rangle \otimes |\psi_2\rangle \otimes \cdots \otimes |\psi_N\rangle \tag{12.22}$$

The numbers before the colons denote which copy of $\mathcal{H}_1$ the respective one-particle state belongs to. Now there are $\binom{N}{2} = N(N-1)/2$ different transposition operators T_{ij}, which operate on the state $|1 : \psi_1; \ldots; N : \psi_N\rangle$ by exchanging the positions i and j,

$$T_{ij}|1 : \psi_1; \ldots; i : \psi_i; \ldots; j : \psi_j; \ldots; N : \psi_N\rangle$$
$$= |1 : \psi_1; \ldots; i : \psi_j; \ldots; j : \psi_i; \ldots; N : \psi_N\rangle. \qquad (12.23)$$

To ensure absolute indistinguishability, nature now requests that each N-particle state is an eigenvector of *all* transposition operators. One can easily show that all of the corresponding eigenvalues τ_{ij} have to be identical. Either $\tau_{ij} = -1$ for all (i, j) or $\tau_{ij} = +1$ for all (i, j). The eigenspace with $\tau_{ij} = -1$ for all (i, j) is denoted $\mathcal{H}_N^{(-)}$, the one with $\tau_{ij} = +1$ for all (i, j) denoted $\mathcal{H}_N^{(+)}$. Boson states live in $\mathcal{H}_N^{(+)}$, fermion states in $\mathcal{H}_N^{(-)}$.

Exercise 12.2
Verify that

$$T_{ni}T_{mj}T_{nm}T_{ni}T_{mj} = T_{ij} \qquad (12.24)$$

Find an equation for the associated eigenvalues and conclude that τ_{ij} has to be the same for all values of (i, j).

Due to their antisymmetry,

$$T_{ij}|1 : \psi_1; \ldots; i : \psi_i; \ldots; j : \psi_j; \ldots; N : \psi_N\rangle$$
$$= -|1 : \psi_1; \ldots; i : \psi_j; \ldots; j : \psi_i; \ldots; N : \psi_N\rangle \qquad (12.25)$$

two fermions can obviously again not occupy the same state.

Now what do the states in $\mathcal{H}_N^{(\pm)}$ look like? To answer this question, we have to find a basis for these two spaces. The remaining states are then linear combinations of those. To determine such a basis, we have to briefly discuss **permutations**. A permutation σ of the numbers $1, 2, \ldots, N$ is a reordering of these numbers, i.e. σ maps each number i to a number $\sigma(i)$, such that among the $\sigma(i)$ each number from 1 to N occurs exactly once. These permutations form a group Σ_N, since permutations can be performed successively, and they also can be inverted. The unit element is the permutation which does not permute anything. Σ_N is generated by the transpositions, i.e. any permutation can be represented as a successive execution of transpositions. A permutation is called **even** if an even number of transpositions is required, otherwise **odd**. The **sign** $p(\sigma)$ of a permutation is $+1$ if the permutation is even, -1 if it is odd. For the Hilbert space $\mathcal{H}_N$, eigenstates of all transpositions are also eigenstates of all permutations (one only needs to apply the transpositions successively). In $\mathcal{H}_N^{(+)}$, the eigenvalue for all permutations is $+1$, whereas in $\mathcal{H}_N^{(-)}$ it is given by $p(\sigma)$.

Let $\{|j\rangle\}$ be a basis of $\mathcal{H}_1$. Assume we have an N-boson state $|\Psi\rangle \in \mathcal{H}_N^{(+)}$ describing one particle in the state $|j_1\rangle$ and $N - 1$ particles in the state $|j_2\rangle$. Then $|j_1\rangle$ must appear in $|\Psi\rangle$ in every position,

$$|\Psi\rangle = \frac{1}{\sqrt{N}} (|1 : j_1; \ldots\rangle + |\ldots; 2 : j_1; \ldots\rangle + \cdots + |\ldots; N : j_1\rangle). \quad (12.26)$$

(The dots inside the terms represent occurrences of j_2-states.) This is because the individual terms are exchanged by permutations; for $|\Psi\rangle$ to be an eigenstate of all permutations, all of these terms must be contained in $|\Psi\rangle$; otherwise some permutation would generate an additional term, and $|\Psi\rangle$ would not be an eigenstate of this permutation. One says the N-boson state has been **symmetrized** over all positions of the one-particle states. The same holds for all possible constellations of one-particle states. It only matters which state occurs how many times. A basis state of $\mathcal{H}_N^{(+)}$ is therefore characterized by its **occupation numbers**, i.e. by the set of numbers $\{n_j\}$, where each n_j denotes how many particles are in the state $|j\rangle$. We denote such an N-boson basis state as $|\{n_j\}\rangle^{(+)}$.

Exercise 12.3

(a) Show by combinatorial considerations that a state $|\{n_j\}\rangle^{(+)}$ contains

$$\frac{N!}{\prod_{j|n_j>1} n_j!} \quad (12.27)$$

terms. Hint: figure out how many permutations there are altogether, and how many of them don't change anything since they exchange equal states.

(b) Confirm the result for a three-particle state with $n_1 = 2$, $n_2 = 1$, i.e. with two particles in the state $|1\rangle$ and one in the state $|2\rangle$. What is the corresponding basis state $|\{n_j\}\rangle^{(+)}$?

For fermions, things are similar, but with two differences

- Due to the exclusion principle, all occupation numbers can take only the values 0 and 1.
- One has to **antisymmetrize** instead of symmetrize. This means that terms which are related via an odd permutation have a relative minus sign.

We denote fermion basis states by $|\{n_j\}\rangle^{(-)}$. Since individual states cannot occur twice, any permutation of a term $|1 : j_1; \ldots; N : j_N\rangle$ generates a different term. Therefore, there are always $N!$ terms in $|\{n_j\}\rangle^{(-)}$. If the fermions occupy the states

$|j_1\rangle, \ldots, |j_N\rangle$, i.e. $n_{j_1} = n_{j_2} = \cdots = n_{j_N} = 1$, all other $n_j = 0$, the corresponding basis state is

$$|\{n_j\}\rangle^{(-)} = \frac{1}{\sqrt{N!}} \sum_\sigma p(\sigma)|1 : j_{\sigma(1)}; 2 : j_{\sigma(2)}; \ldots; N : j_{\sigma(N)}\rangle. \tag{12.28}$$

This reminds us of the expression for a determinant. Indeed one can formally express this state in terms of the so-called **Slater determinant**,

$$|\{n_j\}\rangle^{(-)} = \frac{1}{\sqrt{N!}} \det \begin{pmatrix} 1 : |j_1\rangle & 2 : |j_1\rangle & \cdots & N : |j_1\rangle \\ 1 : |j_2\rangle & 2 : |j_2\rangle & \cdots & N : |j_2\rangle \\ \vdots & \vdots & & \vdots \\ 1 : |j_N\rangle & 2 : |j_N\rangle & \cdots & N : |j_N\rangle \end{pmatrix}. \tag{12.29}$$

At this point, we want to emphasize the importance of the Pauli principle for nature. Among other things, it ensures that each $|nlm\rangle$-state in an atom can be occupied by only two electrons (because of two independent spin states). For the higher elements, the shells are therefore occupied step by step from inside to outside. This is what gives the elements their characteristic properties and makes chemistry thereby possible.

After we have investigated N-particle states and found we have to symmetrize or antisymmetrize over all particles of a species, the question arises why it is still possible to study individual particles. The (anti-)symmetrization constitutes a kind of entanglement, cf. the definition of entanglement in Sect. 2.10. If every particle is entangled with all other particles of the same species, why is it still possible to pick one and plug it into the one-particle Schrödinger equation? This question is not that trivial to respond to. A similar problem exists already for generic kinds of entanglement. In QM, everything is entangled with everything in a quite esoteric way, and it is not always so clear by which criteria a system can be decoupled from the rest of the universe and described in isolation. In many cases, the spatial distance helps (small overlap of wave functions), or the weakness of interactions (small tendency for further entanglement). This also holds for N-particle systems. In Shankar (2011) and Messiah (2014) you can find calculations which demonstrate how small the error is that you produce if you ignore the existence of other particles, given a sufficient spatial separation.

Self-check questions:

1. How does nature ensure that indistinguishable particles cannot be "marked"?
2. What are occupation numbers? How are they used to construct a basis for N-boson and N-fermion Hilbert spaces?
3. What is the difference between orthohelium and parahelium?

12.2 Fock Space

The Fock space $\mathcal{H}^{(\pm)}$ of a given particle species is the direct sum of all its N-particle Hilbert spaces,

$$\mathcal{H}^{(\pm)} = \{|0\rangle\} \oplus \mathcal{H}_1 \oplus \mathcal{H}_2^{(\pm)} \oplus \mathcal{H}_3^{(\pm)} \oplus \cdots \tag{12.30}$$

Here $|0\rangle$ is the zero-particle state, called **vacuum**. So far we were only dealing with Hilbert spaces describing a fixed number of particles. In the Hilbert space of an electron, all kinds of strange things could happen, but at least it was clear that there was *one* electron. In the antisymmetric Hilbert space of two electrons, even stranger things could happen, in particular entanglement, but at least it was clear that there were *two* electrons. In Fock space you can't even rely on that. Now there can be superpositions between states of different particle numbers. If for example

$$|\Psi\rangle = \frac{1}{\sqrt{2}}(|\Psi_1\rangle + |\Psi_2\rangle), \tag{12.31}$$

$$|\Psi_1\rangle = |\psi_1\psi_2\rangle^{(\pm)} \in \mathcal{H}_2^{(\pm)}, \quad |\Psi_1\rangle = |\phi_1\phi_2\phi_3\rangle^{(\pm)} \in \mathcal{H}_3^{(\pm)}, \tag{12.32}$$

then there is a superposition of a two-particle and a three-particle state. In Fock space there are raising and lowering operators, so-called **creation and annihilation operators**, which interfere between the individual N-particle spaces. Let $\{|j\rangle\}$ be a basis of $\mathcal{H}_1$. Then the creation operator $A_j^\dagger$ (the notation is inherited from the harmonic oscillator) maps a subspace $\mathcal{H}_N^{(\pm)}$ to the subspace $\mathcal{H}_{N+1}^{(\pm)}$,

$$A_j^\dagger : \mathcal{H}_N^{(\pm)} \to \mathcal{H}_{N+1}^{(\pm)} \quad \text{for all } N, \tag{12.33}$$

by creating an additional particle in the state $|j\rangle$. The corresponding annihilation operator A_j removes such a particle if it exists. Using the exchange properties of bosons and fermions one can easily show that the creation operators commute in the case of bosons, but anticommute in the case of fermions,

$$\mathcal{H}^{(+)}: \quad A_{j_1}^\dagger A_{j_2}^\dagger = A_{j_2}^\dagger A_{j_1}^\dagger, \qquad \mathcal{H}^{(-)}: \quad A_{j_1}^\dagger A_{j_2}^\dagger = -A_{j_2}^\dagger A_{j_1}^\dagger. \tag{12.34}$$

The Fock space plays an important role in quantum field theory (QFT) where particles are created and annihilated all the time. It is foundational for the concept of **virtual particles**: If you have another look on the second order term in time-dependent perturbation theory, (11.40), you find appearances of H_1 at times t_2 and t_1 (t_2 is the earlier moment, hence the order). In between, there is a state $|j\rangle$ which was "created" by $H_1(t_2)$ and "annihilated" by $H_1(t_1)$. If H_1 contains creation and annihilation operators, then this corresponds to the actual creation and annihilation of a particle; this is what is called a virtual particle.

Nerd's Corner 12.1

Things become even wilder in QFT on curved spacetimes, i.e. in combination with general relativity. This is not yet Quantum Gravity, since the gravitational field (the spacetime geometry) is here only used as a background, as a classical field which is not influenced by the quantum objects under study. This is analogous to our treatment of a charged quantum object in an electromagnetic field. The latter was considered as a classical field which acts on the quantum object via the Hamiltonian operator, without begin influenced in turn by the quantum object. For that reason, what we did was not yet quantum electrodynamics.

So, in QFT on curved spacetimes it happens that it depends on the reference frame to which part of Fock space a state belongs. In flat space we have seen that a rotation or translation of the coordinate system is accompanied by a transformation of the Hilbert space. The wave functions themselves had to be rotated or shifted. The number of particles remained unchanged, of course. For the transformations on curved spacetimes this is no longer true. The most extreme example may be the **evaporation of black holes**: For an observer in an inertial frame at the event horizon there is only vacuum, the zero-particle state. For an observer outside, this vacuum appears as a mixture of N-particle states, streaming out of the black hole.

12.3 Density Operator

Given is a ray of electrons half of which have spin state $|z+\rangle$, the other half $|z-\rangle$. If we consider the ray as a huge N-particle state, we can say that the occupation number of $|z+\rangle$ equals that of $|z-\rangle$, namely $N/2$. Doesn't this contradict our claim that fermion occupation numbers are maximally 1? No, because electron states are tensor products of wave function and spin state. Only the combination has to be unique. As long as two electrons don't have the same wave function, they can occupy the same spin state. Assume the electrons are as wave packets spatially sufficiently separated, so that we can regard them as individual objects (see the remark at the end of Sect. 12.1). At time t one of these electrons runs into a detector which measures its spin (in an arbitrary direction).

How do we determine the probability for the measurement results? We see that we have nested probabilities: At first there is a probability that we got an electron in the state $|z+\rangle$ (or $|z-\rangle$). And then there is the probability to get a certain measurement result for this state. If we measure the spin in i-direction (i stands for x, y or z), then the probability $p(i+)$ for finding the value $+\hbar/2$ is given by

$$p(i+) = p_1|\langle i+|z+\rangle|^2 + p_2|\langle i+|z-\rangle|^2 = \frac{1}{2}|\langle i+|z+\rangle|^2 + \frac{1}{2}|\langle i+|z-\rangle|^2, \quad (12.35)$$

where $p_{1,2}$ are the "outer" probabilities that the electron was in the state $|z\pm\rangle$ before the measurement, and $|\langle i + |z\pm\rangle|^2$ is the "inner" probability that for a given state the measurement yields the result $(i+)$. The expectation value for the S_i-measurement is

$$\langle S_i \rangle = p_1 \langle z + |S_i|z+\rangle + p_2 \langle z - |S_i|z-\rangle. \tag{12.36}$$

Exercise 12.4
Show that in our case for any $i = x, y, z$ the result is $p(i+) = 1/2$ and $\langle S_i \rangle = 0$.

Exercise 12.5
Repeat the calculation for the case that half of the electrons have spin state $|z+\rangle$, the other half spin state $|x+\rangle$.

Can we avoid the nested probabilities by incorporating the outer probabilities into the state? What if we set

$$|\chi\rangle = \frac{1}{\sqrt{2}}(|z+\rangle + |z-\rangle) \tag{12.37}$$

for the spin state? Doesn't this give the same result? Is a mixture of $|z+\rangle$ and $|z-\rangle$ states not the same as a superposition of them? You can easily verify that for $i = y, z$ the result is indeed the same, but not for $i = x$, since $|\chi\rangle = |x+\rangle$, and so $\langle S_x \rangle = +\hbar/2$. Or if we set instead

$$|\chi\rangle = \frac{1}{\sqrt{2}}(|z+\rangle + i|z-\rangle), \tag{12.38}$$

then $\langle S_x \rangle = 0$, but $\langle S_y \rangle$ does not vanish. No matter what we try, a mixture is *not* the same as a superposition, and therefore we are dealing with two different kinds of probabilities. A problem of this kind is in general treated using the **density operator**, also called **density matrix**.

Given a Hilbert space $\mathcal{H}$ with basis $|j\rangle$ and a set of indistinguishable particles whose one-particle Hilbert space $\mathcal{H}_1$ contains the factor $\mathcal{H}$,

$$\mathcal{H}_1 = \mathcal{H} \otimes \mathcal{H}_{\text{rest}}. \tag{12.39}$$

In the example above $\mathcal{H}$ was the space of spin-$\frac{1}{2}$ states and $\mathcal{H}_{\text{rest}}$ the space of wave functions. The particles are w.r.t. $\mathcal{H}$ in n different states $|\psi_k\rangle, k = 1, \ldots, n$, where the ratio of particles occupying such a state is p_k. This means: if one randomly picks up one of the particles, then it is with probability p_k in the $\mathcal{H}$-state $|\psi_k\rangle$. In the example above we had $n = 2$, $|\psi_1\rangle = |z+\rangle$, $|\psi_2\rangle = |z-\rangle$, $p_1 = p_2 = 1/2$. In contrast to many other books we don't assume here that the $|\psi_k\rangle$ are orthogonal to each other. In the second exercise above, for example, $|\psi_1\rangle = |z+\rangle$ and $|\psi_2\rangle = |x+\rangle$ were not orthogonal. In the case $n = 1$ (all particles occupy the same $\mathcal{H}$-state, $p_1 = 1$) one speaks of a **pure state**, otherwise of a **mixture**.

The density operator ρ is defined as

$$\rho = \sum_k p_k \, |\psi_k\rangle\langle\psi_k|. \tag{12.40}$$

We remind ourselves that $|\psi_k\rangle\langle\psi_k|$ is the projection operator on the state $|\psi_k\rangle$. The density operator is thus a sum of projection operators on the occurring states, weighted with the respective probabilities. It is obviously hermitian. We further remind ourselves that the matrix components of an operator A in a given basis are given by $A_{ij} = \langle i|A|j\rangle$, and that the **trace** is a basis-independent property of an operator, which however can within a given basis be determined as the sum of the diagonal entries,

$$\mathrm{tr}(A) = \sum_i A_{ii} = \sum_i \langle i|A|i\rangle. \tag{12.41}$$

For infinite-dimensional Hilbert spaces, the trace is in general not defined, since the sum does not necessarily converge. Even if it converges, it may depend on the order of the terms, it is then no longer basis-independent. We therefore consider the trace of an operator A on an infinite-dimensional Hilbert space as well-defined only if A operates only on a finite-dimensional subspace $\mathcal{H}_A$ of $\mathcal{H}$. This means that there is a subspace $\mathcal{H}_A$ of $\mathcal{H}$ with the following properties:

- In the **orthogonal complement** of $\mathcal{H}_A$ (i.e. in all directions orthogonal to $\mathcal{H}_A$) A vanishes identically.
- $\mathcal{H}_A$ contains Image(A), i.e. all vectors $A|\psi\rangle$ for $|\psi\rangle \in \mathcal{H}$.
- $\mathcal{H}_A$ is chosen minimally, i.e. there is no smaller smaller subspace of $\mathcal{H}$ with the two properties mentioned above.
- $\mathcal{H}_A$ is finite-dimensional.

Then we define the trace of A via restriction on this subspace, $\mathrm{tr}(A) := \mathrm{tr}(A|_{\mathcal{H}_A})$.

The trace of the density operator is

$$\mathrm{tr}(\rho) = \sum_j \langle j|\rho|j\rangle \tag{12.42}$$

$$= \sum_{j,k} p_k \langle j|\psi_k\rangle\langle\psi_k|j\rangle \tag{12.43}$$

$$= \sum_{j,k} p_k \langle\psi_k|j\rangle\langle j|\psi_k\rangle \tag{12.44}$$

$$= \sum_k p_k \langle\psi_k|\psi_k\rangle = \sum_k p_k = 1, \tag{12.45}$$

where we have used $\sum_j |j\rangle\langle j| = 1$. This trace is well-defined, since we have assumed that ρ consists of only *finitely* many states $|\psi_k\rangle$; ρ operates therefore only on the subspace $\mathcal{H}_\rho$ spanned by these n states. So, we found the first important property of the density operator:

$$\text{tr}(\rho) = 1 \tag{12.46}$$

For a pure state, $\rho = |\psi_1\rangle\langle\psi_1|$ and

$$\rho^2 = |\psi_1\rangle\langle\psi_1|\psi_1\rangle\langle\psi_1| = |\psi_1\rangle\langle\psi_1| = \rho. \tag{12.47}$$

The inverse statement also holds: if $\rho^2 = \rho$, then ρ is a projection operator (this is just the abstract definition of a projection operator). A projection operator P projects states into a subspace $\mathcal{H}'$. Inside of this subspace, P equals the unit operator, in the orthogonal complement it vanishes. Its trace is therefore just the dimension d of $\mathcal{H}'$. The operator ρ has a trace of 1, hence the projection can occur only into a one-dimensional subspace, i.e. on a single state, which implies that ρ describes a pure state. So, we found the second important property of the density operator:

$$\rho^2 = \rho \quad \Leftrightarrow \quad \text{pure state} \tag{12.48}$$

The expectation value of an operator A is

$$\langle A \rangle = \sum_k p_k \langle\psi_k|A|\psi_k\rangle \tag{12.49}$$

$$= \sum_{i,j,k} p_k \langle\psi_k|i\rangle\langle i|A|j\rangle\langle j|\psi_k\rangle \tag{12.50}$$

$$= \sum_{i,j} \langle i|A|j\rangle \sum_k p_k \langle j|\psi_k\rangle\langle\psi_k|i\rangle \tag{12.51}$$

$$= \sum_{ij} A_{ij}\rho_{ji} = \text{tr}(A\rho). \tag{12.52}$$

Again, this trace is well-defined.

Exercise 12.6
What is $\mathcal{H}_{A\rho}$? Hint: the dimension d of this subspace obeys $d \leq 2n$, where n is the number of states of which ρ is composed.

So, we found the third important property of the density operator: The expectation value of A is given by the trace of $A\rho$,

$$\langle A \rangle = \text{tr}(A\rho). \tag{12.53}$$

Exercise 12.7
Compute the expectation values from Exercises 12.4 and 12.5 again, this time using the density matrix. In each case, determine the density matrix in the basis $\{|z+\rangle, |z-\rangle\}$ using matrix multiplication,

$$|\alpha(z+) + \beta(z-)\rangle\langle\alpha(z+) + \beta(z-)| = \begin{pmatrix} \alpha \\ \beta \end{pmatrix} \begin{pmatrix} \alpha^* & \beta^* \end{pmatrix} = \begin{pmatrix} \alpha^*\alpha & \beta^*\alpha \\ \alpha^*\beta & \beta^*\beta \end{pmatrix}.$$
(12.54)

Check that $\mathrm{tr}(\rho) = 1$.

Exercise 12.8

(a) Determine the density matrix for the pure state $|x+\rangle$ in the basis $\{|z+\rangle, |z-\rangle\}$ and verify that $\rho^2 = \rho$.
(b) Given the density matrix

$$\rho = \frac{1}{9}\begin{pmatrix} 4 & 4-2i \\ 4+2i & 5 \end{pmatrix}.$$
(12.55)

Show by taking the square that this describes a pure state. Which one? Use (12.54) and choose α to be real. This is always possible, because the state is not changed by a phase rotation.

Finally we investigate the time evolution. We use the die Schrödinger equation and its hermitian conjugate,

$$i\hbar|\dot\psi\rangle = H|\psi\rangle, \quad -i\hbar\langle\dot\psi| = \langle\psi|H.$$
(12.56)

It follows

$$i\hbar\dot\rho = i\hbar\frac{d}{dt}\sum_k p_k|\psi_k\rangle\langle\psi_k|$$
(12.57)

$$= i\hbar\sum_k p_k\left(|\dot\psi_k\rangle\langle\psi_k| + |\psi_k\rangle\langle\dot\psi_k|\right)$$
(12.58)

$$= \sum_k p_k\left(H|\psi_k\rangle\langle\psi_k| - |\psi_k\rangle\langle\psi_k|H\right)$$
(12.59)

$$= [H, \rho].$$
(12.60)

So, we found the fourth important property of the density operator: its time evolution is given by the **von Neumann equation**,

$$i\hbar\dot{\rho} = [H, \rho]. \tag{12.61}$$

In summary:

Density operator

- Definition:

$$\rho = \sum_k p_k |\psi_k\rangle\langle\psi_k| \tag{12.62}$$

- Trace:

$$\mathrm{tr}(\rho) = 1 \tag{12.63}$$

- Pure state:

$$\rho^2 = \rho \quad \Leftrightarrow \quad \text{pure state} \tag{12.64}$$

- Expectation values:

$$\langle A \rangle = \mathrm{tr}(A\rho) \tag{12.65}$$

- Time evolution (von Neumann equation):

$$i\hbar\dot{\rho} = [H, \rho] \tag{12.66}$$

The density operator can not only be applied to particles, but also to larger quantum systems. It can be used to describe ensembles of quantum systems in the sense of statistical mechanics.

Self-check questions:

1. How is the density operator defined and in which situations is it needed?
2. How is it used to determine expectation values?
3. How can you check if the density operator describes a pure state?

References

R. Shankar, *Principles of Quantum Mechanics*, 2nd edn. (Springer, 2011). (Excellent textbook. Two extensive chapters on the path integral)

A. Messiah, *Quantum Mechanics* (Dover Publications, 2014). (A classic and still one of the best textbooks on QM. Particularly excellent regarding relativistic generalizations)

Chapter 13
Path Integral

Abstract We demonstrate how path integrals can be made sense of mathematically.

The path integral is a method to determine the **spatial propagator** $U(\mathbf{r}', t', \mathbf{r}_0, t_0)$ without using the Schrödinger equation. Here the spatial propagator is defined via the normal propagator $U(t, t_0)$ by

$$U(\mathbf{r}', t', \mathbf{r}_0, t_0) = \langle \mathbf{r}' | U(t', t_0) | \mathbf{r}_0 \rangle. \tag{13.1}$$

The norm squared of $U(\mathbf{r}', t', \mathbf{r}_0, t_0)$ is the probability density for finding a quantum object at time t' at the position $\mathbf{r}'$ when it has been at time t_0 at the position $\mathbf{r}_0$. A standard example is the double slit experiment, for example with electrons: $\mathbf{r}_0$ is the position of the electron source which emits an electron at time t_0. What is the probability density for finding the electron at a later time t' at a position $\mathbf{r}'$ on the screen? The answer, leading to the famous interference pattern, is that one has to add up two contributions of the wave functions, where each contribution results from one of the two paths the electron can take, with a phase that depends on this path:

$$U(\mathbf{r}', t', \mathbf{r}_0, t_0) \sim \sum_{\text{paths}} e^{i\,\text{phase(path)}} \tag{13.2}$$

The idea of the path integral is to generalize this summation: a quantum object takes all possible paths to get from $\mathbf{r}_0$ to $\mathbf{r}'$. The phases resulting from each of these paths interfere at $(\mathbf{r}', t')$, where a certain amplitude is left over, namely $U(\mathbf{r}', t', \mathbf{r}_0, t_0)$. In short, the claim made by the theory of the path integral is:

Path integral

$$U(\mathbf{r}', t', \mathbf{r}_0, t_0) = \int \mathcal{D}[\mathbf{r}(t)]\, e^{i\,S[\mathbf{r}(t)]/\hbar} \tag{13.3}$$

© Springer International Publishing Switzerland 2016
J.-M. Schwindt, *Conceptual Basis of Quantum Mechanics*,
Undergraduate Lecture Notes in Physics, DOI 10.1007/978-3-319-24526-3_13

"What the heck does this mean?", you may ask. In the literature you will find some "proofs" why the statement above is equivalent to the Schrödinger equation (see e.g. Shankar (2011)). The proofs all have some pitfalls, their open or hidden assumptions, and are all to be treated with caution. And good authors (Shankar is one of them) point that out. We don't even want to be so ambitious here to show the equivalence or even to calculate something useful with this formalism (for that, Shankar (2011) also gives some nice examples). Here we only want to try to clarify what the right hand side of (13.3) actually means. For that, we restrict ourselves to a single space dimension and replace $\mathbf{r}$ with x,

$$U(x', t', x_0, t_0) = \int \mathcal{D}[x(t)] e^{i S[x(t)]/\hbar}. \tag{13.4}$$

The expression $\int \mathcal{D}[x(t)]$ is the "integral over all paths", which we want to define in a moment by taking three subsequent limits. But before that, we try to make sense of the exponential expression $e^{i S[x(t)]/\hbar}$. Here $S[x(t)]$ is the classical **action**

$$S = \int_{t_0}^{t'} dt\, L(x(t), \dot{x}(t)). \tag{13.5}$$

$L = T - V$ is the Lagrangian function (kinetic minus potential energy), which does not explicitly depend on time here (no time-dependent potential). The action $S[x(t)]$ is a **functional**, i.e. it maps each function $x(t)$ to a number. The function $x(t)$ obeys the boundary conditions $x(t_0) = x_0$ and $x(t') = x'$. The functional $S[x(t)]$ has the property that it is extremal for the **classical path** $x_{cl}(t)$ (**Hamilton's principle**), i.e. for the path which leads from $x(t_0) = x_0$ to $x(t') = x'$ and thereby obeys the classical equations of motions (assuming such a path exists and is unique).

Hamilton's principle is also the reason why S (up to a factor) is such a good candidate for the phase in the path integral (that's at least the common folklore): One can imagine that S oscillates very fast for paths which deviate strongly from the classical one, so that the different phases average to (almost) zero in the summation "over all paths". Only in a small region around the classical path, the variation of S is small, due to the extremum, so that paths in this region interfere constructively, due to their similar phases, and thereby provide the main contribution to the path integral. This is what one expects for quantum theory (according to the folklore). Classical physics is in many respects a good approximation. In particular, expectation values obey the classical equations of motion. Therefore the assumption seems justified that QM only induces "quantum fluctuations" around the classical path. We are not going to discuss here how far this folklore is justified. (Only one remark: the tunnel effect does not fit into the picture. In this case there is no classical path leading through the potential barrier.)

A different more formal motivation for choosing S as the phase in the path integral is given by the connection between QM and the Hamilton-Jacobi formalism, see nerd's corner 3.3. The action of the classical path occurs there as a solution for the phase of the wave function in the classical approximation.

As a little example that will turn out to be useful in a moment, we calculate the action of a free particle ($V = 0$), moving on the classical path, i.e. with constant velocity v, from x_0 to x':

$$S[x(t)] = \int_{t_0}^{t'} dt\, L(x(t)) = \int_{t_0}^{t'} dt\, \frac{m}{2} v^2 \tag{13.6}$$

$$= (t' - t_0)\frac{m}{2}\left(\frac{x' - x_0}{t' - t_0}\right)^2 = \frac{m}{2}\frac{(x' - x_0)^2}{t' - t_0} \tag{13.7}$$

Now we turn to the main problem: the strange expression $\int \mathcal{D}[x(t)]$, the "integral over all paths". Here we have to go a bit further afield. Initially, we restrict the allowed values of x to a finite interval $[-a, a]$ and discretize spacetime for $t \in [t_0, t']$ and $x \in [-a, a]$. That is, we calculate with finitely many spatial points and finitely many points in time, with uniform distances

$$\Delta x = \epsilon_1 = \frac{2a}{N_1}, \quad \Delta t = \epsilon_2 = \frac{t' - t_0}{N_2}. \tag{13.8}$$

We have thus reduced the aforementioned region of two-dimensional spacetime to $(N_1 + 1)(N_2 + 1)$ points. The points in time are enumerated by $t_n = t_0 + n\epsilon_2$, for $n = 0, 1, \ldots, N_2$, in particular $t' = t_{N_2}$. Now we form all possible functions $x(t)$ on this lattice, with boundary conditions $x(t_0) = x_0$ and $x(t_{N_2}) = x_1$. (We assume that x_0 and x' are on the lattice. Otherwise we shift the origin and change a accordingly.) For each moment t_1 to t_{N_2-1}, the x-values can be freely chosen from the $N_1 + 1$ possibilities, we therefore get $(N_1 + 1)^{N_2-1}$ possible "paths" from x_0 to x' we have to sum over. But first we have to define $S[x(t)]$. We denote the function values $x(t_n)$ by x_n. The action $S[x(t)]$ is the sum of the actions $S_n(x_{n-1}, x_n)$ of the individual time slices $[t_{n-1}, t_n]$, with $n = 1, \ldots, N_2$,

$$S[x(t)] = \sum_{n=1}^{N_2} S_n(x_{n-1}, x_n). \tag{13.9}$$

Here we *define* that $S_n(x_{n-1}, x_n)$ results from a uniform motion from x_{n-1} to x_n, and the potential V is evaluated at the point $\frac{x_{n-1}+x_n}{2}$. That is, we set per definition

$$S_n(x_{n-1}, x_n) = \frac{m}{2}\frac{(x_n - x_{n-1})^2}{\epsilon_2} - \epsilon_2 V\left(\frac{x_{n-1} + x_n}{2}\right)^2, \tag{13.10}$$

where we have used the result (13.7) for the kinetic term. The summation over all paths in (13.4) currently reads

$$\sum_{x_1}\sum_{x_2}\cdots\sum_{x_{N_2-1}} e^{iS[x(t)]/\hbar}, \tag{13.11}$$

where each sum has to be performed over the $N_1 + 1$ possible values of x_n, and for S the (13.9) and (13.10) have to be inserted. This is a well-defined expression. Now we proceed to the actual path integral by taking three subsequent limits:

- At first we take $\epsilon_1 \to 0$. Thereby the sums become integrals, and the expression for the path integral now reads

$$\int_{-a}^{a} dx_1 \int_{-a}^{a} dx_2 \cdots \int_{-a}^{a} dx_{N_2-1} \, e^{i S[x(t)]/\hbar}. \tag{13.12}$$

So far no problem, everything is still well-defined.

- Next, the limit $a \to \infty$ shall be taken. Now it gets problematic. The exponential function oscillates faster and faster when some x_n is increased, due to the quadratic term in (13.10). The integrals therefore don't converge. However, such a behavior is not new to us. We remember the formal integral

$$\frac{1}{2\pi} \int_{-\infty}^{\infty} dx \, e^{ikx} = \delta(k). \tag{13.13}$$

This integral also doesn't converge, actually, for $k \neq 0$ (for $k = 0$ it doesn't anyhow), and still we use it successfully. How can that be justified? We can help ourselves by defining the limit such that it "averages" over the integral at infinity. For example, one can use the following definition:

$$\int_{-\infty}^{\infty} dx \, f(x) := \lim_{L \to \infty} \frac{1}{L} \int_{L}^{2L} dL' \int_{-L'}^{L'} dx \, f(x) \tag{13.14}$$

In other words, we nest the actual integral inside a second one, which averages over several limits of the actual one. We demonstrate this for the example $f(x) = e^{ikx}$ with $k \neq 0$. The inner integral is then

$$\int_{-L'}^{L'} dx \, e^{ikx} = \frac{2}{k} \sin(kL'). \tag{13.15}$$

The outer integral

$$\frac{1}{L} \int_{L}^{2L} dL' \frac{2}{k} \sin(kL') \tag{13.16}$$

takes the average over the sine function in the interval $L < L' < 2L$, and this average value converges to 0. With this definition, $\int_{-\infty}^{\infty} dx \, e^{ikx} = 0$ is for $k \neq 0$ well-defined. For functions which are integrable from $-\infty$ to $+\infty$ in the ordinary sense, nothing changes, since

$$\frac{1}{L}\int_L^{2L} dL' \int_{-L'}^{L'} dx\; f(x) \tag{13.17}$$

$$= \int_{-L}^{L} dx\; f(x) + \frac{1}{L}\int_L^{2L} dL' \left[\int_{-L'}^{-L} dx\; f(x) + \int_L^{L'} dx\; f(x) \right].$$

In the limit $L \to \infty$, the first term becomes the ordinary $\int_{-\infty}^{\infty} dx\; f(x)$, and the expression in square brackets converges to zero. The definition (13.14) makes sense. The path integral thus becomes

$$\int_{-\infty}^{\infty} dx_1 \int_{-\infty}^{\infty} dx_2 \cdots \int_{-\infty}^{\infty} dx_{N_2-1}\; e^{i S[x(t)]/\hbar}. \tag{13.18}$$

- Before we turn to the final limit, we remember that we want to obtain (13.4) in the end, an expression for the spatial propagator. It turns out that therefore each integral in (13.18) has to be provided with a constant factor $1/C$, which we don't further specify here, to give the correct result. Another factor has to be applied to the entire expression. The path integral now reads

$$\frac{1}{C} \int_{-\infty}^{\infty} \frac{dx_1}{C} \int_{-\infty}^{\infty} \frac{dx_1}{C} \cdots \int_{-\infty}^{\infty} \frac{dx_{N_2-1}}{C}\; e^{i S[x(t)]/\hbar}. \tag{13.19}$$

- The final limit is $\epsilon_2 \to 0$, i.e. $N_2 \to \infty$. This takes us from discrete to continuous time. This limit should not be taken too soon, for otherwise one gets a meaningless C^∞ in the denominator and infinitely many integrations. The integrations have to be performed *first* and combined with the factors of $1/C$, only then the limit $\epsilon_2 \to 0$ can be taken.

So, we have explained how the right hand side of (13.4) is to be interpreted, i.e. what a path integral is. How to perform calculations with it, is a different question. In the end, it almost always amounts to an expansion around the classical path. For examples we refer you to Shankar (2011). In QFT, path integrals are primarily used as a purely formal tool to derive Feynman diagrams, using a certain heuristics. An interesting aspect is the formal similarity with the partition function of statistical mechanics. For that, we also refer you to the literature.

Reference

R. Shankar, *Principles of Quantum Mechanics*, 2nd edn. (Springer, 2011). (Excellent textbook. Two extensive chapters on the path integral)

Chapter 14
Dirac Equation

Abstract We show how a glance of beauty entered QM, and in what way we live on an infinite sea of particles with negative energy.

Glorious moments of theoretical physics come about when a representative of this guild is able to derive from a few simple but fundamental principles an equation which is friendly approved by nature and which explains a number of until then unexplained phenomena. This was for example the case when Einstein found general relativity. In QM things looked totally different. The theorists were driven by seemingly con-tradictory, absurd observations; only with great effort they were able to bring some order into the chaos, with Heisenberg's matrix mechanics or Schrödinger's wave mechanics; but the interpretation and deeper meaning of the whole thing remained unclear. Only with the Dirac equation, the glory of theory entered QM. With a few simple considerations, Dirac found an equation which at one blow

- provided a relativistic generalization of the QM of the electron,
- explained the electron's spin
- described the interaction between spin and a magnetic field,
- explained the spin-orbit interaction of the hydrogen atom,
- predicted the positron, the first example of an antimatter particle.

In the following we set for simplicity the speed of light $c = 1$, as it is often done in theoretical physics. This means that we measure temporal as well as spatial distances in meters, with the conversion factor provided by $c = 1$, i.e. $1s = 300.000$ km.

In relativistic mechanics, position $\mathbf{r}$ and time t are combined into a four-position $x^\mu = (t, x, y, z)$, similarly momentum $\mathbf{p}$ and energy E into a four-momentum $p^\mu = (E, p_x, p_y, p_z)$. There is an aspect of QM which encourages a relativistic generalization: the association of E with a time derivative (the operator H of energy is in the Schrödinger equation identified with a time derivative) and $\mathbf{p}$ with spatial derivatives fits well into this four-schema.

The nonrelativistic energy of a free particle is $E = p^2/(2m)$, and therefore the Schrödinger equation contains only a first derivative w.r.t. time, but second derivatives w.r.t. space. Relativistically, $E^2 = p^2 + m^2$; on the level of operators we therefore set

$$H^2|\psi\rangle = (P^2 + m^2)|\psi\rangle. \tag{14.1}$$

© Springer International Publishing Switzerland 2016
J.-M. Schwindt, *Conceptual Basis of Quantum Mechanics*,
Undergraduate Lecture Notes in Physics, DOI 10.1007/978-3-319-24526-3_14

With the replacement

$$H \to i\hbar \frac{\partial}{\partial t}, \quad P_i \to -i\hbar \frac{\partial}{\partial x_i}, \tag{14.2}$$

this becomes the **Klein-Gordon equation**

$$\left[\frac{\partial^2}{\partial t^2} - \Delta + \frac{m^2}{\hbar^2} \right] \psi(\mathbf{r}, t) = 0. \tag{14.3}$$

It has the disadvantage that it is second order in time, which goes against a fundamental property of QM, namely that the time evolution of a state can be deduced from the knowledge of the state alone (and the Hamiltonian operator), without knowledge of its first derivative w.r.t. time. (There are further problems with the Klein-Gordon equation, which we don't discuss here; we refer the interested reader to Messiah (2014).) But maybe an equation of the form

$$i\hbar \frac{\partial}{\partial t} |\psi\rangle = H |\psi\rangle \tag{14.4}$$

is still possible? The simplest would be to take the square root of $H^2 = P^2 + m^2$

$$i\hbar \frac{\partial}{\partial t} |\psi\rangle = \sqrt{P^2 + m^2} |\psi\rangle, \tag{14.5}$$

as we have done already for the relativistic correction to the hydrogen atom, and to expand it in powers of P. But this equation violates the spirit of relativity: on the right hand side, spatial derivatives of arbitrary order occur, and the symmetry between space and time is broken. Dirac's goal was to find an equation which is first order in space and time and which implies the Klein-Gordon equation and hence the Lorentz symmetry. Therefore he had to extend the space $\mathcal{H}_\psi$ of wave functions via a tensor product with a d-dimensional space $\mathcal{H}_\zeta$,

$$\mathcal{H} = \mathcal{H}_\zeta \otimes \mathcal{H}_\psi, \tag{14.6}$$

similar to what we have done for the spin. (At this point we assume that we don't know yet anything about spin.) An element of $\mathcal{H}_\zeta$ is called **Dirac spinor**. If one expresses a state $|\psi\rangle$ as a wave function, this has d components, $\psi_a(\mathbf{r}, t)$, $a = 1, \ldots d$. Now Dirac made the following ansatz for the Hamiltonian operator:

$$H = \sum_{i=1}^{3} \alpha_i \otimes P_i + m\beta \otimes \mathbf{1} \tag{14.7}$$

Here α_i and β are $(d \times d)$-matrices acting in $\mathcal{H}_\zeta$. If one can accomplish that

$$\alpha_i^2 = \beta^2 = 1, \quad \alpha_i \alpha_j + \alpha_j \alpha_i = 2\delta_{ij} \mathbf{1}, \quad \alpha_i \beta + \beta \alpha_i = 0, \tag{14.8}$$

then it follows (please verify!)

$$H^2 = 1 \otimes P^2 + m^2 1 \otimes 1, \qquad (14.9)$$

which is the correct relativistic expression, only enhanced by irrelevant unit matrices in $\mathcal{H}_\zeta$. With (14.7) One obtains as a relativistic generalization of the free Schrödinger equation

---

Dirac equation

$$i\hbar \frac{d}{dt} |\psi\rangle = \left(\sum_{i=1}^{3} \alpha_i \otimes P_i + m\beta \otimes 1 \right) |\psi\rangle. \qquad (14.10)$$

In a more modern notation one defines $\gamma^0 = \beta$, $\gamma^i = \beta \alpha_i$, for then one can rewrite (14.10) to

$$\left(i\hbar \sum_{\mu=0}^{3} \gamma^\mu \otimes \frac{\partial}{\partial x^\mu} - m \right) \psi(\mathbf{r}, t) = 0, \qquad (14.11)$$

where $\psi(\mathbf{r}, t)$ is to be understood as a d-dimensional vector in $\mathcal{H}_\zeta$. Since we are interested in energy eigenvalues, the notation (14.10) is more useful to us.

So far everything was hypothetical, for we first have to find matrices with the properties (14.8). It turns out that d has to be at least 4. Then, a possibility is, written in 2×2 blocks,

$$\alpha_i = \begin{pmatrix} 0 & \sigma_i \\ \sigma_i & 0 \end{pmatrix}, \quad \beta = \begin{pmatrix} 1 & 0 \\ 0 & -1 \end{pmatrix} \qquad (14.12)$$

(other possibilities can be obtained via unitary transformations in $\mathcal{H}_\zeta$).

Exercise 14.1

Verify using the known properties of the Pauli matrices σ_i that these matrices α_i, β have the properties required by (14.8).

At this point it appears somewhat disturbing that we have four spinor components. We expected only two! To solve this puzzle, we write down the energy eigenvalue equation,

$$\left(\sum_{i=1}^{3} \alpha_i \otimes P_i + m\beta \otimes 1 \right) |\psi\rangle = E |\psi\rangle. \qquad (14.13)$$

Next, we split the four-component spinor ψ into two two-component parts χ and ξ,

$$\begin{pmatrix} \psi_1 \\ \psi_2 \\ \psi_3 \\ \psi_4 \end{pmatrix} = \begin{pmatrix} \chi_1 \\ \chi_2 \\ \xi_1 \\ \xi_2 \end{pmatrix}. \tag{14.14}$$

Each component is to be understood as a function of space and time. Inserting (14.12) into (14.13), we obtain

$$\begin{pmatrix} (E-m)\mathbf{1} & -\boldsymbol{\sigma}\cdot\mathbf{P} \\ -\boldsymbol{\sigma}\cdot\mathbf{P} & (E+m)\mathbf{1} \end{pmatrix} \begin{pmatrix} \chi \\ \xi \end{pmatrix} = \begin{pmatrix} 0 \\ 0 \end{pmatrix} \tag{14.15}$$

with the definition

$$\boldsymbol{\sigma}\cdot\mathbf{P} = \sum_{i=1}^{3} \sigma_i \otimes P_i. \tag{14.16}$$

Now we have a look at an eigenstate of momentum with $\mathbf{p} = 0$. Since we have a free particle (no potential in the Dirac equation), this is automatically also an energy eigenstate. For the energy eigenvalue we expect $E = m$. But let's keep this open for the moment and plug $\mathbf{p} = 0$ into (14.15):

$$\begin{pmatrix} (E-m)\mathbf{1} & 0 \\ 0 & (E+m)\mathbf{1} \end{pmatrix} \begin{pmatrix} \chi \\ \xi \end{pmatrix} = \begin{pmatrix} 0 \\ 0 \end{pmatrix} \tag{14.17}$$

This gives $(E-m)\chi = 0$ and $(E+m)\xi = 0$, with two possible solutions: $E = m$ and $\xi = 0$, or $E = -m$ and $\chi = 0$. For the expected solution with $E = m$ there are only two possible spinor components, the other two components correspond to strange states with negative energy $E = -m$. Now what is that supposed to mean? One can easily verify that also for $\mathbf{p} \neq 0$ these circumstances are obtained: there are two eigenspinors for the energy $E = \sqrt{p^2 + m^2}$ and two for the energy $E = -\sqrt{p^2 + m^2}$. Only the eigenspinors of positive energy are physically meaningful. The two-dimensional subspace of H_ζ spanned by them is the spinor space we know from nonrelativistic QM.

Dirac interpreted this in the following way: If the states of negative energy were unoccupied, then all electrons would drop down into these states and set free huge amounts of energy. This does not happen, and so they have to be already occupied. Infinitely many electrons fill this "Dirac sea" of negative energies, so that all these states are occupied. If one of the sea electrons receives an energy of $2m$, for example by absorption of a photon, it can jump from $E = -m$ to $E = +m$ and become a "real" electron in this way. In the sea this creates a hole. This hole has properties similar to those of the electron, but with positive charge. It is a piece of antimatter, a positron. In this way Dirac predicted the positron, and a short time later it was discovered. The sea interpretation with its infinitely many sea electrons is still quite adventurous, and was in the mean time replaced by more modern but also more complicated points of view.

Finally we want to derive the interaction between spin and magnetic field as well as the g-factor $g_e = 2$ from the Dirac equation. In the following, instructions are provided; the details of the calculations are left to you as an exercise.

Exercise 14.2

(a) Conclude from (14.15) that

$$\xi = \frac{\sigma \cdot P}{E + m} \chi. \tag{14.18}$$

Estimate for the nonrelativistic case $v \ll 1$ (we are only interested in this case) the relative sizes. We consider here only solutions with positive energy! Make the approximation

$$\xi \approx \frac{\sigma \cdot P}{2m} \chi. \tag{14.19}$$

(b) Plug this result back into (14.15) and obtain for χ in the nonrelativistic approximation the equation

$$\frac{(\sigma \cdot P)(\sigma \cdot P)}{2m} \chi = (E - m)\chi = E_S \chi. \tag{14.20}$$

Here $E_S = E - m$ is the energy as it appears in the Schrödinger equation (the energy associated with the rest mass is missing there).

Now we introduce again an electromagnetic potential A (we don't need the electrostatic one, as we are only interested in a constant magnetic field). Again we replace

$$P \rightarrow P - qA \tag{14.21}$$

and obtain the **Pauli equation**

$$\frac{[\sigma \cdot (P - qA)][\sigma \cdot (P - qA)]}{2m} \chi = (E - m)\chi = E_S \chi. \tag{14.22}$$

(c) Show, using the properties of the Pauli matrices, the general relation

$$(\sigma \cdot U)(\sigma \cdot V) = U \cdot V + i\sigma \cdot (U \times V). \tag{14.23}$$

(d) Why do vectorial operators U not necessarily obey $U \times U = 0$? Show that

$$(P - qA) \times (P - qA) = iq\hbar B, \tag{14.24}$$

where **B** is the magnetic field corresponding to **A**.

(e) Conclude

$$\left[\frac{(\mathbf{P} - q\mathbf{A})^2}{2m} - \frac{q\hbar}{2m}\sigma \cdot \mathbf{B} \right] \chi = E_S \chi. \qquad (14.25)$$

and from that $g_e = 2$.

For a deeper understanding of these topics we recommend (Messiah 2014).

Reference

A. Messiah, *Quantum Mechanics* (Dover Publications, 2014) (A classic and still one of the best textbooks on QM. Particularly excellent regarding relativistic generalizations.)

Solutions to the Exercises

Solution 2.1

(a)

$$A^{(e)}|v^{(e)}\rangle = \begin{pmatrix} 0 & 1 \\ 1 & 0 \end{pmatrix}\begin{pmatrix} 0 \\ 1 \end{pmatrix} = \begin{pmatrix} 1 \\ 0 \end{pmatrix}$$

$$\langle u^{(e)}|A^{(e)} = (1\ 0)\begin{pmatrix} 0 & 1 \\ 1 & 0 \end{pmatrix} = (0\ 1)$$

$$\langle u^{(e)}|A^{(e)}v^{(e)}\rangle = (1\ 0)\begin{pmatrix} 1 \\ 0 \end{pmatrix} = 1 = (0\ 1)\begin{pmatrix} 0 \\ 1 \end{pmatrix} = \langle u^{(e)}A^{(e)}|v^{(e)}\rangle$$

(b)

$$\langle u^{(f)}| = (1\ 0), \quad |v^{(f)}\rangle = \begin{pmatrix} 0 \\ \frac{1}{2} \end{pmatrix}$$

$$A|f_1\rangle = A|e_1\rangle = |e_2\rangle = \frac{1}{2}|f_2\rangle$$

$$A|f_2\rangle = 2A|e_2\rangle = 2|e_1\rangle = 2|f_1\rangle$$

$$\Rightarrow \quad A^{(f)} = \begin{pmatrix} 0 & 2 \\ \frac{1}{2} & 0 \end{pmatrix}$$

$$\Rightarrow \quad A^{(f)}|v^{(f)}\rangle = \begin{pmatrix} 1 \\ 0 \end{pmatrix}, \quad \langle u^{(f)}|A^{(f)} = (0\ 2)$$

© Springer International Publishing Switzerland 2016
J.-M. Schwindt, *Conceptual Basis of Quantum Mechanics*,
Undergraduate Lecture Notes in Physics, DOI 10.1007/978-3-319-24526-3

Scalar product in the f-basis:

$$(0\ 1) \cdot \begin{pmatrix} 0 \\ 1 \end{pmatrix} = \langle f_2 | f_2 \rangle = 4 \quad \Rightarrow \quad (\alpha\ \beta) \cdot \begin{pmatrix} \gamma \\ \delta \end{pmatrix} = \alpha\gamma + 4\beta\delta$$

$$\Rightarrow \quad \langle u^{(e)} | A^{(e)} v^{(e)} \rangle = 1, \quad \langle u^{(e)} A^{(e)} | v^{(e)} \rangle = 4$$

Solution 2.2

It is sufficient to show: $[(AB)^{(e)} | v^{(e)} \rangle]_i = [(A^{(e)} B^{(e)}) | v^{(e)} \rangle]_i$

With $AB|v\rangle = A(B|v\rangle)$ we compute:

$$\left[(AB)^{(e)} | v^{(e)} \rangle \right]_i = \left[A^{(e)}(B^{(e)} | v^{(e)} \rangle) \right]_i$$

$$= \sum_j A_{ij}^{(e)} (B^{(e)} | v^{(e)} \rangle)_j = \sum_j A_{ij}^{(e)} \sum_k B_{jk}^{(e)} | v^{(e)} \rangle_k$$

$$= \sum_k \left(\sum_j A_{ij}^{(e)} B_{jk}^{(e)} \right) | v^{(e)} \rangle_k = \left(A^{(e)} B^{(e)} \right)_{ik} | v^{(e)} \rangle_k$$

$$= \left[(A^{(e)} B^{(e)}) | v^{(e)} \rangle \right]_i$$

Solution 2.4

Characteristic polynomial:

$$\det(\lambda \mathbf{1} - \sigma_y) = \lambda^2 - 1 \quad \Rightarrow \quad \text{Eigenvalues } \pm 1$$

Eigenvalue 1:

$$\begin{pmatrix} 1 & i \\ -i & 1 \end{pmatrix} \begin{pmatrix} \alpha \\ \beta \end{pmatrix} = 0 \quad \rightarrow \quad \beta = i\alpha$$

Eigenvalue -1:

$$\begin{pmatrix} -1 & i \\ -i & -1 \end{pmatrix} \begin{pmatrix} \alpha \\ \beta \end{pmatrix} = 0 \quad \rightarrow \quad \beta = -i\alpha$$

$\Rightarrow$ The eigenspace for $\lambda = 1$ is spanned by $\begin{pmatrix} 1 \\ i \end{pmatrix}$, that of $\lambda = -1$ by $\begin{pmatrix} 1 \\ -i \end{pmatrix}$.

Solution 2.5

(a) The characteristic polynomial is $(\lambda - 1)^2$, giving 1 as the only eigenvalue. The corresponding eigenspace is spanned by $\begin{pmatrix} 1 \\ 0 \end{pmatrix}$, which is only a subspace of $\mathcal{H}$.

(b) The characteristic polynomial is $(\lambda - 1)(\lambda - 2)$. The eigenvalues are 1 and 2, the corresponding eigenspaces are spanned by $\begin{pmatrix} 1 \\ -1 \end{pmatrix}$ and $\begin{pmatrix} 1 \\ 0 \end{pmatrix}$, respectively. They are not orthogonal.

Solution 2.6

$$|y+\rangle = \frac{1}{\sqrt{2}}\begin{pmatrix}1\\i\end{pmatrix}, \quad \frac{1}{\sqrt{2}}\langle y+| = (1, -i),$$

$$|y-\rangle = \frac{1}{\sqrt{2}}\begin{pmatrix}1\\-i\end{pmatrix}, \quad \frac{1}{\sqrt{2}}\langle y+| = (1, i),$$

$$P_{y+} = |y+\rangle\langle y+| = \frac{1}{2}\begin{pmatrix}1 & -i\\i & 1\end{pmatrix}, \quad P_{y-} = |y-\rangle\langle y-| = \frac{1}{2}\begin{pmatrix}1 & i\\-i & 1\end{pmatrix},$$

therefore $P_{y+} + P_{y-} = 1$.

$$p(y+) = \langle v|P_{y+}|v\rangle = \frac{1}{2}\begin{pmatrix}\alpha^* & \beta^*\end{pmatrix}\begin{pmatrix}1 & -i\\i & 1\end{pmatrix}\begin{pmatrix}\alpha\\\beta\end{pmatrix}$$

$$= \frac{1}{2}\begin{pmatrix}\alpha^* & \beta^*\end{pmatrix}\begin{pmatrix}\alpha - i\beta\\i\alpha + \beta\end{pmatrix} = \frac{1}{2}(\alpha^* + i\beta^*)(\alpha - i\beta)$$

and similarly

$$p(y-) = \frac{1}{2}(\alpha^* - i\beta^*)(\alpha + i\beta).$$

In combination we get again

$$p(y+) + p(y-) = \alpha^*\alpha + \beta^*\beta = 1.$$

Solution 2.7
Simply plug $|x+\rangle = \frac{1}{\sqrt{2}}\begin{pmatrix}1\\1\end{pmatrix}$ and $|x-\rangle = \frac{1}{\sqrt{2}}\begin{pmatrix}1\\-1\end{pmatrix}$ into the known expressions for $p(y\pm)$ and $p(z\pm)$. Same for $|y\pm\rangle$ and $|z\pm\rangle$.

Solution 2.8

$$\det(\lambda 1 - P_{x+}) = \det\begin{pmatrix}\lambda - \frac{1}{2} & -\frac{1}{2}\\-\frac{1}{2} & \lambda - \frac{1}{2}\end{pmatrix}$$

$$= (\lambda - \frac{1}{2})^2 - \frac{1}{4} = \lambda^2 - \lambda = \lambda(\lambda - 1),$$

so the eigenvalues are 0 and 1. Similarly for P_{x-}.

Solution 2.10
Just as for σ_y one has

$$\sigma_x^{2n+1} = \sigma_x, \quad \sigma_x^{2n} = 1.$$

It follows

$$U_x(\alpha) = e^{-i\alpha\sigma_x} = \sum_{n=0}^{\infty} \frac{i^n \alpha^n}{n!} \sigma_x^n$$

$$= \begin{pmatrix} \sum_{k=0}^{\infty}(-1)^k \frac{\alpha^{2k}}{(2k)!} & (-i)\sum_{k=0}^{\infty}(-1)^k \frac{\alpha^{2k+1}}{(2k+1)!} \\ (-i)\sum_{k=0}^{\infty}(-1)^k \frac{\alpha^{2k+1}}{(2k+1)!} & \sum_{k=0}^{\infty}(-1)^k \frac{\alpha^{2k}}{(2k)!} \end{pmatrix}$$

$$= \begin{pmatrix} \cos\alpha & -i\sin\alpha \\ -i\sin\alpha & \cos\alpha \end{pmatrix}.$$

The result for $U_z(\alpha)$ can be read directly from σ_z, since σ_z is diagonal.

Solution 2.11
The result follows using

$$U_x^{\dagger}(\alpha) = \begin{pmatrix} \cos\alpha & i\sin\alpha \\ i\sin\alpha & \cos\alpha \end{pmatrix}, \quad U_y^{\dagger}(\alpha) = \begin{pmatrix} \cos\alpha & \sin\alpha \\ -\sin\alpha & \cos\alpha \end{pmatrix},$$

$$U_z^{\dagger}(\alpha) = \begin{pmatrix} e^{i\alpha} & 0 \\ 0 & e^{-i\alpha} \end{pmatrix}.$$

Solution 2.13
With

$$U_x(\frac{\pi}{4}) = \frac{1}{\sqrt{2}} \begin{pmatrix} 1 & -i \\ -i & 1 \end{pmatrix}$$

one derives

$$U_x(\frac{\pi}{4})|y+\rangle = |z+\rangle, \quad U_x(\frac{\pi}{4})|y-\rangle = -i|z-\rangle,$$

$$U_x(\frac{\pi}{4})|z+\rangle = |y-\rangle, \quad U_x(\frac{\pi}{4})|z-\rangle = -i|y+\rangle.$$

The phase factor i is irrelevant. With arbitrary α one finds

$$U_x(\alpha)|x+\rangle = \frac{1}{\sqrt{2}} \begin{pmatrix} \cos\alpha - i\sin\alpha \\ \cos\alpha - i\sin\alpha \end{pmatrix} = e^{-i\alpha}|x+\rangle$$

and similarly

$$U_x(\alpha)|x-\rangle = e^{i\alpha}|x-\rangle.$$

With

$$U_z(\frac{\pi}{4}) = \frac{1}{\sqrt{2}} \begin{pmatrix} 1-i & 0 \\ 0 & 1+i \end{pmatrix}$$

one derives

$$U_z(\tfrac{\pi}{4})|x+\rangle = \frac{1-i}{\sqrt{2}}|y-\rangle, \quad U_z(\tfrac{\pi}{4})|x-\rangle = \frac{1-i}{\sqrt{2}}|y+\rangle,$$

$$U_z(\tfrac{\pi}{4})|y+\rangle = \frac{1-i}{\sqrt{2}}|x+\rangle, \quad U_z(\tfrac{\pi}{4})|y-\rangle = \frac{1-i}{\sqrt{2}}|x-\rangle.$$

Directly from the matrix $U_z(\alpha)$ one can read off

$$U_z(\alpha)|z\pm\rangle = e^{\mp i\alpha}|z\pm\rangle.$$

Solution 2.14
After the matrix multiplication one has to apply the identities

$$\cos^2\alpha - \sin^2\alpha = \cos(2\alpha), \qquad 2\sin\alpha\cos\alpha = \sin(2\alpha).$$

Solution 2.16
The observable a takes on the values a_i. In the corresponding orthonormal basis $\{|e_i\rangle\}$ of eigenvectors, $A^{(e)}$ is diagonal with the values a_i on the diagonal. The observable b takes the values $b_i = \sum_n \alpha_n a_i^n$. The eigenvectors are the same. Therefore $B^{(e)}$ is diagonal with values b_i on the diagonal. This is the matrix of the operator $\sum_n \alpha_n A^n$.

Solution 2.17
From the product rule follows

$$\frac{d}{dt}e^A = \frac{d}{dt}A + \frac{1}{2}\frac{d}{dt}A^2 + \frac{1}{6}\frac{d}{dt}A^3 + \cdots$$

$$= \dot{A} + \frac{1}{2}(\dot{A}A + A\dot{A}) + \frac{1}{6}(\dot{A}A^2 + A\dot{A}A + A^2\dot{A}) + \cdots,$$

where we used the notation $\dot{A}$ for $\frac{d}{dt}A$. Only if A commutes with $\dot{A}$, one can pull $\dot{A}$ to the front of the products, and it follows

$$\frac{d}{dt}e^A = \dot{A} + \dot{A}A + \frac{1}{2}\dot{A}A^2 + \cdots$$

$$= \dot{A}(1 + A + \frac{1}{2}A^2 + \cdots) = \dot{A}e^A.$$

With

$$\frac{d}{dt}\int_{t_0}^t dt'\, H(t') = H(t),$$

(2.168) is a solution of (2.165) if and only if $\frac{d}{dt}e^A = \dot{A}e^A$ can be applied, that is, if $H(t)$ commutes with $\int_{t_0}^t dt'\, H(t')$. In general this is not the case.

Solution 2.18

Exchange the names of the integration variables in the second term, $t_1 \leftrightarrow t_2$. Then it remains to show:

$$\int_{t_0}^{t} dt_1 \int_{t_0}^{t_1} dt_2 \, H(t_1)H(t_2) = \int_{t_0}^{t} dt_2 \int_{t_2}^{t} dt_1 \, H(t_1)H(t_2)$$

The integrals are identical: In both cases one integrates over the area of the triangle with vertices (t_0, t_0), (t, t_0) and (t, t) in the (t_1, t_2)-plane. You should draw it to verify this.

Solution 2.19

For the example of $[\sigma_x, \sigma_y]$ (the other cases are analogous):

$$[\sigma_x, \sigma_y] = \sigma_x \sigma_y - \sigma_y \sigma_x$$

$$= \begin{pmatrix} 0 & 1 \\ 1 & 0 \end{pmatrix} \begin{pmatrix} 0 & -i \\ i & 0 \end{pmatrix} - \begin{pmatrix} 0 & -i \\ i & 0 \end{pmatrix} \begin{pmatrix} 0 & 1 \\ 1 & 0 \end{pmatrix}$$

$$= \begin{pmatrix} i & 0 \\ 0 & -i \end{pmatrix} - \begin{pmatrix} -i & 0 \\ 0 & i \end{pmatrix} = 2i \begin{pmatrix} 1 & 0 \\ 0 & -1 \end{pmatrix} = 2i\sigma_z$$

Solution 2.20

Similar to Eq. (2.182).

Solution 2.22

$$\mathrm{Sp}(AB) = \sum_i (AB)_{ii} = \sum_{i,j} A_{ij} B_{ji}$$

$$= \sum_{i,j} B_{ji} A_{ij} = \sum_j (BA)_{jj} = \mathrm{Sp}(BA)$$

$$\Rightarrow \quad \mathrm{Sp}[A, B] = \mathrm{Sp}(AB) - \mathrm{Sp}(BA) = 0$$

But $\lambda \mathbf{1}$ has the trace λd, where d is the dimension of the Hilbert space.

Solution 2.23

The pairs of eigenvalues $(1, 1)$, $(1, -1)$, $(-1, 1)$ and $(-1, -1)$ of A and B have the corresponding eigenvectors

$$\frac{1}{\sqrt{2}} \begin{pmatrix} 1 \\ 1 \\ 0 \\ 0 \end{pmatrix}, \quad \frac{1}{\sqrt{2}} \begin{pmatrix} 1 \\ -1 \\ 0 \\ 0 \end{pmatrix}, \quad \frac{1}{\sqrt{2}} \begin{pmatrix} 0 \\ 0 \\ 1 \\ 1 \end{pmatrix}, \quad \frac{1}{\sqrt{2}} \begin{pmatrix} 0 \\ 0 \\ 1 \\ -1 \end{pmatrix}.$$

Solution 2.24

The pairs of eigenvalues mentioned in the previous exercise characterize their respective states completely, since the corresponding eigenspaces are one-dimensional.

Solution 2.27

With Eq. (2.161) we have:

$$U(t,0) = e^{-i\omega t \sigma_z} = \begin{pmatrix} e^{-i\omega t} & 0 \\ 0 & e^{i\omega t} \end{pmatrix}$$

$$\Rightarrow (S_x)_H = U^\dagger(t,0)\frac{\hbar}{2}\sigma_x U(t,0) = \frac{\hbar}{2}\begin{pmatrix} 0 & e^{2i\omega t} \\ e^{-2i\omega t} & 0 \end{pmatrix}$$

$$(S_y)_H = U^\dagger(t,0)\frac{\hbar}{2}\sigma_y U(t,0) = \frac{\hbar}{2}\begin{pmatrix} 0 & -i\,e^{2i\omega t} \\ i\,e^{-2i\omega t} & 0 \end{pmatrix}$$

Expectation value in the Heisenberg picture:

$$\langle s_x(t)\rangle_v = \langle x+|(S_x)_H(t)|x+\rangle$$

$$= \frac{\hbar}{4}(1\ 1)\begin{pmatrix} 0 & e^{2i\omega t} \\ e^{-2i\omega t} & 0 \end{pmatrix}\begin{pmatrix} 1 \\ 1 \end{pmatrix}$$

$$= \frac{\hbar}{4}(e^{2i\omega t} + e^{-2i\omega t}) = \frac{\hbar}{2}\cos(2\omega t)$$

Expectation value in the Schrödinger picture:

$$\langle s_x(t)\rangle_v := \langle s_x\rangle_{v(t)} = \langle v(t)|S_x|v(t)\rangle$$

$$= \frac{\hbar}{4}(e^{i\omega t}\ e^{-i\omega t})\begin{pmatrix} 0 & 1 \\ 1 & 0 \end{pmatrix}\begin{pmatrix} e^{-i\omega t} \\ e^{i\omega t} \end{pmatrix}$$

$$= \frac{\hbar}{4}(e^{2i\omega t} + e^{-2i\omega t}) = \frac{\hbar}{2}\cos(2\omega t)$$

Solution 2.28

We assume that

$$\mathbf{w} = (a_1\mathbf{e}_1^{(1)} + a_2\mathbf{e}_2^{(1)}) \otimes (b_1\mathbf{e}_1^{(2)} + b_2\mathbf{e}_2^{(2)})$$

$$= a_1b_1\,\mathbf{e}_1^{(1)} \otimes \mathbf{e}_1^{(2)} + a_1b_2\,\mathbf{e}_1^{(1)} \otimes \mathbf{e}_2^{(2)} + a_2b_1\,\mathbf{e}_2^{(1)} \otimes \mathbf{e}_1^{(2)} + a_2b_2\,\mathbf{e}_2^{(1)} \otimes \mathbf{e}_2^{(2)}.$$

Comparing with (2.254) yields:

$$a_1b_1 = -a_2b_2 = 1, \qquad a_1b_2 = a_2b_1 = 0$$

The second set of equations requires two coefficients to be zero, which is in contradiction with the first set of equations.

Solution 2.29

$$\sigma_x \otimes \sigma_x = \begin{pmatrix} 0 & 0 & 0 & 1 \\ 0 & 0 & 1 & 0 \\ 0 & 1 & 0 & 0 \\ 1 & 0 & 0 & 0 \end{pmatrix}, \qquad \sigma_x \otimes \sigma_y = \begin{pmatrix} 0 & 0 & 0 & -i \\ 0 & 0 & i & 0 \\ 0 & -i & 0 & 0 \\ i & 0 & 0 & 0 \end{pmatrix},$$

$$\sigma_y \otimes \sigma_x = \begin{pmatrix} 0 & 0 & 0 & -i \\ 0 & 0 & -i & 0 \\ 0 & i & 0 & 0 \\ i & 0 & 0 & 0 \end{pmatrix}, \qquad \sigma_y \otimes \sigma_y = \begin{pmatrix} 0 & 0 & 0 & -1 \\ 0 & 0 & 1 & 0 \\ 0 & 1 & 0 & 0 \\ -1 & 0 & 0 & 0 \end{pmatrix}.$$

Solution 2.30

$$[A^{(1)}, B^{(2)}](\mathbf{u} \otimes \mathbf{v}) = (A \otimes \mathbf{1})(\mathbf{1} \otimes B)(\mathbf{u} \otimes \mathbf{v}) - (\mathbf{1} \otimes B)(A \otimes \mathbf{1})(\mathbf{u} \otimes \mathbf{v})$$
$$= (A \otimes \mathbf{1})(\mathbf{u} \otimes B\mathbf{v}) - (\mathbf{1} \otimes B)(A\mathbf{u} \otimes \mathbf{v})$$
$$= A\mathbf{u} \otimes B\mathbf{v} - A\mathbf{u} \otimes B\mathbf{v} = 0$$

Solution 2.31
The claim follows from

$$|x+, x+\rangle = \frac{1}{2}\begin{pmatrix} 1 \\ 1 \\ 1 \\ 1 \end{pmatrix}, \qquad |x-, x-\rangle = \frac{1}{2}\begin{pmatrix} 1 \\ -1 \\ -1 \\ 1 \end{pmatrix},$$

$$|y+, y-\rangle = \frac{1}{2}\begin{pmatrix} 1 \\ -i \\ i \\ 1 \end{pmatrix}, \qquad |y-, y+\rangle = \frac{1}{2}\begin{pmatrix} 1 \\ i \\ -i \\ 1 \end{pmatrix}.$$

Solution 2.32
Already known:

$$P_{z+}^{(1)} = \mathrm{diag}(1, 1, 0, 0), \qquad P_{z+}^{(1)} = \mathrm{diag}(0, 0, 1, 1)$$

Still required:

$$P_{x+}^{(2)} = \mathbf{1} \otimes |x+\rangle\langle x+| = \mathbf{1} \otimes \frac{1}{2}\begin{pmatrix} 1 & 1 \\ 1 & 1 \end{pmatrix} = \frac{1}{2}\begin{pmatrix} 1 & 1 & 0 & 0 \\ 1 & 1 & 0 & 0 \\ 0 & 0 & 1 & 1 \\ 0 & 0 & 1 & 1 \end{pmatrix}$$

$$P_{x-}^{(2)} = \mathbf{1} \otimes |x-\rangle\langle x-| = \mathbf{1} \otimes \frac{1}{2} \begin{pmatrix} 1 & -1 \\ -1 & 1 \end{pmatrix} = \frac{1}{2} \begin{pmatrix} 1 & -1 & 0 & 0 \\ -1 & 1 & 0 & 0 \\ 0 & 0 & 1 & -1 \\ 0 & 0 & -1 & 1 \end{pmatrix}$$

We calculate the conditional probability for the spin of system 1 to be measured positive in z-direction, if the spin of system 2 was measured positive in x-direction:

$$P_{x+(2)}(z+^{(1)}) = \frac{p(z+, x+)}{p(x+^{(2)})} = \frac{\langle w | P_{z+}^{(1)} P_{x+}^{(2)} | w \rangle}{\langle w | P_{x+}^{(2)} | w \rangle}$$

The denominator evaluates to

$$\frac{1}{4} (1\ 0\ 0\ 1) \begin{pmatrix} 1 & 1 & 0 & 0 \\ 1 & 1 & 0 & 0 \\ 0 & 0 & 1 & 1 \\ 0 & 0 & 1 & 1 \end{pmatrix} \begin{pmatrix} 1 \\ 0 \\ 0 \\ 1 \end{pmatrix} = \frac{1}{2}.$$

The numerator is

$$\frac{1}{4} (1\ 0\ 0\ 1) \begin{pmatrix} 1 & 0 & 0 & 0 \\ 0 & 1 & 0 & 0 \\ 0 & 0 & 0 & 0 \\ 0 & 0 & 0 & 0 \end{pmatrix} \begin{pmatrix} 1 & 1 & 0 & 0 \\ 1 & 1 & 0 & 0 \\ 0 & 0 & 1 & 1 \\ 0 & 0 & 1 & 1 \end{pmatrix} \begin{pmatrix} 1 \\ 0 \\ 0 \\ 1 \end{pmatrix} = \frac{1}{4}.$$

As a result, one gets $P_{x+(2)}(z+^{(1)}) = \frac{1}{2}$. All other conditional probabilities are obtained similarly. The correlation is

$$\langle w | \sigma_z \otimes \sigma_x | w \rangle = \frac{1}{2} (1\ 0\ 0\ 1) \begin{pmatrix} 0 & 1 & 0 & 0 \\ 1 & 0 & 0 & 0 \\ 0 & 0 & 0 & -1 \\ 0 & 0 & -1 & 0 \end{pmatrix} \begin{pmatrix} 1 \\ 0 \\ 0 \\ 1 \end{pmatrix} = 0.$$

Solution 2.33

A spin in direction $\frac{1}{\sqrt{2}}(\mathbf{e}_x + \mathbf{e}_z)$ is measured with the operator $S_{xz} = \frac{\hbar}{2}\sigma_{xz}$, where

$$\sigma_{xz} := \frac{1}{\sqrt{2}}(\sigma_x + \sigma_z) = \frac{1}{\sqrt{2}} \begin{pmatrix} 1 & 1 \\ 1 & -1 \end{pmatrix}.$$

The correlation is

$$\langle w | \sigma_z \otimes \sigma_{xz} | w \rangle = \frac{1}{2\sqrt{2}} (1\ 0\ 0\ 1) \begin{pmatrix} 1 & 1 & 0 & 0 \\ 1 & -1 & 0 & 0 \\ 0 & 0 & -1 & -1 \\ 0 & 0 & -1 & 1 \end{pmatrix} \begin{pmatrix} 1 \\ 0 \\ 0 \\ 1 \end{pmatrix} = \frac{1}{\sqrt{2}}$$

and is thus between 0 and 1.

Solution 3.1
The proof is obtained by induction. Assume $\{|f_1\rangle, \ldots, |f_k\rangle$ is known to be orthonormal. It is then sufficient to show orthogonality for $k + 1$. For $j \leq k$ we have:

$$\langle f_j | \tilde{f}_{k+1} \rangle = \langle f_j | e_{k+1} \rangle - \sum_{i=1}^{k} \langle f_j | f_i \rangle \langle f_i | e_{k+1} \rangle$$

$$= \langle f_j | e_{k+1} \rangle - \sum_{i=1}^{k} \delta_{ij} \langle f_i | e_{k+1} \rangle$$

$$= \langle f_j | e_{k+1} \rangle - \langle f_j | e_{k+1} \rangle = 0$$

Solution 3.3

$$P_0(x) = \frac{1}{\sqrt{2}}, \qquad P_2(x) = \frac{\sqrt{5}}{2\sqrt{2}}(3x^2 - 1)$$

$$\langle P_0 | e_4 \rangle = \int_{-1}^{1} dx \, \frac{1}{\sqrt{2}} x^4 = \frac{\sqrt{2}}{5}$$

$$\langle P_2 | e_4 \rangle = \int_{-1}^{1} dx \, \frac{\sqrt{5}}{2\sqrt{2}}(3x^2 - 1)x^4 = \frac{\sqrt{5}}{2\sqrt{2}} \left(\frac{6}{7} - \frac{2}{5} \right) = \frac{4}{7}\sqrt{25}$$

$$|\tilde{P}_4\rangle = |e_4\rangle - |P_0\rangle\langle P_0 | e_4 \rangle - |P_2\rangle\langle P_2 | e_4 \rangle$$

$$\Rightarrow \quad \tilde{P}_4(x) = x^4 - \frac{1}{5} - \frac{2}{7}(3x^2 - 1) = x^4 - \frac{6}{7}x^2 + \frac{3}{35}$$

$$\langle \tilde{P}_4 | \tilde{P}_4 \rangle = \int_{-1}^{1} dx \left[x^8 - \frac{12}{7}x^6 + \left(\frac{6}{5 \cdot 7} + \frac{36}{7^2} \right) x^4 - \frac{36}{5 \cdot 7^2}x^2 + \frac{9}{5^2 \cdot 7^2} \right]$$

$$= \frac{2}{3^2} - \frac{24}{7^2} + \frac{444}{5^2 \cdot 7^2} - \frac{72}{3 \cdot 5 \cdot 7^2} + \frac{18}{5^2 \cdot 7^2} = \frac{128}{3^2 \cdot 5^2 \cdot 7^2}$$

$$\Rightarrow \quad P_4(x) = \left(\frac{128}{3^2 \cdot 5^2 \cdot 7^2} \right)^{-1/2} \left(x^4 - \frac{6}{7}x^2 + \frac{3}{35} \right)$$

$$= \frac{3 \cdot 5 \cdot 7}{8\sqrt{2}} \left(x^4 - \frac{6}{7}x^2 + \frac{3}{35} \right)$$

$$\Rightarrow \quad x P_3(x) = \frac{5\sqrt{7}}{2\sqrt{2}} \left(x^4 - \frac{3}{5}x^2 \right) = \frac{4}{3\sqrt{7}} P_4(x) + \frac{3}{\sqrt{35}} P_2(x)$$

Solution 3.4
The calculation yields

$$A_{(4)}^{-1} X_{(4)}^{(e)} A_{(4)} = \begin{pmatrix} 0 & \frac{1}{\sqrt{3}} & 0 & \frac{\sqrt{7}}{2} \\ \frac{1}{\sqrt{3}} & 0 & \frac{2}{\sqrt{15}} & 0 \\ 0 & \frac{2}{\sqrt{15}} & 0 & -\sqrt{\frac{7}{5}} \\ 0 & 0 & \frac{3}{\sqrt{35}} & 0 \end{pmatrix} \neq X_{(4)}^{(P)}.$$

Here, the indicator (4) denotes the restriction of the matrices onto $\mathrm{Pol}_3([-1, 1], \mathbb{R})$. The deviation results from the fact that $\mathrm{Pol}_3([-1, 1], \mathbb{R})$ is not closed under the action of X. $X|e_3\rangle = |e_4\rangle$ contains portions of $|P_0\rangle$ and $|P_2\rangle$ which are not accounted for in the backtransformation with $A_{(4)}^{-1}$ and are therefore being lost.

Solution 3.5
Each of these properties is preserved under the addition of two functions and under multiplication by a real number.

Solution 3.6

$$\langle v_{2n-1}|X|v_0\rangle = \int_{-1}^{1} dx \, \frac{1}{\sqrt{2}} x \, \sin(n\pi x)$$

$$= \int_{-1}^{1} dx \, \frac{1}{\sqrt{2}} \frac{1}{n\pi} \cos(n\pi x) + \frac{1}{\sqrt{2}\,n\pi} [x \, \cos(n\pi x)]_{-1}^{1}$$

$$= 0 + \frac{1}{\sqrt{2}\,n\pi} \cdot 2 \cdot (-1)^n = \frac{(-1)^n \sqrt{2}}{n\pi}.$$

The coefficients $\langle v_{2n}|X|v_0\rangle$ vanish, because they represent integrals over odd functions.

Solution 3.7
For $k \in \mathbb{Z}\backslash\{0\}$ one has:

$$\int_{-1}^{1} dx \, e^{ik\pi x} = \frac{1}{ik\pi} \left[e^{ik\pi x} \right]_{-1}^{1} = 0$$

It follows:

$$\langle w_0|w_{2n}\rangle = \langle w_0|w_{2n-1}\rangle = \langle w_{2n}|w_{2n'-1}\rangle = 0$$

$$\langle w_{2n}|w_{2n'}\rangle = \langle w_{2n-1}|w_{2n'-1}\rangle = \delta_{nn'}$$

Solution 3.8

$$\langle v_{2n}|v_{2n-1}\rangle = \frac{i}{2}\left[\langle w_{2n-1}|w_{2n-1}\rangle - \langle w_{2n}|w_{2n}\rangle\right] = 0$$

For $n \neq n'$ one reads off:

$$\langle v_{2n}|v_{2n'}\rangle = \langle v_{2n-1}|v_{2n'-1}\rangle = \langle v_{2n}|v_{2n'-1}\rangle = 0$$

For then all $|w_i\rangle$ appearing on the right hand side of (3.91) and (3.92) are different.

Solution 3.9
Because of the block diagonal shape the following calculation is sufficient:

$$\begin{pmatrix} \frac{i}{\sqrt{2}} & \frac{1}{\sqrt{2}} \\ -\frac{i}{\sqrt{2}} & \frac{1}{\sqrt{2}} \end{pmatrix} \begin{pmatrix} 0 & -n\pi \\ n\pi & 0 \end{pmatrix} \begin{pmatrix} -\frac{i}{\sqrt{2}} & \frac{i}{\sqrt{2}} \\ \frac{1}{\sqrt{2}} & \frac{1}{\sqrt{2}} \end{pmatrix} = \begin{pmatrix} -in\pi & 0 \\ 0 & in\pi \end{pmatrix}$$

Solution 3.10

$$\tilde{\psi}(p) = \frac{\sqrt{\sigma}}{\sqrt{\hbar}\pi^{1/4}} e^{-\frac{\sigma^2(p-p_0)^2}{2\hbar^2}}$$

$$\langle X\rangle_\psi = \langle\psi|X|\psi\rangle = i\hbar\int_{-\infty}^{\infty} dp\,\tilde{\psi}^*(p)\frac{d}{dp}\tilde{\psi}(p)$$

$$= \frac{i\sigma}{\sqrt{\pi}}\int_{-\infty}^{\infty} dp\left(-\frac{\sigma^2}{\hbar^2}\right)(p-p_0)e^{-\frac{\sigma^2(p-p_0)^2}{\hbar^2}}$$

$$= 0$$

$$\langle X^2\rangle_\psi = \langle\psi|X^2|\psi\rangle = -\hbar^2\int_{-\infty}^{\infty} dp\,\tilde{\psi}^*(p)\frac{d^2}{dp^2}\tilde{\psi}(p)$$

$$= -\frac{\hbar\sigma}{\sqrt{\pi}}\int_{-\infty}^{\infty} dp\left[\frac{\sigma^4}{\hbar^4}(p-p_0)^2 - \frac{\sigma^2}{\hbar^2}\right]e^{-\frac{\sigma^2(p-p_0)^2}{\hbar^2}}$$

$$= -\frac{\hbar\sigma}{\sqrt{\pi}}\left[\frac{\sqrt{\pi}}{2}\left(\frac{\hbar}{\sigma}\right)^3\frac{\sigma^4}{\hbar^4} - \sqrt{\pi}\frac{\hbar}{\sigma}\frac{\sigma^2}{\hbar^2}\right]$$

$$= \frac{\sigma^2}{2}$$

Solution 3.11

$$\psi(x) = \frac{1}{\sqrt{\sigma}\pi^{1/4}} e^{i\frac{p_0}{\hbar}x} e^{-\frac{x^2}{2\sigma^2}}$$

$$\langle P \rangle_\psi = \langle \psi | P | \psi \rangle = -i\hbar \int_{-\infty}^{\infty} dx \, \psi^*(x) \frac{d}{dx} \psi(x)$$

$$= -\frac{i\hbar}{\sigma\sqrt{\pi}} \int_{-\infty}^{\infty} dx \left(\frac{ip_0}{\hbar} - \frac{x}{\sigma^2} \right) e^{-\frac{x^2}{\sigma^2}}$$

$$= \frac{p_0}{\sigma\sqrt{\pi}} \int_{-\infty}^{\infty} dx \, e^{-\frac{x^2}{\sigma^2}}$$

$$= p_0$$

$$\langle P^2 \rangle_\psi = \langle \psi | P^2 | \psi \rangle = -\hbar^2 \int_{-\infty}^{\infty} dx \, \psi^*(x) \frac{d^2}{dx^2} \psi(x)$$

$$= -\frac{\hbar^2}{\sigma\sqrt{\pi}} \int_{-\infty}^{\infty} dx \left[\left(\frac{ip_0}{\hbar} - \frac{x}{\sigma^2} \right)^2 - \frac{1}{\sigma^2} \right] e^{-\frac{x^2}{\sigma^2}}$$

$$= -\frac{\hbar^2}{\sigma\sqrt{\pi}} \int_{-\infty}^{\infty} dx \left[\frac{x^2}{\sigma^4} - \frac{p_0^2}{\hbar^2} - \frac{1}{\sigma^2} \right] e^{-\frac{x^2}{\sigma^2}}$$

$$= -\frac{\hbar^2}{\sigma\sqrt{\pi}} \left[\frac{1}{\sigma^4} \frac{\sqrt{\pi}}{2} \sigma^3 - \sqrt{\pi} \sigma \left(\frac{p_0^2}{\hbar^2} + \frac{1}{\sigma^2} \right) \right]$$

$$= p_0^2 + \frac{\hbar^2}{2\sigma^2}$$

Solution 3.12

$$\tilde{\psi}(p) = \frac{\sqrt{\sigma}}{\sqrt{\hbar} \pi^{1/4}} e^{-\frac{\sigma^2 (p-p_0)^2}{2\hbar^2}} e^{-\frac{ipx_0}{\hbar}}$$

$$\langle X \rangle_\psi = \langle \psi | X | \psi \rangle = i\hbar \int_{-\infty}^{\infty} dp \, \tilde{\psi}^*(p) \frac{d}{dp} \tilde{\psi}(p)$$

$$= \frac{i\sigma}{\sqrt{\pi}} \int_{-\infty}^{\infty} dp \left[\left(\frac{\sigma^2}{\hbar^2} \right) (p - p_0) - \frac{ix_0}{\hbar} \right] e^{-\frac{\sigma^2 (p-p_0)^2}{\hbar^2}}$$

$$= x_0$$

$$\langle X^2 \rangle_\psi = \langle \psi | X^2 | \psi \rangle = -\hbar^2 \int_{-\infty}^{\infty} dp \, \tilde{\psi}^*(p) \frac{d^2}{dp^2} \tilde{\psi}(p)$$

$$= -\frac{\hbar\sigma}{\sqrt{\pi}} \int_{-\infty}^{\infty} dp \left\{ \left[\left(-\frac{\sigma^2}{\hbar^2} \right) (p - p_0) - \frac{ix_0}{\hbar} \right]^2 - \frac{\sigma^2}{\hbar^2} \right\} e^{-\frac{\sigma^2 (p-p_0)^2}{\hbar^2}}$$

$$= -\frac{\hbar\sigma}{\sqrt{\pi}} \int_{-\infty}^{\infty} dp \left(\frac{\sigma^4}{\hbar^4} (p - p_0)^2 - \frac{x_0^2}{\hbar^2} - \frac{\sigma^2}{\hbar^2} \right) e^{-\frac{\sigma^2 (p-p_0)^2}{\hbar^2}}$$

$$= -\frac{\hbar\sigma}{\sqrt{\pi}}\left[\frac{\sqrt{\pi}}{2}\frac{\hbar^3}{\sigma^3}\frac{\sigma^4}{\hbar^4} - \frac{\sqrt{\pi}\,\hbar}{\sigma}\left(\frac{x_0^2}{\hbar^2} + \frac{\sigma^2}{\hbar^2}\right)\right]$$

$$= x_0^2 + \frac{\sigma^2}{2}$$

The uncertainty is unchanged, $(\Delta X)_\psi = \frac{\sigma}{\sqrt{2}}$. The values of $\langle P\rangle_\psi$, $\langle P^2\rangle_\psi$ and $(\Delta P)_\psi$ are unchanged, because the phase cancels in the multiplication.

Solution 3.14

$$(X_i P_j - P_j X_i)\psi(\mathbf{r}) = -i\hbar\left(x_i\frac{d}{dx_j}\psi(\mathbf{r}) - \frac{d}{dx_j}(x_i\psi(\mathbf{r}))\right)$$

$$= -i\hbar\left[x_i\frac{d}{dx_j}\psi(\mathbf{r}) - x_i\frac{d}{dx_j}\psi(\mathbf{r}) - \left(\frac{d}{dx_j}x_i\right)\psi(\mathbf{r})\right]$$

$$= i\hbar\delta_{ij}\,\psi(\mathbf{r})$$

Solution 3.16
The identities for the Poisson brackets are obviously fulfilled, in particular

$$\{x', p'\} = \{p, -x\} = \{x, p\} = 1.$$

Therefore the wave function can be written in the form $\tilde{\psi}(x') = \tilde{\psi}(p)$. The corresponding operators act in the following way:

$$X'\tilde{\psi}(x') = x'\tilde{\psi}(x') \quad \Rightarrow \quad P\tilde{\psi}(p) = p\,\tilde{\psi}(p)$$

$$P'\tilde{\psi}(x') = -i\hbar\frac{d}{dx'}\tilde{\psi}(x') \quad \Rightarrow \quad -X\tilde{\psi}(p) = -i\hbar\frac{d}{dp}\tilde{\psi}(p)$$

Solution 3.18

$$\mathbf{r}^{(1)} = \mathbf{r}_S - \frac{m^{(2)}}{m^{(1)} + m^{(2)}}\mathbf{r}_R, \qquad \mathbf{r}^{(2)} = \mathbf{r}_S + \frac{m^{(1)}}{m^{(1)} + m^{(2)}}\mathbf{r}_R$$

$$\frac{\partial}{\partial x_{Si}} = \frac{\partial x_i^{(1)}}{\partial x_{Si}}\frac{\partial}{\partial x_i^{(1)}} + \frac{\partial x_i^{(2)}}{\partial x_{Si}}\frac{\partial}{\partial x_i^{(2)}} = \frac{\partial}{\partial x_i^{(1)}} + \frac{\partial}{\partial x_i^{(2)}}$$

$$\frac{\partial}{\partial x_{Ri}} = \frac{\partial x_i^{(1)}}{\partial x_{Ri}}\frac{\partial}{\partial x_i^{(1)}} + \frac{\partial x_i^{(2)}}{\partial x_{Ri}}\frac{\partial}{\partial x_i^{(2)}} = -\frac{m^{(2)}}{m^{(1)} + m^{(2)}}\frac{\partial}{\partial x_i^{(1)}} + \frac{m^{(1)}}{m^{(1)} + m^{(2)}}\frac{\partial}{\partial x_i^{(2)}}$$

Insertion into (3.289) and comparison with (3.287) and (3.288) gives the required result.

Solution 5.1

(a)

$$\int_{-\infty}^{\infty} dx \int_{-\infty}^{\infty} dy\, e^{-(x^2+y^2)} = \int_0^{\infty} d\rho \int_0^{2\pi} d\phi\, \rho e^{-\rho^2} = 2\pi \int_0^{\infty} d\rho\, \rho e^{-\rho^2}$$

$$= \pi \int_0^{\infty} du\, e^{-u} = -\pi \left[e^{-u} \right]_0^{\infty} = \pi$$

(b) If a is complex, after substitution the path of integration is a sloped line in the complex plane. Connect this line to the real axis via two vertical lines at $u = \pm R$ (where R is arbitrary). The integral along the resulting contour vanishes by the Residue theorem. For $R \to \infty$ the integral along the vertical pieces vanishes. The integral along the sloped line is therefore equal to the integral along the real axis.

(c)

$$\int_{-\infty}^{\infty} dy\, y e^{-\frac{(y-y_0)^2}{a}} = \int_{-\infty}^{\infty} du\, (u + y_0) e^{-\frac{u^2}{a}}$$

$$= \int_{-\infty}^{\infty} du\, y_0\, e^{-\frac{u^2}{a}} = \sqrt{\pi a}\, y_0$$

(d) Partial integration yields:

$$\int_{-\infty}^{\infty} du\, u^2 e^{-u^2} = \frac{1}{2} \int_{-\infty}^{\infty} du\, u \left(2u\, e^{-u^2} \right) = \frac{1}{2} \int_{-\infty}^{\infty} du\, e^{-u^2} = \frac{\sqrt{\pi}}{2}$$

It follows:

$$\int_{-\infty}^{\infty} dy\, y^2 e^{-\frac{(y-y_0)^2}{a}} = \int_{-\infty}^{\infty} du\, \sqrt{a}(\sqrt{a}\, u + y_0)^2 e^{-u^2}$$

$$= \sqrt{a} \int_{-\infty}^{\infty} du\, (au^2 + y_0^2)\, e^{-u^2}$$

$$= \sqrt{a} \left(y_0^2 \sqrt{\pi} + a \int_{-\infty}^{\infty} du\, u^2 e^{-u^2} \right)$$

$$= \sqrt{\pi a} \left(\frac{a}{2} + y_0^2 \right)$$

Solution 5.2

$$\tilde{\psi}(p, t) = \frac{\sqrt{\sigma}}{\sqrt{\hbar}\pi^{1/4}} e^{-\frac{\sigma^2(p-p_0)^2}{2\hbar^2} - i\frac{p^2}{2m\hbar} t}$$

$$\langle X \rangle_\psi = \langle \psi | X | \psi \rangle = i\hbar \int_{-\infty}^{\infty} dp\, \tilde{\psi}^*(p) \frac{d}{dp} \tilde{\psi}(p)$$

$$= \frac{i\sigma}{\sqrt{\pi}} \int_{-\infty}^{\infty} dp \left(-\frac{\sigma^2}{\hbar^2}(p - p_0) - \frac{ipt}{m\hbar} \right) e^{-\frac{\sigma^2(p-p_0)^2}{\hbar^2}}$$

$$= \frac{i\sigma}{\sqrt{\pi}} \int_{-\infty}^{\infty} dp \left(-\frac{ipt}{m\hbar} \right) e^{-\frac{\sigma^2(p-p_0)^2}{\hbar^2}}$$

$$= \frac{\sigma t}{m\hbar\sqrt{\pi}} \int_{-\infty}^{\infty} dp \, p \, e^{-\frac{\sigma^2(p-p_0)^2}{\hbar^2}}$$

$$= \frac{p_0 t}{m}$$

The first term in the second line doesn't appear afterwards, since it is an odd function in $(p - p_0)$ whose integral vanishes. In the end we used (5.3).

$$\langle X^2 \rangle_\psi = \langle \psi | X^2 | \psi \rangle = -\hbar^2 \int_{-\infty}^{\infty} dp \, \tilde{\psi}^*(p) \frac{d^2}{dp^2} \tilde{\psi}(p)$$

$$= -\frac{\hbar\sigma}{\sqrt{\pi}} \int_{-\infty}^{\infty} dp \left\{ \left[-\frac{\sigma^2}{\hbar^2}(p - p_0) - \frac{ipt}{m\hbar} \right]^2 - \frac{\sigma^2}{\hbar^2} - \frac{it}{m\hbar} \right\} e^{-\frac{\sigma^2(p-p_0)^2}{\hbar^2}}$$

$$= -\frac{\hbar\sigma}{\sqrt{\pi}} \int_{-\infty}^{\infty} dq \left\{ \left[-\frac{\sigma^2}{\hbar^2}q - \frac{i(q + p_0)t}{m\hbar} \right]^2 - \frac{\sigma^2}{\hbar^2} - \frac{it}{m\hbar} \right\} e^{-\frac{\sigma^2 q^2}{\hbar^2}}$$

$$= -\frac{\hbar\sigma}{\sqrt{\pi}} \int_{-\infty}^{\infty} dq \left\{ \left[\frac{\sigma^4}{\hbar^4} - \frac{t^2}{m^2\hbar^2} + \frac{2i\sigma^2 t}{m\hbar^3} \right] q^2 - \frac{p_0^2 t^2}{m^2\hbar^2} - \frac{\sigma^2}{\hbar^2} - \frac{it}{m\hbar} \right\} e^{-\frac{\sigma^2 q^2}{\hbar^2}}$$

$$= -\hbar\sigma \left[\frac{\hbar^3}{2\sigma^3} \left(\frac{\sigma^4}{\hbar^4} - \frac{t^2}{m^2\hbar^2} + \frac{2i\sigma^2 t}{m\hbar^3} \right) - \frac{\hbar}{\sigma} \left(\frac{p_0^2 t^2}{m^2\hbar^2} + \frac{\sigma^2}{\hbar^2} + \frac{it}{m\hbar} \right) \right]$$

$$= \frac{p_0^2 t^2}{m^2} + \frac{\sigma^2}{2} + \frac{\hbar^2 t^2}{2\sigma^2 m^2}$$

Solution 5.3
The exponent is brought into a convenient form by completing the square:

$$i\frac{p}{\hbar}x - \frac{\sigma^2(p - p_0)^2}{2\hbar^2} - i\frac{p^2}{2m\hbar}t$$

$$= -\frac{it\hbar + m\sigma^2}{2m\hbar^2}p^2 + \frac{i\hbar x + \sigma^2 p_0}{\hbar^2}p - \frac{\sigma^2 p_0^2}{2\hbar^2}$$

$$= \frac{it\hbar + m\sigma^2}{2m\hbar^2} \left(-p^2 + p\frac{2m\hbar^2}{it\hbar + m\sigma^2}\frac{i\hbar x + \sigma^2 p_0}{\hbar^2} \right) - \frac{\sigma^2 p_0^2}{2\hbar^2}$$

$$= \frac{it\hbar + m\sigma^2}{2m\hbar^2} \left[-\left(p - \frac{m\hbar^2}{it\hbar + m\sigma^2}\frac{i\hbar x + \sigma^2 p_0}{\hbar^2} \right)^2 \right]$$

$$+ \frac{m\hbar^2}{2(it\hbar + m\sigma^2)} \left(\frac{i\hbar x + \sigma^2 p_0}{\hbar^2} \right)^2 - \frac{\sigma^2 p_0^2}{2\hbar^2}$$

It follows:

$$
\frac{\sqrt{\sigma}}{\sqrt{2}\,\hbar\pi^{3/4}} \int_{-\infty}^{\infty} dp\, \exp\left(i\frac{p}{\hbar}x - \frac{\sigma^2(p-p_0)^2}{2\hbar^2} - i\frac{p^2}{2m\hbar}t\right)
$$

$$
= \frac{\sqrt{\sigma}}{\sqrt{2}\,\hbar\pi^{1/4}} \sqrt{\frac{2m\hbar^2}{it\hbar + m\sigma^2}}\, \exp\left(\frac{m(i\hbar x + \sigma^2 p_0)^2}{2\hbar^2(it\hbar + m\sigma^2)} - \frac{\sigma^2 p_0^2}{2\hbar^2}\right)
$$

$$
= \frac{\sqrt{\sigma}}{\pi^{1/4}} \frac{1}{\sigma^2 + i\frac{\hbar t}{m}}\, \exp\left[-\frac{\left(x - \frac{i\sigma^2 p_0}{\hbar}\right)^2}{2\left(\sigma^2 + i\frac{\hbar t}{m}\right)} - \frac{\sigma^2 p_0^2}{2\hbar^2}\right]
$$

For the equivalence with (5.13) it remains to be shown:

$$
-\frac{\left(x - \frac{i\sigma^2 p_0}{\hbar}\right)^2}{2\left(\sigma^2 + i\frac{\hbar t}{m}\right)} - \frac{\sigma^2 p_0^2}{2\hbar^2} = -\frac{\left(x - \frac{p_0 t}{m}\right)^2}{2\left(\sigma^2 + i\frac{\hbar t}{m}\right)} + i\frac{p_0}{\hbar}\left(x - \frac{p_0 t}{2m}\right)
$$

This is verified by multiplying both sides with $2\left(\sigma^2 + i\frac{\hbar t}{m}\right)$, multiplying out the brackets and comparing the terms.

Solution 5.4
$\psi \sim e^{i(kx - Et/\hbar)}$ $\Rightarrow$ wavefronts of constant phase move to the right for growing t.

Solution 5.5
For simplicity we set $a_I^{(1)} = 1$. Then the four solutions are:

$$
a_I^{(1)} = 1, \quad b_I^{(1)} = \frac{k_I - k_{II}}{k_I + k_{II}}, \quad a_{II}^{(1)} = \frac{2k_I}{k_I + k_{II}}, \quad b_{II}^{(1)} = 0
$$

$$
a_I^{(2)} = \frac{k_I - k_{II}}{k_I + k_{II}}, \quad b_I^{(2)} = 1, \quad a_{II}^{(2)} = 0, \quad b_{II}^{(2)} = \frac{2k_I}{k_I + k_{II}}
$$

$$
a_I^{(3)} = 0, \quad b_I^{(3)} = \frac{2k_{II}}{k_I + k_{II}}, \quad a_{II}^{(3)} = \frac{k_{II} - k_I}{k_I + k_{II}}, \quad b_{II}^{(3)} = 1
$$

$$
a_I^{(4)} = \frac{2k_{II}}{k_I + k_{II}}, \quad b_I^{(4)} = 0, \quad a_{II}^{(4)} = 1, \quad b_{II}^{(4)} = \frac{k_{II} - k_I}{k_I + k_{II}}
$$

It is easy to check:

$$
\text{Lsg } 3 = \frac{k_I + k_{II}}{2k_I} \text{Lsg } 2 + \frac{k_{II} - k_I}{2k_I} \text{Lsg } 1
$$

$$\text{Lsg } 4 = \frac{k_I + k_{II}}{2k_I} \text{ Lsg } 1 + \frac{k_{II} - k_I}{2k_I} \text{ Lsg } 2$$

Solution 5.8

$$j_0 = \frac{\hbar k_I |a_I|^2}{m}, \quad j_d = \frac{\hbar k_I |a_{III}|^2}{m}$$

$$
\begin{aligned}
T &= \frac{|j_d|}{|j_0|} = \left| \frac{a_{III}}{a_I} \right| \\
&= \frac{16z^2}{\left[(z+1)^2 e^{-2ik_{II}x_0} - (z-1)^2 e^{2ik_{II}x_0}\right]\left[(z+1)^2 e^{2ik_{II}x_0} - (z-1)^2 e^{-2ik_{II}x_0}\right]} \\
&= \frac{16z^2}{(z+1)^4 + (z-1)^4 - 2(z^2-1)^2 \cos(4k_{II}x)} \\
&= \frac{16z^2}{(z+1)^4 + (z-1)^4 - 2(z^2-1)^2 \left(1 - 2\sin^2(2k_{II}x)\right)} \\
&= \frac{4z^2}{4z^2 + (z^2-1)^2 \sin^2(2k_{II}x_0)}
\end{aligned}
$$

In the second last line we used:

$$\cos 2\phi = \cos^2 \phi - \sin^2 \phi = \cos^2 \phi + \sin^2 \phi - 2\sin^2 \phi = 1 - 2\sin^2 \phi$$

The reflexion coefficient follows from $R = 1 - T$.

Solution 5.9
The numerator of R vanishes for $k_{II} = \frac{n\pi}{2x_0}$ with integer n. The corresponding energies are

$$E = \frac{\hbar^2 k_{II}^2}{2m} + V_0 = \frac{\hbar^2 n^2 \pi^2}{8mx_0^2} + V_0$$

Solution 5.10
The continuity conditions are identical up to the replacement $ik_{II} \to \kappa_{II}$. This replacement runs through the entire calculation and implies

$$iz \to w, \quad z^2 \to -w^2,$$

$$\sin(2k_{II}x_0) \to \sin(-2i\kappa_{II}x_0) = -i\sinh(2\kappa_{II}x_0),$$

$$\sin^2(2k_{II}x_0) \to -\sinh^2(2\kappa_{II}x_0).$$

Solution 6.2
(a)

$$i\hbar \left(A_x A_y^\dagger - A_x^\dagger A_y \right)$$
$$= i\hbar \frac{1}{2\hbar} \left(-iX P_y + i P_x Y - iX P_y + i P_x Y \right)$$
$$= X P_y - Y P_x = L$$

(b) to (d) are obtained by inserting the definitions of A_x and A_y.

(e)

$$A_R^\dagger = \frac{1}{\sqrt{2}} \left(A_x^\dagger + i A_y^\dagger \right)$$
$$= \frac{1}{\sqrt{2}} \sqrt{\frac{m\omega}{2\hbar}} \left[x - \frac{\hbar}{m\omega} \frac{\partial}{\partial x} + i \left(y - \frac{\hbar}{m\omega} \frac{\partial}{\partial y} \right) \right]$$
$$= \sqrt{\frac{m\omega}{4\hbar}} \left[(x + iy) - \frac{\hbar}{m\omega} \left(\frac{\partial}{\partial x} + i \frac{\partial}{\partial y} \right) \right]$$
$$= \sqrt{\frac{m\omega}{4\hbar}} \left[e^{i\phi} r - \frac{\hbar}{m\omega} \left(\cos\phi \frac{\partial}{\partial r} - \frac{1}{r} \sin\phi \frac{\partial}{\partial \phi} + i \sin\phi \frac{\partial}{\partial r} + \frac{i}{r} \cos\phi \frac{\partial}{\partial \phi} \right) \right]$$
$$= \sqrt{\frac{m\omega}{4\hbar}} \left[e^{i\phi} r - \frac{\hbar}{m\omega} \left((\cos\phi + i\sin\phi) \frac{\partial}{\partial r} + \frac{i}{r} (\cos\phi + i\sin\phi) \frac{\partial}{\partial \phi} \right) \right]$$
$$= \sqrt{\frac{m\omega}{4\hbar}} e^{i\phi} \left[r - \frac{\hbar}{m\omega} \left(\frac{\partial}{\partial r} + \frac{i}{r} \frac{\partial}{\partial \phi} \right) \right]$$

$A_L^\dagger$ equivalently, only replacing $i \to -i$.
(f) and (g) are obtained directly by insertion.

Solution 7.1
The sum of three non-negative integers adding up to n can be symbolically represented via delimiters, for example $7 = 4 + 1 + 2$ can be written like this:

$$\cdots \cdot | \cdot | \cdots$$

That's $n + 2$ symbols, 2 of them delimiters. There are $\binom{n+2}{2}$ possibilities for their arrangement.

Solution 7.2

$$[A, L_z] = \frac{1}{2m} \left[P_x^2 + P_y^2 + P_z^2, X P_y - Y P_x \right]$$
$$= \frac{1}{2m} \left([P_x^2, X] P_y - [P_y^2, Y] P_x \right)$$

$$= \frac{1}{2m} \left(P_x[P_x, X]P_y + [P_x, X]P_x P_y - P_y[P_y, Y]P_x - [P_y, Y]P_y P_x \right)$$

$$= \frac{1}{2m} \left(-2i\hbar P_x P_y + 2i\hbar P_y P_x \right) = 0$$

Similarly L_x und L_y.

Solution 7.3

For simplicity we suppress the quantum number α. From (7.61) we get

$$L_+|l, -l\rangle = \hbar\sqrt{(2l) \cdot 1} \, |l, -l + 1\rangle$$

$$L_+|l, -l + 1\rangle = \hbar\sqrt{(2l - 1) \cdot 2} \, |l, -l + 2\rangle$$

etc., so

$$L_+^{l+m}|l, -l\rangle = \hbar^{l+m} \sqrt{(2l)(2l - 1) \cdots (l - m + 1)} \sqrt{(l + m)!} \, |l, m\rangle$$

$$= \sqrt{\frac{(2l)!(l + m)!}{(l - m)!}} \, |l, m\rangle.$$

Similarly L_-.

Solution 7.4

For $|z\pm\rangle$ we have $l = \frac{1}{2}$, $m = \pm\frac{1}{2}$. We therefore have to show

$$S_+|z-\rangle = \hbar|z+\rangle, \quad S_+|z+\rangle = 0, \quad S_-|z-\rangle = 0, \quad S_-|z+\rangle = \hbar|z-\rangle.$$

This follows directly from

$$S_+ = \frac{\hbar}{2}(\sigma_x + i\sigma_y) = \hbar \begin{pmatrix} 0 & 1 \\ 0 & 0 \end{pmatrix}, \quad S_- = \frac{\hbar}{2}(\sigma_x - i\sigma_y) = \hbar \begin{pmatrix} 0 & 0 \\ 1 & 0 \end{pmatrix}.$$

Solution 7.7

$$L^2 \left(\sin^l \theta \, e^{-il\phi} \right)$$

$$= -\hbar^2 \left(\frac{\partial^2}{\partial\theta^2} + \cot\theta \frac{\partial}{\partial\theta} + \frac{1}{\sin^2\theta} \frac{\partial^2}{\partial\phi^2} \right) \left(\sin^l \theta \, e^{-il\phi} \right)$$

$$= -\hbar^2 \left[\left(\frac{\partial}{\partial\theta} + \cot\theta \right) l \sin^{l-1}\theta \cos\theta + \sin^{l-2}\theta(-l^2) \right] e^{-il\phi}$$

$$= -\hbar^2 \left[l(l - 1) \sin^{l-2}\theta \cos^2\theta - l \sin^l\theta + l \sin^{l-2}\theta \cos^2\theta - l^2 \sin^{l-2}\theta \right] e^{-il\phi}$$

$$= -\hbar^2 l \sin^{l-2}\theta \left[l \cos^2\theta - \sin^2\theta - l \right] e^{-il\phi}$$

$$= -\hbar^2 l \sin^{l-2}\theta \left[l\cos^2\theta - \sin^2\theta - l(\cos^2\theta + \sin^2\theta) \right] e^{-il\phi}$$
$$= \hbar^2 l(l+1) \sin^l\theta\, e^{-il\phi}$$

Solution 7.10
Equation (7.80) gives:

$$\Delta \frac{1}{r} = \frac{1}{r} \frac{\partial^2}{\partial r^2} \left(r\frac{1}{r} \right) = \frac{1}{r} \frac{\partial^2}{\partial r^2} 1 = 0$$

Determination of the gradient:

$$\frac{\partial}{\partial x} \frac{1}{r} = \frac{\partial}{\partial x} \frac{1}{\sqrt{x^2 + y^2 + z^2}} = -\frac{x}{\sqrt{x^2 + y^2 + z^2}^3} = -\frac{x}{r^3}$$

etc., so

$$\nabla \frac{1}{r} = -\frac{\mathbf{r}}{r^3}.$$

It follows

$$\int_S d\mathbf{S} \cdot \left(\nabla \frac{1}{r} \right) = \int_{-1}^{1} d\cos\theta \int_0^{2\pi} d\phi\, \mathbf{e}_r \cdot (-\mathbf{r}) \frac{1}{r^3} = -4\pi,$$

for on the unit sphere one has $\mathbf{e}_r \cdot (-\mathbf{r}) = -1$ and $\frac{1}{r^3} = 1$.

Solution 7.11
The case $n = 1$ is obvious. Assume the claim is fulfilled for n.

$$\sum_{l=0}^{n} (2l+1) = \sum_{l=0}^{n-1} (2l+1) + (2n+1) = n^2 + 2n + 1 = (n+1)^2$$

Solution 8.1
Simply plug ψ_{in} and ψ_{sc} into (3.239).

Solution 8.2
Equation (7.80) implies for $r > 0$:

$$(\Delta + k^2) \frac{e^{\pm ikr}}{r} = \frac{1}{r} \frac{\partial^2}{\partial r^2} e^{\pm ikr} + k^2 \frac{e^{\pm ikr}}{r}$$
$$= -k^2 \frac{e^{\pm ikr}}{r} + k^2 \frac{e^{\pm ikr}}{r} = 0$$

The δ^3 term follows from (7.151).

Solution 8.3
Set

$$U_{nl}(r) = e^{i(kr - \gamma \ln kr)}$$

and ignore terms of order r^{-2} during the differentiation:

$$\frac{d}{dr} U_{nl}(r) = \left(ik - i\frac{\gamma}{r} \right) e^{i(kr - \gamma \ln kr)}$$

$$\frac{d^2}{dr^2} U_{nl}(r) = \left(ik - i\frac{\gamma}{r} \right)^2 e^{i(kr - \gamma \ln kr)} = \left(-k^2 + 2\frac{\gamma}{k} \right) e^{i(kr - \gamma \ln kr)}$$

Plugging this into (8.39) we get

$$\left[\frac{\hbar^2 k^2}{2m} - \frac{\hbar^2 \gamma k}{rm} + \frac{g}{r} \right] e^{i(kr - \gamma \ln kr)} = \frac{\hbar^2 k^2}{2m} e^{i(kr - \gamma \ln kr)}.$$

This equation is fulfilled if $\gamma = \frac{gm}{k\hbar^2}$.

Solution 8.4

$$
\begin{aligned}
f(\mathbf{q}) &= -\frac{mg}{2\pi\hbar^2} \int_0^\infty dr \int_{-1}^1 d\cos\theta \int_0^{2\pi} d\phi\, r^2\, e^{iqr\cos\theta} \frac{e^{-\beta r}}{r} \\
&= -\frac{mg}{\hbar^2} \int_0^\infty dr \int_{-1}^1 d\cos\theta\, r\, e^{iqr\cos\theta} e^{-\beta r} \\
&= -\frac{mg}{\hbar^2} \int_0^\infty dr\, \frac{1}{iq} e^{-\beta r} \left[e^{iqr\cos\theta} \right]_{-1}^1 \\
&= -\frac{mg}{iq\hbar^2} \int_0^\infty dr\, e^{-\beta r} \left[e^{iqr} - e^{-iqr} \right] \\
&= -\frac{mg}{iq\hbar^2} \left(\frac{1}{-\beta + iq} \left[e^{(-\beta + iq)r} \right]_0^\infty - \frac{1}{-\beta - iq} \left[e^{(-\beta - iq)r} \right]_0^\infty \right) \\
&= -\frac{mg}{iq\hbar^2} \left(\frac{1}{\beta - iq} - \frac{1}{\beta + iq} \right) \\
&= -\frac{2mg}{\hbar^2} \frac{1}{\beta^2 + q^2}
\end{aligned}
$$

Solution 9.1
Equations (9.3) and (9.4) yield

$$S_+|z-\rangle = \sqrt{2}\hbar\,|z0\rangle, \quad S_+|z0\rangle = \sqrt{2}\hbar\,|z+\rangle,$$

$$S_-|z+\rangle = \sqrt{2}\hbar\,|z0\rangle, \quad S_-|z0\rangle = \sqrt{2}\hbar\,|z-\rangle$$

and therefore

$$S_+ = \sqrt{2}\hbar \begin{pmatrix} 0 & 1 & 0 \\ 0 & 0 & 1 \\ 0 & 0 & 0 \end{pmatrix}, \quad S_- = \sqrt{2}\hbar \begin{pmatrix} 0 & 0 & 0 \\ 1 & 0 & 0 \\ 0 & 1 & 0 \end{pmatrix}.$$

Now follow the same steps as in the case of Spin-$\frac{1}{2}$.

Solution 9.2

(a)

$$\begin{aligned}
[J_i, J_j] &= [J_{1i} \otimes \mathbf{1} + \mathbf{1} \otimes J_{2i}, \ J_{1j} \otimes \mathbf{1} + \mathbf{1} \otimes J_{2j}] \\
&= [J_{1i} \otimes \mathbf{1}, \ J_{1j} \otimes \mathbf{1}] + [\mathbf{1} \otimes J_{2i}, \ \mathbf{1} \otimes J_{2j}] \\
&= [J_{1i}, J_{1j}] \otimes \mathbf{1} + \mathbf{1} \otimes [J_{2i}, J_{2j}] \\
&= i\hbar \sum_k \epsilon_{ijk}(J_{1k} \otimes \mathbf{1} + \mathbf{1} \otimes J_{2k}) \\
&= i\hbar \sum_k \epsilon_{ijk} J_k
\end{aligned}$$

In the third line, (9.25) was used. $[\mathbf{J}^2, J_i] = 0$ follows as before from $[J_i, J_j]$.

(b)

$$\begin{aligned}
\mathbf{J}^2 &= \sum_i (J_{1i} \otimes \mathbf{1} + \mathbf{1} \otimes J_{2i})^2 \\
&= \sum_i J_{1i}^2 \otimes \mathbf{1} + \mathbf{1} \otimes J_{2i}^2 + 2J_{1i} \otimes J_{2i} \\
&= \mathbf{J}_1^2 + \mathbf{J}_2^2 + 2\sum_i J_{1i} \otimes J_{2i}
\end{aligned}$$

$$\begin{aligned}
[\mathbf{J}^2, \mathbf{J}_1^2] &= \left[\mathbf{J}_1^2 \otimes \mathbf{1} + \mathbf{1} \otimes \mathbf{J}_2^2 + 2\sum_i J_{1i} \otimes J_{2i}, \ \mathbf{J}_1^2 \otimes \mathbf{1} \right] \\
&= 0 + 0 + 2\sum_i [J_{1i}, \mathbf{J}_1^2] \otimes J_{2i} = 0
\end{aligned}$$

$$\begin{aligned}
[J_i, \mathbf{J}_1^2] &= [J_{1i} \otimes \mathbf{1} + \mathbf{1} \otimes J_{2i}, \ \mathbf{J}_1^2 \otimes \mathbf{1}] \\
&= [J_{1i}, \mathbf{J}_1^2] \otimes \mathbf{1} + 0 = 0
\end{aligned}$$

$$\begin{aligned}
[\mathbf{J}^2, J_{1z}] &= \left[\mathbf{J}_1^2 \otimes \mathbf{1} + \mathbf{1} \otimes \mathbf{J}_2^2 + 2\sum_i J_{1i} \otimes J_{2i}, \ J_{1z} \otimes \mathbf{1} \right] \\
&= 0 + 0 + 2\sum_i [J_{1i}, J_{1z}] \otimes J_{2i} \\
&= 2i\hbar(-J_{1y} \otimes J_{2x} + J_{1x} \otimes J_{2y})
\end{aligned}$$

Similarly J_2.

Solution 9.3

The proof goes by induction over j_2. For $j_2 = 0$ the relation is obviously fulfilled.
Induction step $j_2 \to j_2 + 1$:

$$\sum_{j=j_1-j_2-1}^{j_1+j_2+1} (2j+1) = \sum_{j=j_1-j_2}^{j_1+j_2} (2j+1) + 2(j_1 - j_2 - 1) + 1 + 2(j_1 + j_2 + 1) + 1$$

$$= (2j_1 + 1)(2j_2 + 1) + 4j_1 + 2$$

$$= (2j_1 + 1)[2(j_2 + 1) + 1]$$

Solution 9.5

$$T(\exp(i\lambda_1))T(\exp(i\lambda_2))$$

$$= \begin{pmatrix} \cos\lambda_1 & \sin\lambda_1 \\ -\sin\lambda_1 & \cos\lambda_1 \end{pmatrix} \begin{pmatrix} \cos\lambda_2 & \sin\lambda_2 \\ -\sin\lambda_2 & \cos\lambda_2 \end{pmatrix}$$

$$= \begin{pmatrix} \cos\lambda_1 \cos\lambda_2 - \sin\lambda_1 \sin\lambda_2 & \cos\lambda_1 \sin\lambda_2 + \sin\lambda_1 \cos\lambda_2 \\ -\cos\lambda_1 \sin\lambda_2 - \sin\lambda_1 \cos\lambda_2 & \cos\lambda_1 \cos\lambda_2 - \sin\lambda_1 \sin\lambda_2 \end{pmatrix}$$

$$= \begin{pmatrix} \cos(\lambda_1 + \lambda_2) & \sin(\lambda_1 + \lambda_2) \\ -\sin(\lambda_1 + \lambda_2) & \cos(\lambda_1 + \lambda_2) \end{pmatrix}$$

$$= T(\exp(i(\lambda_1 + \lambda_2)))$$

$$= T(\exp(i\lambda_1)\exp(i\lambda_2))$$

Compare the calculation of $U_y(\alpha)$ in Sect. 2.6.

Solution 10.1

$$h = \frac{1}{2m}\left(\mathbf{p}^2 - \frac{2q}{c}\mathbf{p}\cdot\mathbf{A} + \frac{q^2}{c^2}\mathbf{A}^2\right) + q\phi$$

$$\dot{p}_i = -\frac{\partial}{\partial x_i}h$$

$$= -\frac{1}{2m}\left(-\frac{2q}{c}\sum_j p_j\frac{\partial}{\partial x_i}A_j + 2\frac{q^2}{c^2}\sum_j A_j\frac{\partial}{\partial x_i}A_j\right) - q\frac{\partial}{\partial x_i}\phi$$

$$= \frac{q}{mc}\sum_j\left(p_j - \frac{q}{c}A_j\right)\frac{\partial}{\partial x_i}A_j - q\frac{\partial}{\partial x_i}\phi$$

$$\dot{x}_i = \frac{\partial}{\partial p_i}h = \frac{1}{m}\left(p_i - \frac{q}{c}A_i\right)$$

$$m\ddot{x}_i = \dot{p}_i - \frac{q}{c}\dot{A}_i$$

$$= \frac{q}{mc}\sum_j\left(p_j - \frac{q}{c}A_j\right)\frac{\partial}{\partial x_i}A_j - q\frac{\partial}{\partial x_i}\phi - \frac{q}{c}\left(\frac{\partial}{\partial t}A_i + \sum_j \dot{x}_j\frac{\partial}{\partial x_j}A_i\right)$$

$$= q\left(-\frac{1}{c}\frac{\partial}{\partial t}A_i - \frac{\partial}{\partial x_i}\phi + \frac{1}{c}\sum_j(\dot{x}_j\frac{\partial}{\partial x_i}A_j - \dot{x}_j\frac{\partial}{\partial x_j}A_i)\right)$$

Solution 10.2

The new current density is obtained with the replacement

$$-i\hbar\nabla\psi \rightarrow -i\hbar\nabla\psi - \frac{q}{c}A\,\psi$$

and the complex conjugate replacement

$$i\hbar\nabla\psi^* \rightarrow i\hbar\nabla\psi^* - \frac{q}{c}A\,\psi^*.$$

Solution 10.3

The additional terms $\sim \frac{\partial\lambda}{\partial t}$ resulting from time derivation and ϕ-transformation are identical:

$$i\hbar\frac{\partial}{\partial t}\psi \rightarrow i\hbar\frac{\partial}{\partial t}\left(e^{i\lambda}\psi\right) = i\hbar e^{i\lambda}\frac{\partial}{\partial t}\psi - \hbar\frac{\partial\lambda}{\partial t}e^{i\lambda}\psi$$

$$q\phi\psi \rightarrow q\phi e^{i\lambda}\psi - \frac{q}{c}\frac{\partial\chi}{\partial t}e^{i\lambda}\psi = q\phi e^{i\lambda}\psi - \hbar\frac{\partial\lambda}{\partial t}e^{i\lambda}\psi$$

The additional terms from $\nabla e^{i\lambda}$ and A' cancel each other:

$$\left(-i\hbar\nabla - \frac{q}{c}A\right)\psi \rightarrow \left(-i\hbar\nabla - \frac{q}{c}A'\right)\left[e^{i\lambda}\psi\right]$$

$$= e^{i\lambda}\left(-i\hbar\nabla + \hbar(\nabla\lambda) - \frac{q}{c}A - \frac{q}{c}(\nabla\chi)\right)\psi$$

$$= e^{i\lambda}\left(-i\hbar\nabla - \frac{q}{c}A\right)\psi$$

It follows:

$$\left(-i\hbar\nabla - \frac{q}{c}A'\right)^2\left[e^{i\lambda}\psi\right] = \left(-i\hbar\nabla - \frac{q}{c}A'\right)e^{i\lambda}\left(-i\hbar\nabla - \frac{q}{c}A\right)\psi$$

$$= e^{i\lambda}\left(-i\hbar\nabla - \frac{q}{c}A\right)^2\psi$$

The Schrödinger equation is therefore also valid for the transformed ψ. There is only an additional factor $e^{i\lambda}$, but all additive extra terms cancel each other.

Solution 10.4

The gauge transformations imply

$$\nabla \chi = \mathbf{C}, \qquad \frac{\partial}{\partial t}\chi = -c\eta$$

$$\Rightarrow \chi = \mathbf{C}\cdot\mathbf{r} - c\eta t + a$$

with an arbitrary constant a.

$$\Rightarrow \psi' = \exp\left[i\left(-\frac{q}{\hbar}\eta t - \frac{q}{\hbar c}\mathbf{C}\cdot\mathbf{r} + \frac{qa}{\hbar c}\right)\right]\psi$$

From the Ehrenfest equations we get

$$\frac{d}{dt}\langle\mathbf{r}\rangle = \frac{1}{m}\left(\langle\mathbf{p}\rangle - \frac{q}{c}\mathbf{A}\right).$$

After the transformation $\langle\mathbf{p}\rangle = \langle\psi|-i\hbar\nabla|\psi\rangle$ is increased by $\frac{q}{c}\mathbf{C}$, but $\frac{q}{c}\mathbf{A}$ is increased by the same value. The changes cancel each other, and the velocity does not depend on the gauge.

Solution 10.5

For vectors we have the identity

$$\mathbf{a}\times(\mathbf{b}\times\mathbf{c}) = \mathbf{b}(\mathbf{a}\cdot\mathbf{c}) - \mathbf{c}(\mathbf{a}\cdot\mathbf{b}).$$

If the nabla operator is involved, one has to take care of the order. In components we get

$$[\nabla\times(\mathbf{r}\times\mathbf{B})]_i = \sum_j\left(\frac{\partial}{\partial x_j}x_i\right)B_j - \left(\frac{\partial}{\partial x_j}x_j\right)B_i$$

$$= \sum_j \delta_{ij}B_j - \delta_{jj}B_i = B_i - 3B_i = -2B_i.$$

Second equation:

$$\nabla\cdot(\mathbf{r}\times\mathbf{B}) = \sum_i\frac{\partial}{\partial x_i}(\mathbf{r}\times\mathbf{B})_i = \sum_{ijk}\epsilon_{ijk}\left(\frac{\partial}{\partial x_i}x_j\right)B_k = 0$$

Solution 11.1

$$H_0 = \frac{\mathbf{P}^2}{2m} - \frac{e^2}{r}, \qquad H_1 = eE_0 Z$$

$$[H_0, H_1] = \frac{eE_0}{2m}[P_z^2, Z] = \frac{eE_0}{2m}(P_z[P_z, Z] + [P_z, Z]P_z) = -\frac{ieE_0\hbar}{m}P_z$$

$$[L_x, Z] = [YP_z - ZP_y, Z] = -i\hbar Y$$

$$\text{similarly}: \quad [L_y, Z] = i\hbar X, \quad [L_y, X] = -i\hbar Z, \quad [L_x, Y] = i\hbar Z$$

$$[L_x, X] = [L_y, Y] = [L_z, Z] = 0$$

$$\Rightarrow [\mathbf{L}^2, H_1] = eE_0[L_x^2 + L_y^2, Z]$$
$$= eE_0\left(L_x[L_x, Z] + [L_x, Z]L_x + L_y[L_y, Z] + [L_y, Z]L_y\right)$$
$$= i\hbar eE_0\left(-L_xY - YL_x + L_yX + XL_y\right)$$
$$= i\hbar eE_0\left(-2L_xY + [L_x, Y] + 2XL_y + [L_y, X]\right)$$
$$= i\hbar eE_0\left(-2L_xY + i\hbar Z + 2XL_y - i\hbar Z\right)$$
$$= 2i\hbar eE_0\left(XL_y - L_xY\right)$$

Solution 11.2

$$\zeta = \int d^3r \, \psi_{2,1,0}^*(\mathbf{r})\psi_{2,0,0}(\mathbf{r})z$$
$$= \int_0^\infty dr \int_{-1}^1 d\cos\theta \int_0^{2\pi} d\phi \, r^2 \frac{1}{32\pi a^3}\frac{r}{a}\left(2 - \frac{r}{a}\right)e^{-r/a}\cos\theta \, r\cos\theta$$
$$= \frac{1}{16a^4}\int_0^\infty dr \left(2r^4 - \frac{r^5}{a}\right)e^{-r/a}\int_{-1}^1 d\cos\theta \, \cos^2\theta$$
$$= \frac{1}{24a^4}\left[\int_0^\infty dr \, 2r^4 e^{-r/a} - \int_0^\infty dr \, \frac{r^5}{a}e^{-r/a}\right]$$
$$= \frac{1}{24a^4}\left[\int_0^\infty dr \, 48\, a^4 e^{-r/a} - \int_0^\infty dr \, 120\, a^4 e^{-r/a}\right]$$
$$= -3\int_0^\infty dr \, e^{-r/a} = -3a$$

The first term in the second last line was obtained after applying partial integration four times, the second term after five times.

Solution 11.3

$$\frac{-\lambda^2}{\hbar^2}\int_0^t dt_1 \int_{t_0}^{t_1} dt_2 \, \langle f|U_0(t, t_1)H_1(t_1)U_0(t_1, t_2)H_1(t_2)U_0(t_2, 0)|i\rangle \, e^{i\omega_{fi}t}$$

$$= \frac{-\lambda^2}{\hbar^2} \int_0^t dt_1 \int_{t_0}^{t_1} dt_2 \sum_{j,k,l,m} \langle f|U_0(t,t_1)|j\rangle \langle j|H_1(t_1)|k\rangle \langle k|U_0(t_1,t_2)|l\rangle$$

$$\cdot \langle l|H_1(t_2)|m\rangle \langle m|U_0(t_1,0)|i\rangle \, e^{i\omega_f t}$$

$$= \frac{-\lambda^2}{\hbar^2} \int_0^t dt_1 \int_{t_0}^{t_1} dt_2 \sum_{j,k,l,m} \langle j|H_1(t_1)|k\rangle \langle l|H_1(t_2)|i\rangle$$

$$\cdot \, e^{-i\omega_f(t-t_1)} \delta_{fj} \, e^{-i\omega_k(t_1-t_2)} \delta_{kl} \, e^{-i\omega_i t_2} \delta_{mi} \, e^{i\omega_f t}$$

$$= \frac{-\lambda^2}{\hbar^2} \int_0^t dt_1 \int_{t_0}^{t_1} dt_2 \sum_k \langle f|H_1(t_1)|k\rangle \langle k|H_1(t_2)|i\rangle \, e^{i(\omega_f-\omega_k)t_1} \, e^{i(\omega_k-\omega_i)t_2}$$

Solution 12.1

$$\dim \mathcal{H}_2^{(+)} = \frac{n(n+1)}{2}, \qquad \dim \mathcal{H}_2^{(-)} = \frac{n(n-1)}{2}$$

The number of dimensions is the number of possibilities to choose two basis vectors of $\mathcal{H}_1$. In one case, the same basis vector can be chosen twice, in the other case not.

Solution 12.2

One can check the equation by following the positions of the states on the left hand side, e.g. $j \to m \to n \to i$. For the eigenvalues we get

$$\tau_{ni}^2 \tau_{mj}^2 \tau_{mn} = \tau_{ij} \quad \Rightarrow \quad \tau_{mn} = \tau_{ij}$$

for arbitrary i, j, m, n.

Solution 12.3

(a) For $n_j > 1$ there are $n_j!$ permutations, exchanging only equal states. For each such j the number of possible permutations therefore has to be divided by $n_j!$.
(b) The number of terms in the sum is $\frac{3!}{2!} = 3$:

$$|\{n_j\}\rangle^{(+)} = \frac{1}{\sqrt{3}} (|1:2, 2:1, 3:1\rangle + |1:1, 2:2, 3:1\rangle + |1:1, 2:1, 3:2\rangle)$$

Solution 12.5

$$p(x+) = \frac{1}{2}|\langle x + |z+\rangle|^2 + \frac{1}{2}|\langle x + |x+\rangle|^2 = \frac{1}{4} + \frac{1}{2} = \frac{3}{4},$$

$$p(y+) = \frac{1}{2}|\langle y + |z+\rangle|^2 + \frac{1}{2}|\langle y + |x+\rangle|^2 = \frac{1}{4} + \frac{1}{4} = \frac{1}{2},$$

$$p(z+) = \frac{1}{2}|\langle z + |z+\rangle|^2 + \frac{1}{2}|\langle z + |x+\rangle|^2 = \frac{1}{2} + \frac{1}{4} = \frac{3}{4}$$

$$\langle S_x \rangle = \frac{\hbar}{2} \left(\frac{1}{2} \langle z + | \sigma_x | z + \rangle + \frac{1}{2} \langle x + | \sigma_x | x + \rangle \right) = \frac{\hbar}{2} \left(\frac{1}{2} + 0 \right) = \frac{\hbar}{4}$$

$$\text{similarly} : \langle S_y \rangle = 0, \quad \langle S_z \rangle = \frac{\hbar}{4}$$

Solution 12.6

$\mathcal{H}_\rho$ is the space spanned by the n states ρ consists of. $A\mathcal{H}_\rho$ is the image of $A|_{\mathcal{H}_\rho}$, i.e. the restriction of A on $\mathcal{H}_\rho$. Then $\mathcal{H}_{A\rho}$ is the space spanned by $\mathcal{H}_\rho$ and $A\mathcal{H}_\rho$, consisting of all linear combinations of vectors in $\mathcal{H}_\rho$ and $A\mathcal{H}_\rho$. Since $\mathcal{H}_\rho$ and $A\mathcal{H}_\rho$ are each not more than n-dimensional, $\mathcal{H}_{A\rho}$ can be maximally $2n$-dimensional.

Solution 12.7
(i)

$$\rho = \frac{1}{2} |z+\rangle\langle z + | + \frac{1}{2} |z-\rangle\langle z - | = \frac{1}{2} \begin{pmatrix} 1 & 0 \\ 0 & 1 \end{pmatrix}$$

$$\langle S_i \rangle = \text{Sp} \left(\frac{\hbar}{2} \sigma_i \frac{1}{2} \mathbf{1} \right) = \frac{\hbar}{4} \text{Sp}(\sigma_i) = 0$$

(ii)

$$\rho = \frac{1}{2} |z+\rangle\langle z + | + \frac{1}{2} |x+\rangle\langle x + | = \frac{1}{4} \begin{pmatrix} 3 & 1 \\ 1 & 1 \end{pmatrix}$$

$$\langle S_x \rangle = \frac{\hbar}{8} \text{Sp} \left(\sigma_x \begin{pmatrix} 3 & 1 \\ 1 & 1 \end{pmatrix} \right) = \frac{\hbar}{8} \text{Sp} \begin{pmatrix} 1 & 1 \\ 3 & 1 \end{pmatrix} = \frac{\hbar}{4}$$

$$\langle S_y \rangle = \frac{\hbar}{8} \text{Sp} \left(\sigma_y \begin{pmatrix} 3 & 1 \\ 1 & 1 \end{pmatrix} \right) = \frac{\hbar}{8} \text{Sp} \begin{pmatrix} -i & -i \\ 3i & i \end{pmatrix} = 0$$

$$\langle S_z \rangle = \frac{\hbar}{8} \text{Sp} \left(\sigma_x \begin{pmatrix} 3 & 1 \\ 1 & 1 \end{pmatrix} \right) = \frac{\hbar}{8} \text{Sp} \begin{pmatrix} 3 & 1 \\ -1 & -1 \end{pmatrix} = \frac{\hbar}{4}$$

Solution 12.8
(b)

$$\alpha^2 = \frac{4}{9} \implies \alpha = \frac{2}{3}, \quad \alpha\beta = \frac{4}{9} + \frac{2}{9}i \implies \beta = \frac{2}{3} + \frac{i}{3}$$

$$\implies |\psi\rangle = \frac{2}{3} |z+\rangle + \left(\frac{2}{3} + \frac{i}{3} \right) |z-\rangle$$

Solution 14.1
With

$$\{\sigma_i, \sigma_j\} := \sigma_i \sigma_j + \sigma_j \sigma_i = 2\delta_{ij} \mathbf{1}$$

(cf. Exercise 2.26) one gets

$$\alpha_i^2 = \begin{pmatrix} 0 & \sigma_i \\ \sigma_i & 0 \end{pmatrix}^2 = \begin{pmatrix} \sigma_i^2 & 0 \\ 0 & \sigma_i^2 \end{pmatrix} = 1$$

$$\alpha_i \alpha_j + \alpha_j \alpha_i = \begin{pmatrix} \sigma_i \sigma_j + \sigma_j \sigma_i & 0 \\ 0 & \sigma_i \sigma_j + \sigma_j \sigma_i \end{pmatrix} = 2\delta_{ij} 1$$

$$\alpha_i \beta + \beta \alpha_i = \begin{pmatrix} 0 & -\sigma_i \\ \sigma_i & 0 \end{pmatrix} + \begin{pmatrix} 0 & \sigma_i \\ -\sigma_i & 0 \end{pmatrix} = 0.$$

Solution 14.2

(c) We need the relation

$$\sigma_i \sigma_j = \delta_{ij} 1 + i \sum_k \epsilon_{ijk} \sigma_k,$$

which can be easily derived from

$$\{\sigma_i, \sigma_j\} = 2\delta_{ij} 1, \qquad [\sigma_i, \sigma_j] = 2i \sum_k \epsilon_{ijk} \sigma_k.$$

Using this we obtain

$$\begin{aligned}
(\sigma \cdot \mathbf{U})(\sigma \cdot \mathbf{V}) &= \sum_{i,j} (\sigma_i \otimes U_i)(\sigma_j \otimes V_j) \\
&= \sum_{i,j} (\sigma_i \sigma_j) \otimes (U_i V_j) \\
&= \sum_{i,j} \delta_{ij} 1 \otimes (U_i V_j) + \sum_{i,j,k} i\epsilon_{ijk} \sigma_k \otimes (U_i V_j) \\
&= \sum_i 1 \otimes U_i V_i + \sum_k i\sigma_k \otimes (\mathbf{U} \times \mathbf{V})_k \\
&= \mathbf{U} \cdot \mathbf{V} + i\sigma \cdot (\mathbf{U} \times \mathbf{V}).
\end{aligned}$$

(d) $\mathbf{U} \times \mathbf{U}$ is not necessarily zero, since the commutator $U_i U_j - U_j U_i$ does not necessarily vanish. But since derivatives with respect to different coordinates commute, we have

$$\mathbf{P} \times \mathbf{P} = 0.$$

For $\mathbf{Q} := (\mathbf{P} - q\mathbf{A}) \times (\mathbf{P} - q\mathbf{A})$ follows:

$$\begin{aligned}
Q_i |\psi\rangle &= [(\mathbf{P} - q\mathbf{A}) \times (\mathbf{P} - q\mathbf{A})]_i \, |\psi\rangle \\
&= -q \, [\mathbf{P} \times \mathbf{A} + \mathbf{A} \times \mathbf{P}]_i \, |\psi\rangle
\end{aligned}$$

$$= i\hbar q \left[\nabla \times \mathbf{A} + \mathbf{A} \times \nabla \right] |\psi\rangle$$

$$= i\hbar q \sum_{j,k} \epsilon_{ijk} \left(\frac{\partial}{\partial x_j} A_k + A_j \frac{\partial}{\partial x_k} \right) |\psi\rangle$$

$$= i\hbar q \sum_{j,k} \epsilon_{ijk} \left(\frac{\partial}{\partial x_j} A_k - A_k \frac{\partial}{\partial x_j} \right) |\psi\rangle$$

$$= i\hbar q \sum_{j,k} \epsilon_{ijk} \left(\frac{\partial}{\partial x_j} (A_k |\psi\rangle) - A_k \frac{\partial}{\partial x_j} |\psi\rangle \right)$$

$$= i\hbar q \sum_{j,k} \epsilon_{ijk} \left(\frac{\partial}{\partial x_j} A_k \right) |\psi\rangle$$

$$= i\hbar q B_i |\psi\rangle$$

In the fifth line $\epsilon_{ikj} = -\epsilon_{ijk}$ was used.

(e) One gets (14.25) by plugging the results of (c) and (d) into the Pauli equation. The result $g_e = 2$ is obtained when we compare with H_{mag}:

$$H_{\text{mag}} = -\mathbf{B} \cdot \mathbf{M} = -\frac{q g_e}{2m} \mathbf{B} \cdot \mathbf{S} = -\frac{q g_e \hbar}{4m} \boldsymbol{\sigma} \cdot \mathbf{B}$$

Index

© Springer International Publishing Switzerland 2016   345
J.-M. Schwindt, *Conceptual Basis of Quantum Mechanics*,
Undergraduate Lecture Notes in Physics, DOI 10.1007/978-3-319-24526-3

Printed in the United States
By Bookmasters